Naphthalenediimide and its Congeners
From Molecules to Materials

Monographs in Supramolecular Chemistry

Series Editors:
Philip Gale, *The University of Sydney, Australia*
Jonathan Steed, *Durham University, UK*

Titles in this Series:
1: Cyclophanes
2: Calixarenes
3: Crown Ethers and Cryptands
4: Container Molecules and Their Guests
5: Membranes and Molecular Assemblies: The Synkinetic Approach
6: Calixarenes Revisited
7: Self-assembly in Supramolecular Systems
8: Anion Receptor Chemistry
9: Boronic Acids in Saccharide Recognition
10: Calixarenes: An Introduction, 2nd Edition
11: Polymeric and Self Assembled Hydrogels: From Fundamental Understanding to Applications
12: Molecular Logic-based Computation
13: Supramolecular Systems in Biomedical Fields
14: Synthetic Receptors for Biomolecules: Design Principles and Applications
15: Polyrotaxane and Slide-Ring Materials
16: Boron: Sensing, Synthesis and Supramolecular Self-Assembly
17: Porous Polymers: Design, Synthesis and Applications
18: Pillararenes
19: Supramolecular Chemistry at Surfaces
20: Aromatic Interactions: Frontiers in Knowledge and Application
21: Naphthalenediimide and its Congeners: From Molecules to Materials

How to obtain future titles on publication:
A standing order plan is available for this series. A standing order will bring delivery of each new volume immediately on publication.

For further information please contact:
Book Sales Department, Royal Society of Chemistry, Thomas Graham House, Science Park, Milton Road, Cambridge, CB4 0WF, UK
Telephone: +44 (0)1223 420066, Fax: +44 (0)1223 420247
Email: booksales@rsc.org
Visit our website at http://www.rsc.org/Shop/Books/

Naphthalenediimide and its Congeners
From Molecules to Materials

Edited by

G. Dan Pantoş
University of Bath, UK
Email: g.d.pantos@bath.ac.uk

ROYAL SOCIETY OF CHEMISTRY

THE QUEEN'S AWARDS
FOR ENTERPRISE:
INTERNATIONAL TRADE
2013

Monographs in Supramolecular Chemistry No. 21

Print ISBN: 978-1-84973-922-1
PDF eISBN: 978-1-78262-138-6
EPUB eISBN: 978-1-78801-119-8
ISSN: 1368-8642

A catalogue record for this book is available from the British Library

The Royal Society of Chemistry is a charity, registered in England and Wales, Number 207890, and a company incorporated in England by Royal Charter (Registered No. RC000524), registered office: Burlington House, Piccadilly, London W1J 0BA, UK, Telephone: +44 (0) 207 4378 6556.

Visit our website at www.rsc.org/books

Printed in the United Kingdom by CPI Group (UK) Ltd, Croydon, CR0 4YY, UK

Preface

Naphthalenediimide and its congeners are not new molecules, however, over the past 15–20 years, their chemistry has developed rapidly. The first mention of naphthalenediimide was in 1887 by Bamberger and Philip. By comparison, pyromellitimide is younger, being first described in 1914 by Meyer and Steiner, while the youngest member is perylenediimide, which was first reported in a German patent in 1919. The larger family of rylene dyes are ever-expanding and increase in complexity despite synthetic challenges. A number of excellent reviews have been published over the years that give an overview of parts of this field, however, when I planned this volume, there had not been a book exclusively devoted to these recent advances. Each chapter in this book is written by experts in the field and aimed at covering the most recent advances in a coherent manner. The content is as varied as the areas in which the naphthalenediimides and their congeners have found applications. Therefore, we have covered topics from organic photovoltaics, anion-slides, and DNA binders to building blocks for complex molecular topologies. The chemistry is rich and we are yet to discover all its secrets. I hope this book reflects the exciting and dynamic chemistry of naphthalenediimide and its congeners and will be a source of inspiration for future generations of PhD students and postdoctoral researchers. On a personal note, I can say that having worked in this field for over a decade, I find it as fascinating as ever because every single time I consider that a certain NDI project has exhausted its novelty, a new observation or discovery leads us to an even more exciting research area.

I must finish with some acknowledgments: I would like to thank the two researchers that introduced me to the chemistry of naphthalenediimides: Prof. Jeremy Sanders who gave me the opportunity to work in this area, and Prof. Brent Iverson who indirectly made me aware of these molecules during

Monographs in Supramolecular Chemistry No. 21
Naphthalenediimide and its Congeners: From Molecules to Materials
Edited by G. Dan Pantoş

Published by the Royal Society of Chemistry, www.rsc.org

my PhD years. I would also like to thank all the students, postdocs and colleagues that have worked with me over the years; they have been an endless source of inspiration and enthusiasm. I would also like to thank the Royal Society of Chemistry books team, led by Rowan Frame, for their endless patience with a project that suffered a number of delays.

Dan Pantoş
University of Bath

To Anca, Tessa and Tricia
without whom nothing makes sense

Contents

Monographs in Supramolecular Chemistry No. 21
Naphthalenediimide and its Congeners: From Molecules to Materials
Edited by G. Dan Pantoş

Published by the Royal Society of Chemistry, www.rsc.org

CHAPTER 1

Supramolecular Chemistry of Naphthalenediimide and its Congeners

GILES M. PRENTICE, LIAM EMMETT, VIJAY LUXAMI AND G. DAN PANTOŞ*

Department of Chemistry, University of Bath, Claverton Down, Bath BA2 7AY, UK
*Email: g.d.pantos@bath.ac.uk

1.1 Introduction

1,4,5,8-Naphthalenediimides (NDI) represent a class of aromatic compounds with an electron-deficient π-system whose electronic, redox and spectroscopic properties have led to their extensive study in the field of supramolecular and materials chemistry. Owing to their planar aromatic nature, naphthalenediimides can exhibit aromatic π-stacking and van der Waals interactions. The electron density on most aromatic rings create a quadrupole moment with a partial negative charge above and below the face and a partial positive charge around the periphery. When electron-deficient NDI is paired with complementary electron-rich species, it adopts an optimum face-centered stacking often referred to as a "donor–acceptor" complex. Different spectroscopic studies have shown that association of π-donor and π-acceptor molecules is more favourable than the self-association of π-donor–donor and π-acceptor–acceptor scaffolds.[1,2] The π system of the NDI core is amenable to anion–π interactions and has been exploited

Monographs in Supramolecular Chemistry No. 21
Naphthalenediimide and its Congeners: From Molecules to Materials
Edited by G. Dan Pantoş

Published by the Royal Society of Chemistry, www.rsc.org

extensively in the construction of artificial anion channels.[3,4] The carbonyl oxygen atoms have been shown to interact with metal cations and form a multitude of hydrogen bonding arrays, while the electron withdrawing nature of the imide groups installs a significant partial positive charge on the aromatic C–Hs, which allows them to participate in non-classical C–H···O hydrogen bonds.[5] The chemistry of NDIs has developed rapidly in the past 15–20 years due to this multitude of supramolecular handles present on the NDI core and their straightforward synthesis.[6–9] A number of reviews have covered parts of this field[10,11] and this chapter will highlight a few of the main areas in which NDIs and their congeners have driven supramolecular chemistry forward.

Iverson *et al.* reported aedamers (aromatic electron donor–acceptor) as the first foldamers to make use of aromatic π–π interactions in water to direct folding. The aedamers exploited the π-acidity of NDI to form the presumed pleated structures based on the complementary face-to-face association of alternating an electron-rich "donor", 1,5-dialkoxynaphthalene (DN) and an electron-deficient "acceptor", NDI.[1,12,13] Aspartic acid residues were linked in between aromatic moieties to facilitate solubility in water (Figure 1.1). These types of NDI-intercalators form strong complexes with DNA, which dissociate extremely slowly, corresponding to a half-life of 16 days.[14]

The effect on folding of the D–A interaction was investigated further by Iverson *et al.* in a study where water-soluble trimers were synthesised with different sequences of NDI and 1,5-DN (Figure 1.2).[13] Again, it was found that the folding always yielded a conformation in which face-centred stacking was observed between the NDI and 1,5-DN, which further reinforces

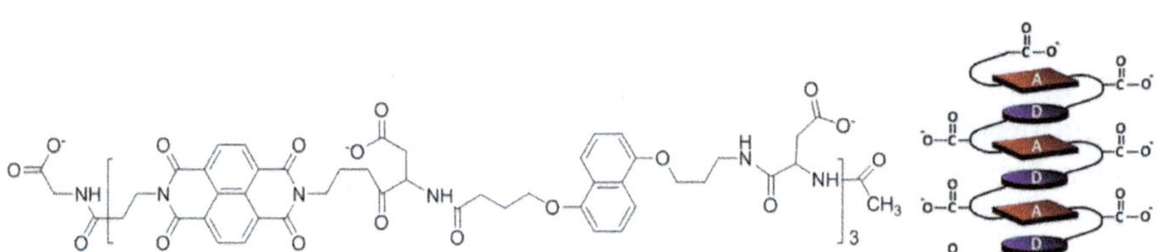

1

Figure 1.1 An NDI–DN aedamer structure and the pleated cartoon representation.[12]

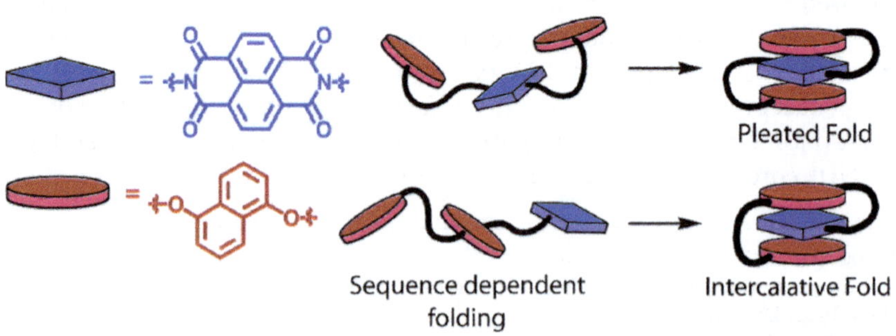

Figure 1.2 Folding in NDI–DN trimers.[13]

the directionality of NDI–DN D–A interactions as shown in Iverson and co-workers' previous work.

1.2 Molecules with Complex Topologies

NDI and BDI have been used extensively in catenanes and rotaxanes. Their properties as acceptors and relative ease of electrochemical reduction have led to a number of templated syntheses of catenanes, as well as properties such as bistability and even potential molecular machines.

Sanders and co-workers in 1998 synthesised a variety of NDI, BDI and DN [2]-catenanes by use of D–A interactions.[15] Crown ethers containing 1,5-DNs were synthesised, where terminally alkyne appended NDIs or BDIs were then coupled using $CuCl_2$ in the presence of these ethers (Figure 1.3).

The formation of [2]-catenanes was obtained as a diimide bound between the DNs in the macrocycle affording the interlocked structure in yields between 29 and 52% for BDI and NDI respectively. The stronger binding of the NDI to the DN when compared to a BDI–DN interaction was confirmed by a competition experiment. A series of modifications on this catenane were later developed by the Sanders group, which showed the versatility of the building blocks used in this approach.

Firstly, a hybrid butadiyne crown ether macrocycle was synthesised and catenation by the original method could be prevented by the use of $Co_2(CO)_6$ to deform the macrocycle by binding to alkynes. This process was found to be reversible by removal of the $Co_2(CO)_6$. It was also possible to assemble the catenane in a single step with $CuCl_2$-catalysed coupling of the alkyne-appended DN and BDI previously used (Figure 1.4).[16–18]

Post synthetic modification of the butadiyne linkers was also performed by hydrogenation by Pd/C (5%). The crown ether was also altered with dialkoxydibenzoates and dialkoxybenzoates in the place of one of the DN moieties; this still allowed for the synthesis of the [2]-catenanes as previously described.[17] Following these modifications, Sanders and co-workers proposed a supramolecular protection of dialkoxybenzoates using Sn porphyrins, which can be used to assemble larger supramolecular assemblies.[19]

Sanders and co-workers then further developed this work synthesising [2]-catenanes with the same aromatic moieties. However, in this case, they utilised a Mitsunobu alkylation between a tetraethylene glycol appended NDI and BDI (Figure 1.5).[20] As in the previous example, the DN crown ether acts as a template with the NDI bound in between the DN moieties (Figure 1.5).

Crown ether containing DNs with NDI and BDI were again used in the synthesis of controllable [2]-rotaxanes (Figure 1.6).[21,22] In this system, a number of electron-deficient stations are installed along the axle of the rotaxane on which the DN macrocycle binds. Modulating the strength of the interaction at each site can control the position of the ring on the axle. This was achieved by two different acceptors, NDI and BDI, installed in the axle; all-NDI and all-BDI [2]-rotaxanes were also synthesised.

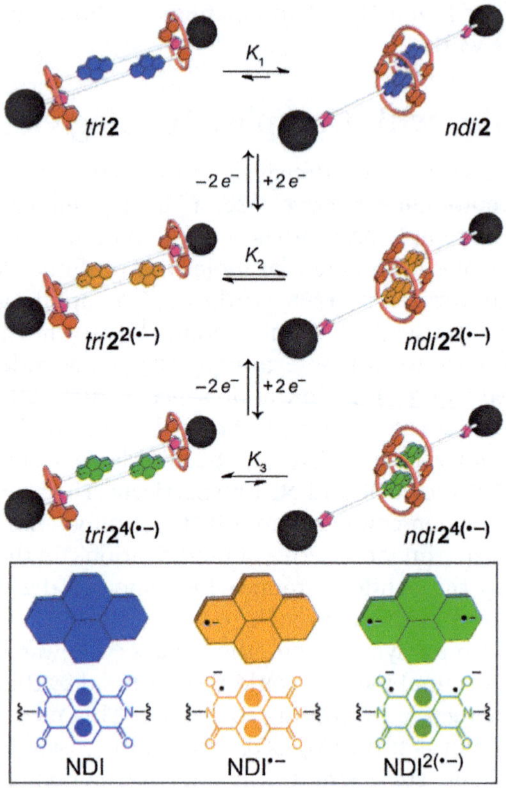

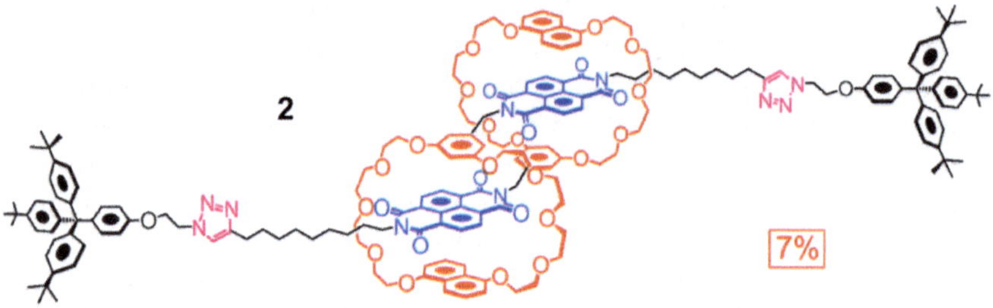

Figure 1.3 A donor–acceptor [2]-catenane synthesised by Sanders *et al.*[15]

While the NDI station is the preferred binding site due to its larger π surface, its properties can be modulated electrochemically or chemically, leading to a switching of the rotaxane conformation. The redox-induced switching was performed by cyclic voltammetry in CH_2Cl_2. Firstly, the NDI site is deactivated by electrochemical reduction (−0.74 V) and the BDI site becomes the preferable acceptor, however, further electrochemical reduction (−1.06 V) can deactivate this site leading to no donor–acceptor interactions. Interestingly, with the all-NDI and BDI, each site was found to have a different reduction potential.

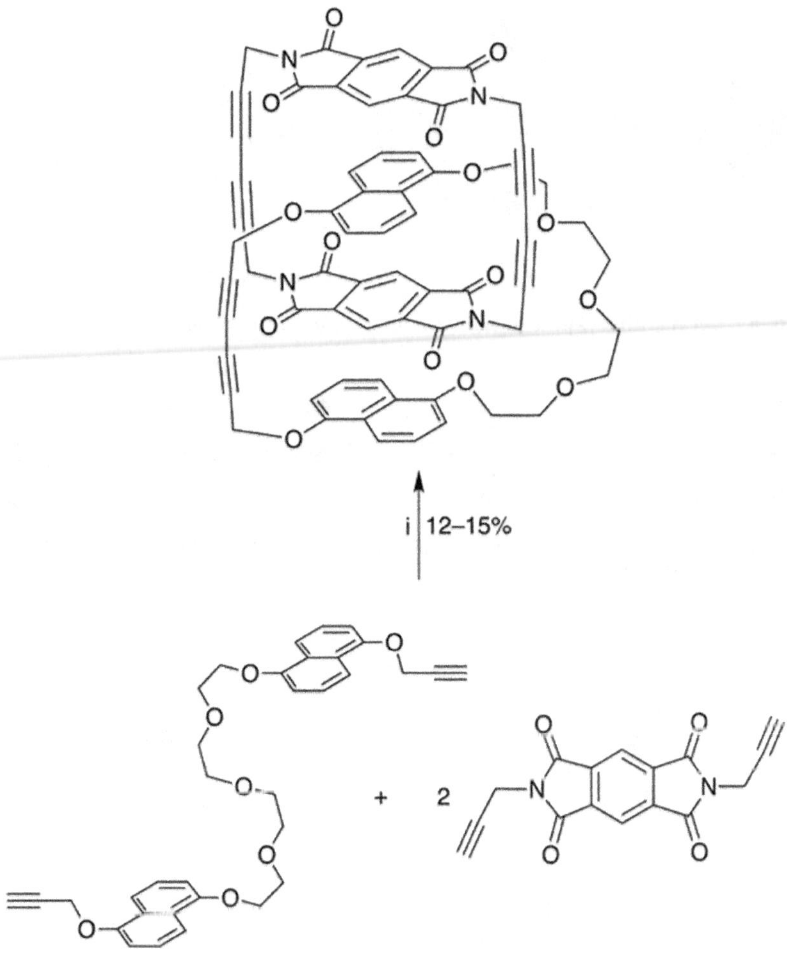

Figure 1.4 Synthesis through an initial step of self-assembly.[18]

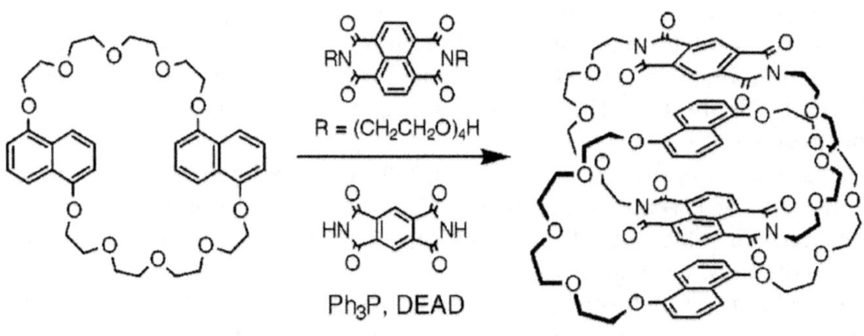

Figure 1.5 An NDI–BDI–DN [2]-catenane prepared by Sanders *et al.*[20]

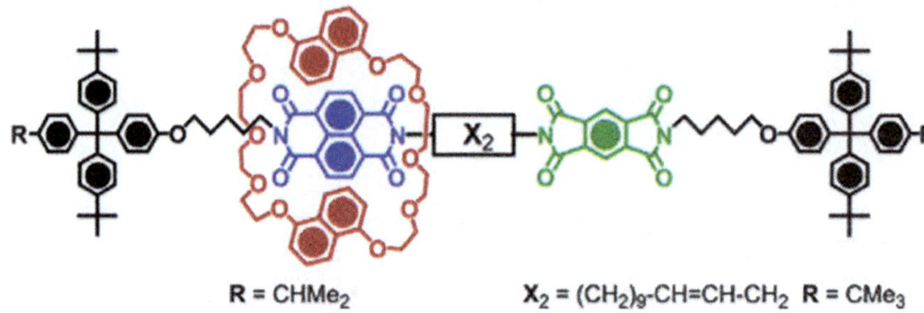

Figure 1.6 A bistable [2]-rotaxane by Sanders and Stoddart *et al.*[21]

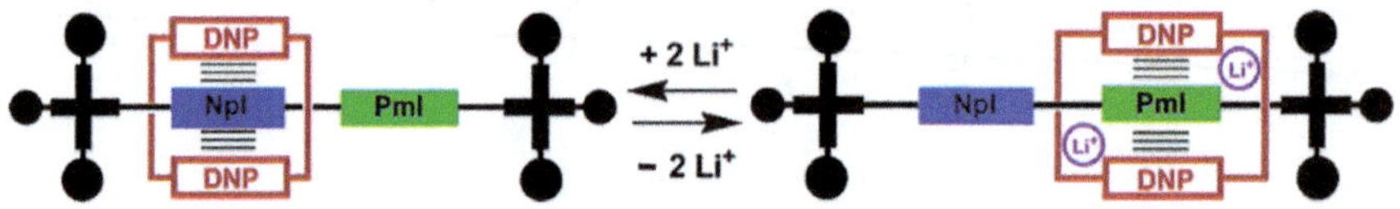

Figure 1.7 Cation-triggered switching in a bistable [2]-rotaxane. NpI = NDI, PmI = BDI, DNP = DN.[21]

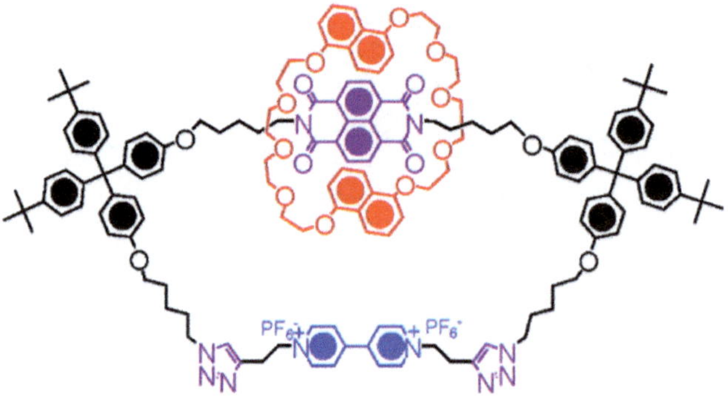

Figure 1.8 An NDI-based bistable [2]-catenane synthesised by Stoddart *et al.*[23]

Chemical switching could be obtained by addition of lithium cations as in the presence of lithium ions, the DN crown ether sits on the BDI as the interaction between the ether loop and the cations is stronger than in the NDI (Figure 1.7). This process was observed by ^{1}H NMR and UV-vis spectroscopy.[21]

This work was further extended in 2008 when Stoddart and co-workers synthesised a [2]-catenane composed of the same DN crown ether with two acceptor sites: an NDI and 1,4-bipyridinium separated by two tetra-arylmethane units (Figure 1.8).[23]

At ambient temperatures, it was found the DN crown ether encircled the NDI, however, when heated to 70 °C and observed by ^{1}H NMR for 60 h, it was found that a mixture of conformers existed in a 60 : 40 ratio for the DN to encircle the 1,4-bipyridinium *vs.* the NDI site. The mixture of conformers was also confirmed by cyclic voltammetry. Further examples have been synthesised by Stoddart and co-workers in which switching between an NDI and triazole using redox methods is achieved.

Recently, the Stoddart group has also used NDI as part of a thermally and electrochemically switchable donor acceptor daisy chain rotaxane in the design of an artificial molecular machine.[24] This used a DN crown ether macrocycle but in comparison to the previous example, it was covalently linked to an NDI axle (Figure 1.9). As in previous examples, electrochemical reduction of the NDI causes the DN to preferentially bind onto another site, in this case the triazole, after a two-electron reduction; one-electron reduction allowed for a mixture of conformers. Thermal switching was studied by VT ^{1}H-NMR and its dynamics further elucidated by computer

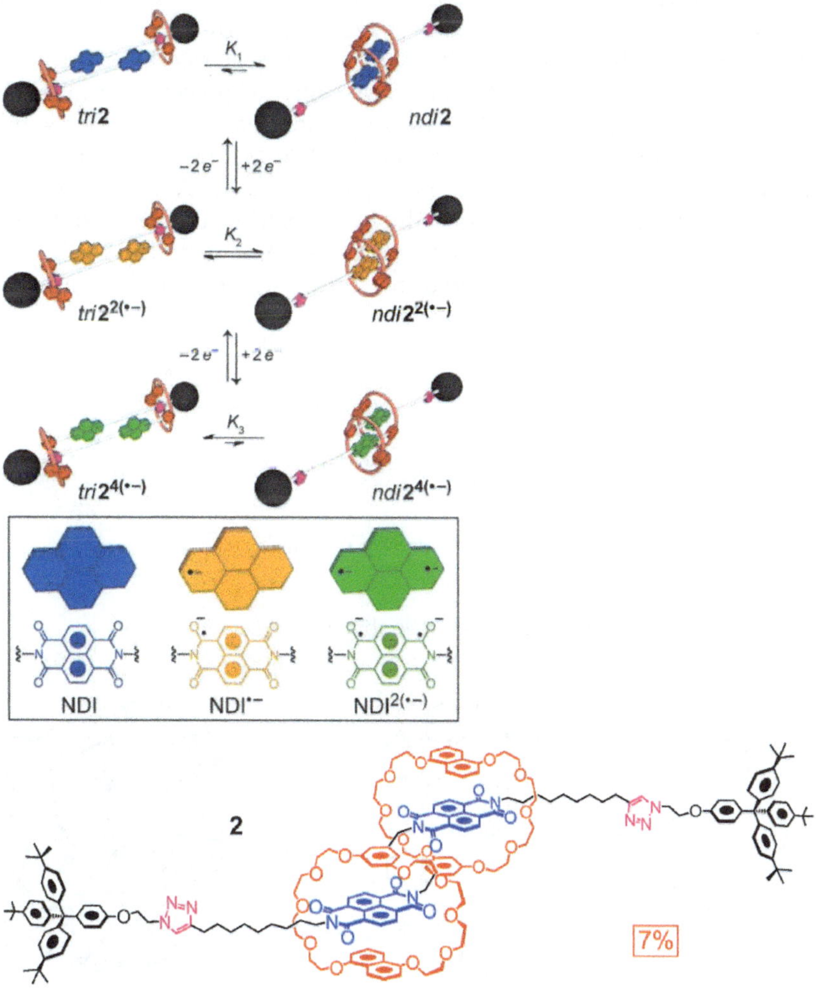

Figure 1.9 The Stoddart [c2]daisy chain rotaxane.[24]

simulation, which indicated the rotaxane was bistable at low temperatures and monostable at high temperatures.

Colquhoun and co-workers have proposed a number of D–A systems that utilise NDI and BDI as part of a cyclophane in non-aqueous media for molecular recognition. Almost all examples were exclusively studied in 6 : 1 $CHCl_3$–hexafluoro-propan-2-ol. Firstly, in 2002,[25] two cyclophanes containing NDI or BDI were synthesised (Figure 1.10). When added to a variety of electron-rich polycyclic aromatics, these formed charge transfer complexes with stronger binding constants observed as the π-surface and the ability for hydrogen bonding increased. This is seen in the case of pyrene, which had a binding constant of 800 M^{-1} in comparison to the 2300 M^{-1} of 1-hydroxypyrene as determined by UV-vis spectroscopy using a dilution technique. The cyclophanes' cavity was also found to contract by 14% from 8.19 A to 7.08 A between the unbound and bound in the case of the NDI cyclophane, which was determined by X-ray crystallography.

This work was then extended in 2003 with the synthesis of "tweezer" complexes (Figure 1.11), which were based on the previous cyclophanes' interactions with 2-pyrenes linked by an isophthalamide, exhibiting high complementarity with very high K_a values (24 000 M^{-1}) with the NDI cyclophane while only 9200 M^{-1} for the BDI (determined by UV-vis spectroscopy using a dilution technique).[26] Such high association is due to both D–A interactions and hydrogen bonding and the complementarity of the substrate, which was further supported by the crystal structure of the complexes.

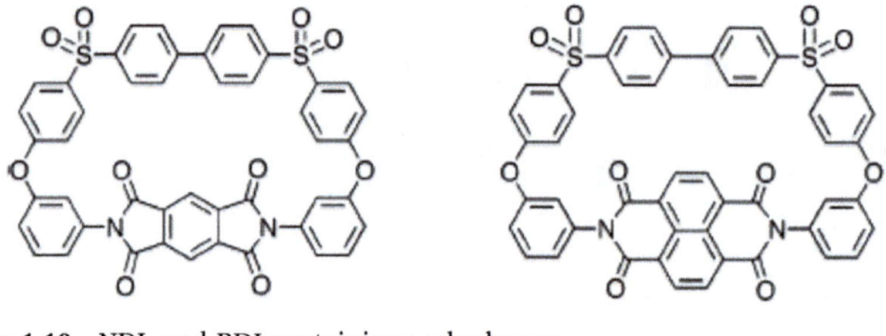

Figure 1.10 NDI- and BDI-containing cyclophanes.

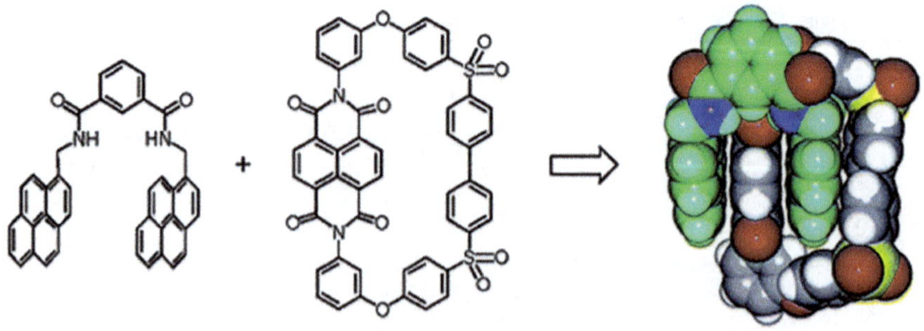

Figure 1.11 Tweezer structure by Colquhoun *et al.*[26]

The two cyclophanes spontaneously complex in their cavity DNs as well as other polycyclic aromatics.[27] The complexation has been studied using spectroscopic techniques in solution and a large number of crystal structures of these complexes have been resolved. Another molecular recognition system, proposed by Shinmyozu and co-workers, used a cyclophane composed of BDI and DN units, which was able to recognise polymethoxybenzenes by charge transfer as well as hydrogen bonding and van der Waals interactions.[28]

The donor–acceptor interactions can be utilised in dynamic combinatorial chemistry (DCC). DCC is a powerful tool for the study of complex molecular networks and systems generally under thermodynamic control.[29] In DCC, simple molecular building blocks are bound together by noncovalent or reversible covalent bonds, generating a complex mixture of interconverting products. The thermodynamically controlled composition of the mixture at equilibrium is known as a dynamic combinatorial library (DCL). DCL responds to various stimuli such as temperature, pH, electric field or chemicals, which can drive the constituents to reorganise in order to minimise the total free energy of the system.

Sanders *et al.* demonstrated that DCC could pave the way for unprecedented stacking of D and A units. The building blocks for these studies comprised a central flat aromatic surface with two-cysteine decorated hydrophilic side arms (Figure 1.12). The aromatic surface was made up of electron-deficient NDI as "A" and electron-rich DN as "D". Here, the aromatic cores engage in D–A interactions, which are augmented by the hydrophobic effect, while the cysteine provides thiol functionalities for reversible disulfide exchange and carboxylic acid groups for water solubility. These DCC building blocks led to different catenanes upon atmospheric oxidation in solution under specific conditions.[30]

The dynamic combinatorial approach has been used to synthesise donor–acceptor [2]-catenanes in an aqueous medium, by employing dialkoxynaphthalenes (DNs) and NDIs as the electron-donor and acceptor functionalities respectively. The efficiency of the assembly process was enhanced by manipulating the library equilibrium either by addition of a templating guest, changing the stoichiometry of the building blocks, or by increasing the overall library concentration and the ionic strength of the solvent by addition of an inert salt.[31] The results of the experiments were unusual and unexpected, as the catenanes possessed either DAAD or DADD

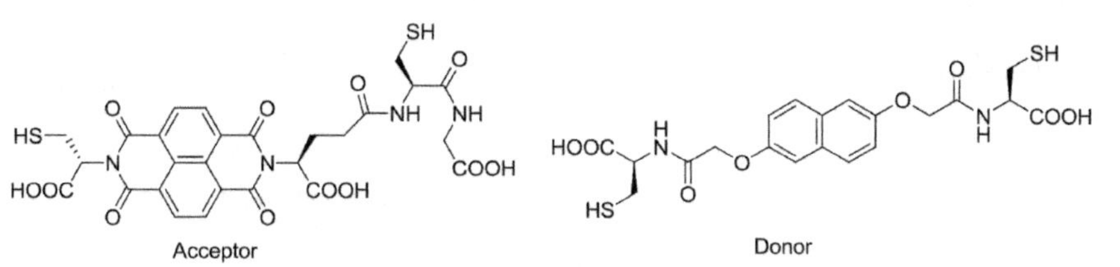

Figure 1.12 Typical NDI and DN building blocks used in DCC.

stacking sequences in the π units of their structures instead of the expected DADA stack (Figure 1.13).[32]

The formation of catenanes and their constitution was found to be dependent on subtle differences in the geometry, dimension, and flexibility of donor building blocks. The "short"-DN building blocks used in these studies led to a decisional flowchart (Figure 1.14) that was used to describe the DCL behavior of these π-rich isomers. The catenane formation was predicted by the presence in the library of both the heterodimer (for the DAAD structures) and donor homodimer (for DADD). The presence of an acceptor unit, which formed a tight homodimer, did not allow threading through the cavity required for the formation of [2]-catenanes.

Further investigation into the mechanism of the [2]-catenane formation pathways under similar library conditions revealed a stepwise assembly process, with donor–acceptor interactions driving and directing their assembly alongside hydrophobic effects (Figure 1.15). In pure water, DCLs composed of donors and acceptors did not lead to the efficient formation of catenanes (yields of under 5% in the best cases, while in the majority of cases, no catenanes were observed). In high salt concentration libraries, the

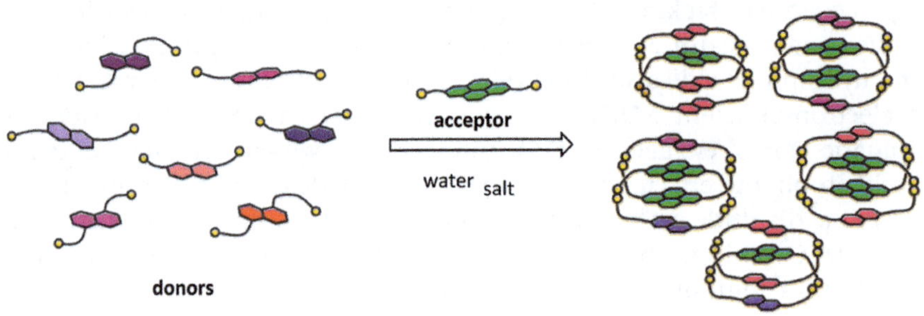

Figure 1.13 A schematic representation of a DCL containing various DNs and NDI building blocks.[32]

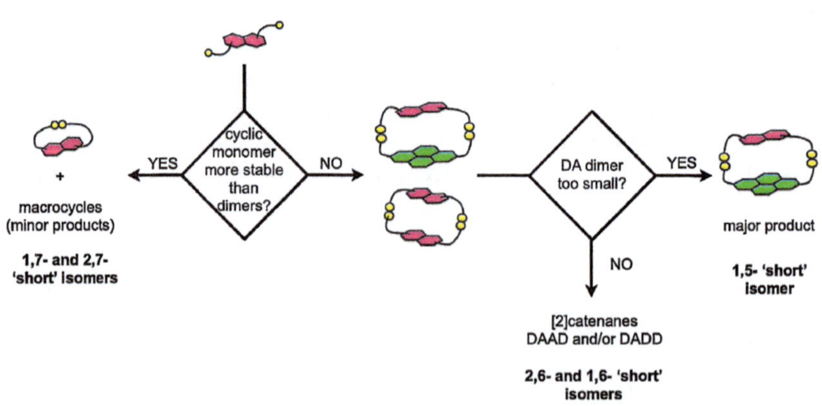

Figure 1.14 Decisional flowchart used to describe the DCL behavior of π-rich DN isomers in the presence of NDI.[32]

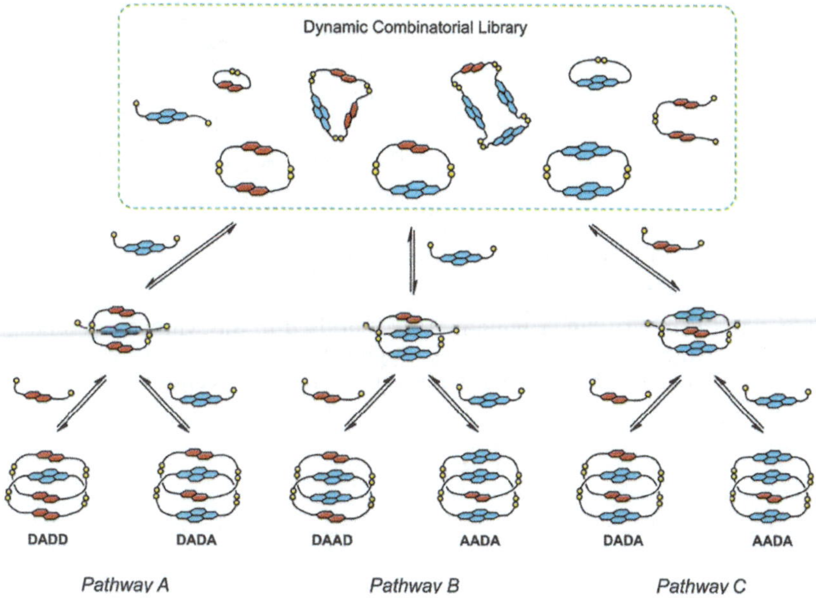

Figure 1.15 Formation pathways for [2]-catenanes assembled in DCLs under similar library conditions.[33]

increased polarity of the medium promoted the decrease of solvent-exposed hydrophobic surfaces and led to a threaded species, which was the first intermediate for catenane formation. The threaded complex formed through pathway I, II, or III, upon linking with a fourth building block (either donor or acceptor) completed the catenane formation. Thus, each pathway led to a characteristic pair of catenanes: DADD and DADA for pathway I, a DAAD and AADA catenane from pathway II, and a DADA and AADA catenane from pathway III.[33]

Sanders and Pantoş *et al.* have synthesised water soluble [3]-catenanes based on NDI cores linked by an alkyl chain of various lengths (2–9 carbons) and terminated with cysteine residues (Figure 1.16). Dynamic combinatorial libraries (DCLs) of these compounds in the presence 2,6-DNs led to the discovery of donor–acceptor [3]-catenanes.[34]

The driving forces behind the selection of a catenane forming pathway were, as in the case of the [2]-catenanes, the donor–acceptor interactions and the hydrophobic effect. Manipulation of the linker length, chirality, kinetic and thermodynamic parameters was found to allow selective synthesis of a range of [2]- and [3]-catenanes. For example, when the linker length of the NDI was under six carbons and contained an even number of atoms, [3]-catenanes were shown to form, while linkers with odd numbers of atoms and under six carbons were found to form macrocycles and unconventional catenanes, which highlighted the remarkably high specificity of the "odd–even" effect in DCLs (Figure 1.17). While the [3]-catenane was the thermodynamic product of odd numbered linkers under six carbons, some libraries

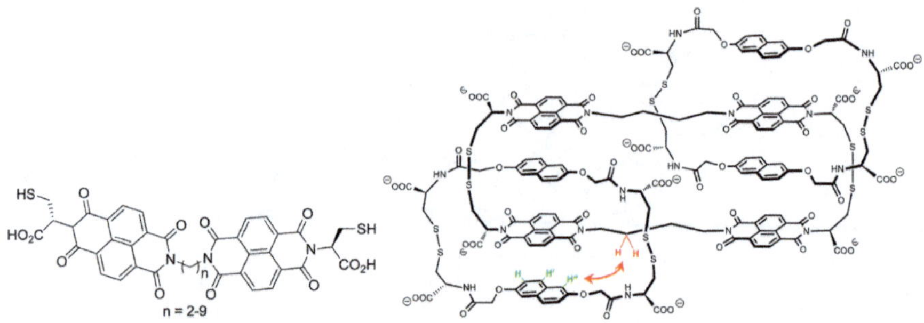

Figure 1.16 Left: NDI dimer building blocks utilised in the synthesis of molecules with complex topologies. Right: [3]-catenane indicating the nOe contacts that allowed the determination of its conformation.

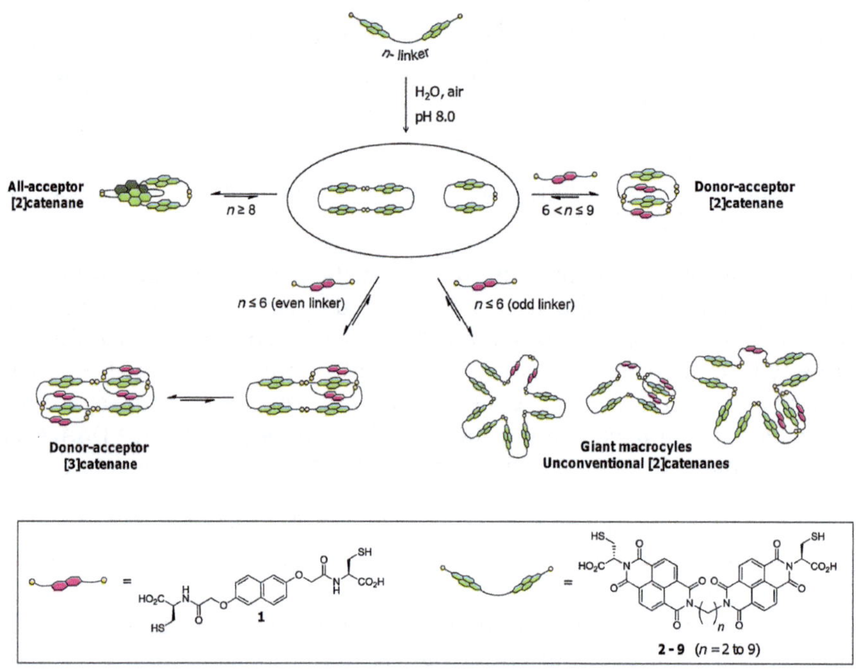

Figure 1.17 Schematic representation of the outcome of donor–acceptor DCLs containing NDI dimers and DNs.[35]

were fully oxidised before they reached thermodynamic equilibrium and only kinetic products were observed. Addition of fresh thiol was found to help this, as in the case of the five-carbon linker, where it was possible to form the [3]-catenane as one of the major species in the DCL. While the [3]-catenane's initial yield was a trace amount, it could be increased by addition of 1 M $NaNO_3$, which is unsurprising as it increases the polarity of the DCL and thus the contribution of the hydrophobic effect to the self-assembly.[35]

Spermine acts as a template in the synthesis of the [3]-catenanes in the four-carbon linker DCL by interacting with the negative charges of the cysteine carboxylate groups to increase the yield 40% from <5% in the un-templated library. Above six carbons, the "odd–even" effect did not manifest itself and these di-NDI building blocks containing longer alkyl chains were found to lead to the formation of [2]-catenanes.

Further, Sanders *et al.* have studied the DCL behaviors of DA building blocks that contain a central core of a flat hydrophobic surface of NDI or DN and flexible hydrophilic thiol side-chains.[36] The size and flexibility of building blocks have a crucial role in the type of products, DCL distributions and library response to templates. It has been demonstrated that in addition to the use of electronically complementary guests, controlling hydrophobic effects through solvent ionic strength is an effective approach for perturbing the relative stability and manipulating the product distribution of the DA library members in these DCLs. The building blocks can effectively be incorporated into new receptors by templates in DCLs, generally combine rigid and flexible subunits, and incorporate recognition moieties within their structure. It was concluded that large, flexible receptors are more responsive to guests and therefore bind more strongly.

The formation of an all-acceptor [2]-catenane is driven by the intercalation of hydrophobic alkyl chains between the electron deficient units. The libraries prepared from short-linker building blocks ($n < 7$) were dominated by a cyclic dimer. With an increase in the chain length to $n = 8$–9, new interlocked species corresponding to an all-acceptor [2]-catenane were formed.

In a recent[37] development, Sanders and Pantoş *et al.* have studied in detail the DCL containing a NDI dimer connected by a hexyl chain (Figure 1.18). The library contained four tetramers that have been identified as a trefoil knot, figure eight knot, Solomon link and the trivial macrocycle. This was a particularly remarkable result as it represented the first synthesis of a figure eight knot.

The structures of the figure eight knot and the Solomon link are quite similar as both contain two stacks of four NDI cores; the difference between these topological isomers is the disulfide connectivity (Figure 1.19).

A remarkable development of the all-acceptor NDI-based DCLs was the synthesis of a trefoil knot. A building block composed of three hydrophobic electron-deficient NDI units (Figure 1.20) connected by hydrophilic amino acids (L-β-amino-alanine) with limited flexibility lead to formation of a trimeric macrocycle, the trefoil knot, at exceptionally short HPLC retention time. The formation of this knot was heavily influenced by the polarity of the medium as its formation is driven exclusively by the hydrophobic effect. The reversible nature of disulfide bonds allowed the equilibrium to be shifted towards its formation from a library containing the cyclic monomer and dimer (Figure 1.20). Further, choice of amino acid chirality was crucial as all-L and all-D building blocks form the mirror image trefoil knots while a D/L building block prevented the knot formation, thus inferring that the point chirality of the building block influences the topological chirality of the trefoil knot.[38]

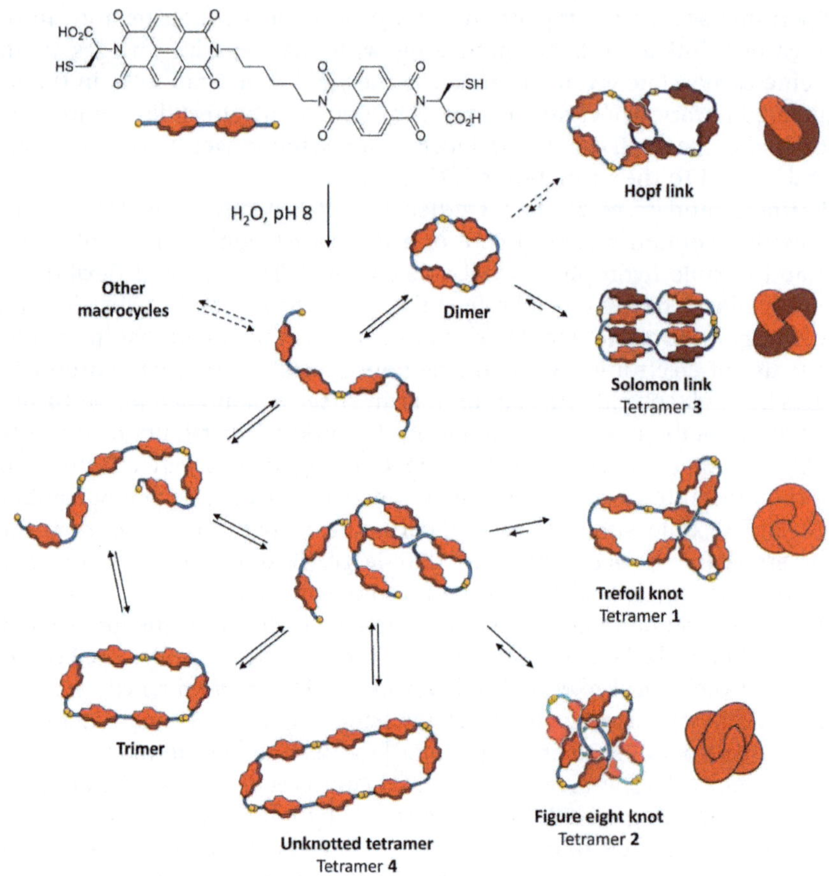

Figure 1.18 Topological isomers that can form in a DCL from a hexyl-linked NDI dimer.[37]

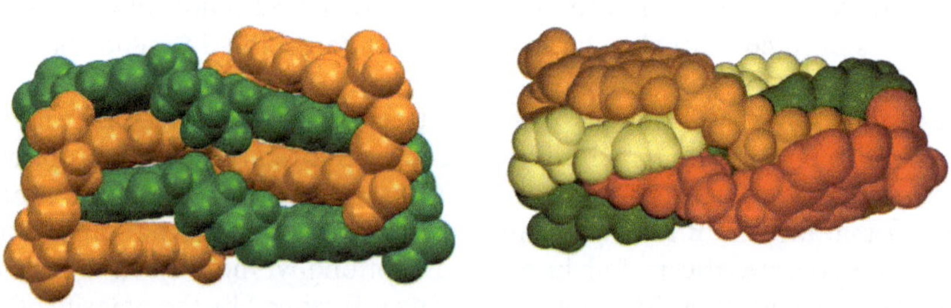

Figure 1.19 Molecular modeling (PM7) structures of the Solomon link (left) and figure eight knot (right).

Li *et al.* have synthesised a perylenediimide (PDI) macrocyclic dimer and a concatenated dimer–dimer ring (a [2]-catenane) from dynamic self-assembly of monomeric bis-*N*,*N*′-(2-(2-(2-(2-thioacetylethoxy)ethoxy)ethoxy)-ethyl)perylenediimide. The monocyclic ring closure and the concatenation

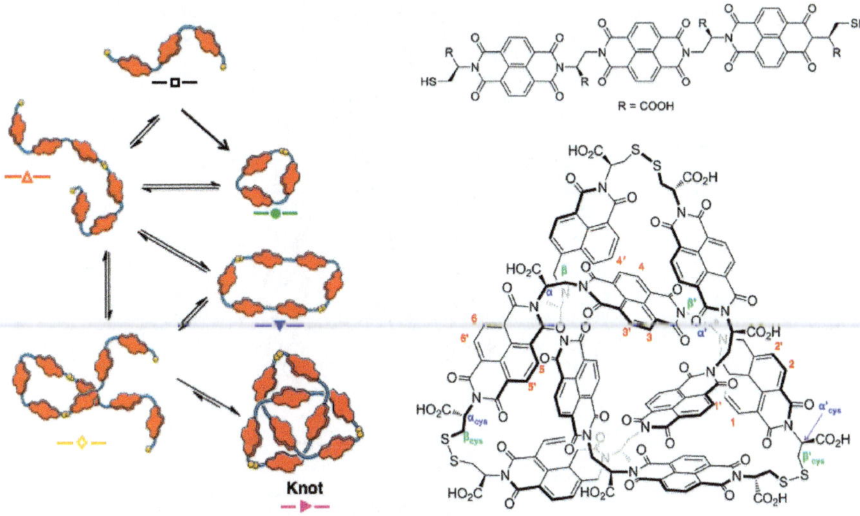

Figure 1.20 Left: the equilibria and species present in the DCL leading to the quasi quantitative assembly of the trefoil knot; right: the structure of the building block and the resulting trefoil knot.[38]

were accomplished through formation of disulfide bonds, which was readily triggered by air oxidisation under basic deacetylation conditions. The perylene cyclic dimer and the [2]-catenane were characterised using both structural methods (NMR, mass spectroscopy) and photophysical measurements (UV-vis spectroscopy). Kinetic analyses offered informative insights into reaction pathways and possible mechanisms, which lead to the formation of complex catenanes (Figure 1.21).[39]

Further, Li *et al.* have used unique model oligomers containing exactly two or four PDI units:[40] linear foldamers **LIN2** and **LIN4**, monocyclic complement **CYC2**, and concatenated foldable rings ([2]-catenane) **CAT4** to study this system further (Figure 1.22). Linear, cyclic, and concatenated foldamers revealed that photoabsorption and excitation induced unfolding and refolding, generating colourful spectral switching from one wavelength to another. Foldamer architectures dictated the unfolding and refolding processes, and hence their spectral dynamics. The linear tetramer exhibited active frame-to-frame spectral switching accompanying dramatic changes in colours, but a concatenated tetramer displayed a multicoloured composite spectrum with little or no spectral switching. Excited state dynamics caused spectral switching: an electronically decoupled PDI monomer emits green fluorescence while electronically coupled PDI π-stacks emit red fluorescence, with longer π-stacks emitting redder fluorescence. Single molecule fluorescence studies on chromophoric foldamers revealed that the maximum domain length was delocalised across just four π-stacked PDI dyes and no new *pure* colour can be found for oligomers beyond the tetramer. The range of fluorescent colours in π-stacks was a function of the number of chromophores only up to the tetramer.

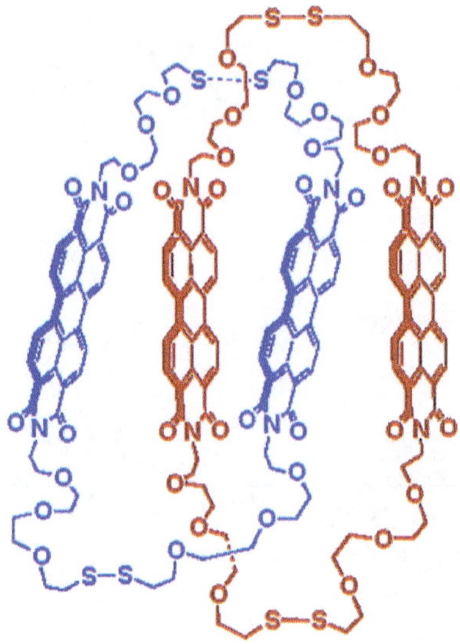

Figure 1.21 Structure of the all-acceptor PDI [2]-catenane.[39]

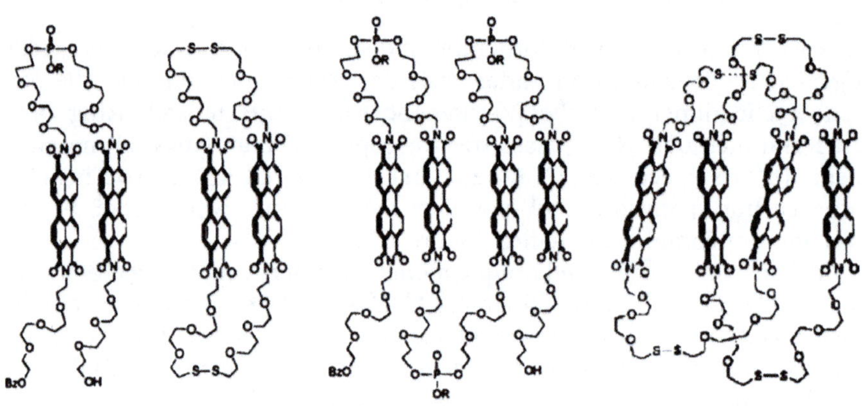

Figure 1.22 Linear foldamers **LIN2** and **LIN4**, monocyclic **CYC2**, and [2]-catenane **CAT4.**[40]

Bay substitutions twist the perylene unit dihedrally out of the plane, with the dihedral angle tunable from 0 to 37°. Li examined dynamic self-assembly between bay-substituted monomers to broadly establish the role of the dihedral twist angle in molecular encoding. Remarkably, monomers with different twist angles preferentially self-assemble into segregated nanostructures even in the presence of other twisted units, thus revealing their unique molecular encryption. The results revealed that when two species co-self-assemble then molecular codes become redundant and the same or very similar reaction pathways are followed. The imprinted twist-angle

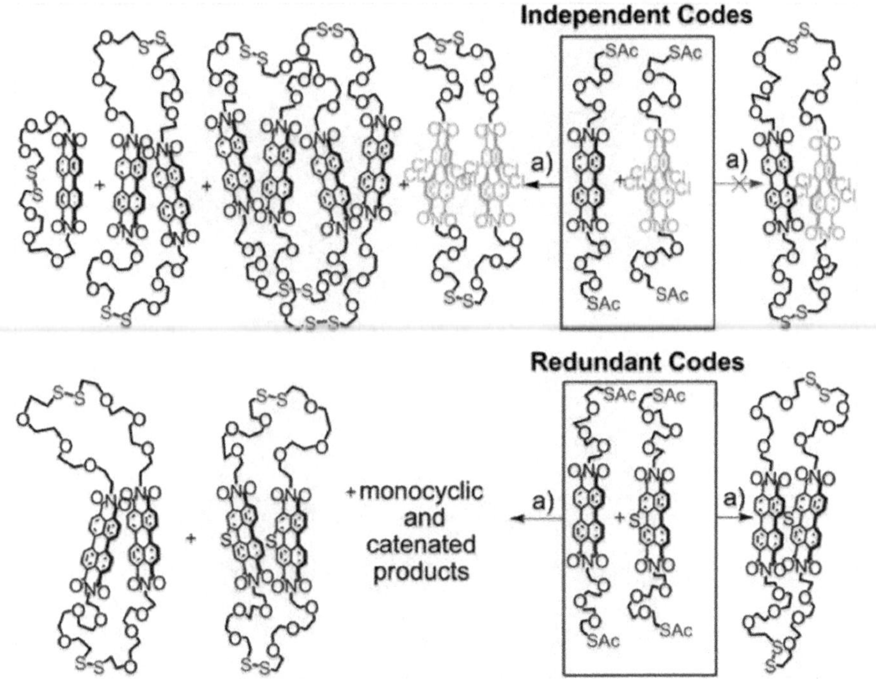

Figure 1.23 Reaction between independently coded monomers and redundant coded monomers leading to catenated and macrocyclic structures.[41]

(structural architecture) determines the thermodynamic parameters and directs the solvophobic interactions, thereby establishing a useful molecular code (Figure 1.23).[41]

1.3 Clathrates and Rigid Molecules

Clathrate compounds have been used for studying molecular channels, organic networks, molecular recognition and stereoselective reactions.[42] Electron-deficient NDI molecules have been used as acceptor moieties in molecular recognition in inclusion complexes. Yamaguchi and co-workers have shown that the most important factor governing guest-inclusion ability between stable rotamers, *anti-* and *syn-N,N'*-bis-(2-*tert*-butylphenyl) NDI was the conformation of the NDI host. To generate a stable host–guest complex, flexible changes to the host's geometry are required.[43]

The conformers used to study this, *anti*NDI and *syn*NDI, were chosen for the electron deficient naphthalene core, a π-system that allows for strong aromatic interactions with electron-rich aromatic compounds (Figure 1.24). The high energy rotational barrier of the N–C single bonds linking the bulky *N*-(2-substituted phenyl) groups to the NDI core allows the isolation of each rotamer, with the conformation of the groups dictating the available space above the π-system.

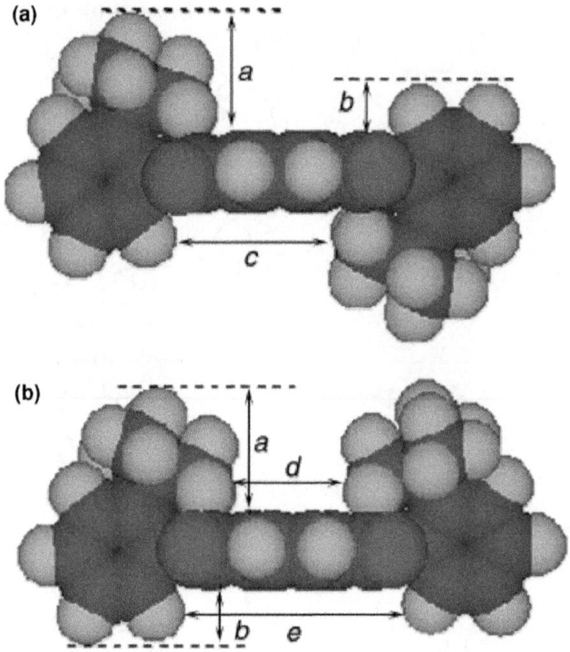

Figure 1.24 Space-filled models of *anti*NDI (a) and *syn*NDI (b).[43]

The *anti*NDI isomer has space on both sides of the naphthalene core suitable for the inclusion of an aromatic ring. The *syn*NDI isomer, due to the bulky substituents being situated on the same side, only has the ability for guest inclusion on the face opposite to the functional groups (Figure 1.24).

The guest inclusion was studied by crystallisation of the *anti*NDI isomer with a variety of aromatic molecules from chloroform solutions. Using ^{1}H NMR and elemental analysis, Yamaguchi *et al.* were able to show that stable host–guest complexes were able to form in host–guest ratios of 1 : 1, 1 : 2 and 1 : 4. When the same procedure was applied to the *syn*NDI isomer, no guest inclusion was seen.

The intermolecular interaction between the *anti*NDI isomer and the range of guests was further investigated using X-ray diffraction of the single crystals of the host–guest complexes. The main interactions observed in the complexation of various guest molecules by the *anti*NDI isomer were NH···O=C or OH···O=C hydrogen bonds and/or charge transfer complexation between the electron-rich and electron-deficient molecules. The host–guest combination is, overall, more rigid and stable under the conditions studied than that of the *anti*NDI isomer alone. In the case of the *syn*NDI isomer, due to the faces of the molecule being unavailable for guest inclusion, a rigid pinwheel array of the *syn*NDI isomer is generated, despite the presence of other aromatic species. With this work, Yamaguchi and co-workers have shown that guest inclusion into a host can be changed by rotational isomerisation with H-bonding and charge transfer driving the formation of clathrates.[43]

Gawronski and co-workers have studied the formation and aggregation of helical and non-helical structures in solution.[44] They utilised the donor–acceptor properties of BDI and NDI substituted with *trans*-1,2-diaminocyclohexane moieties to form dyads and triads. In solution, these molecules exhibit conformational isomerization, which generates two stable foldamers: the triads form foldamers, *C* and *S*, of opposite helicity and the dyads form foldamers *M* and *P* (Figure 1.25).

PM3 and DFT b3lyp/6-31g(d) calculations were used to identify the lowest energy conformations of the dyads and triads. Surprisingly, both *C* and *S* conformers showed a similar lowest energy state with no preferential bias for either conformer.

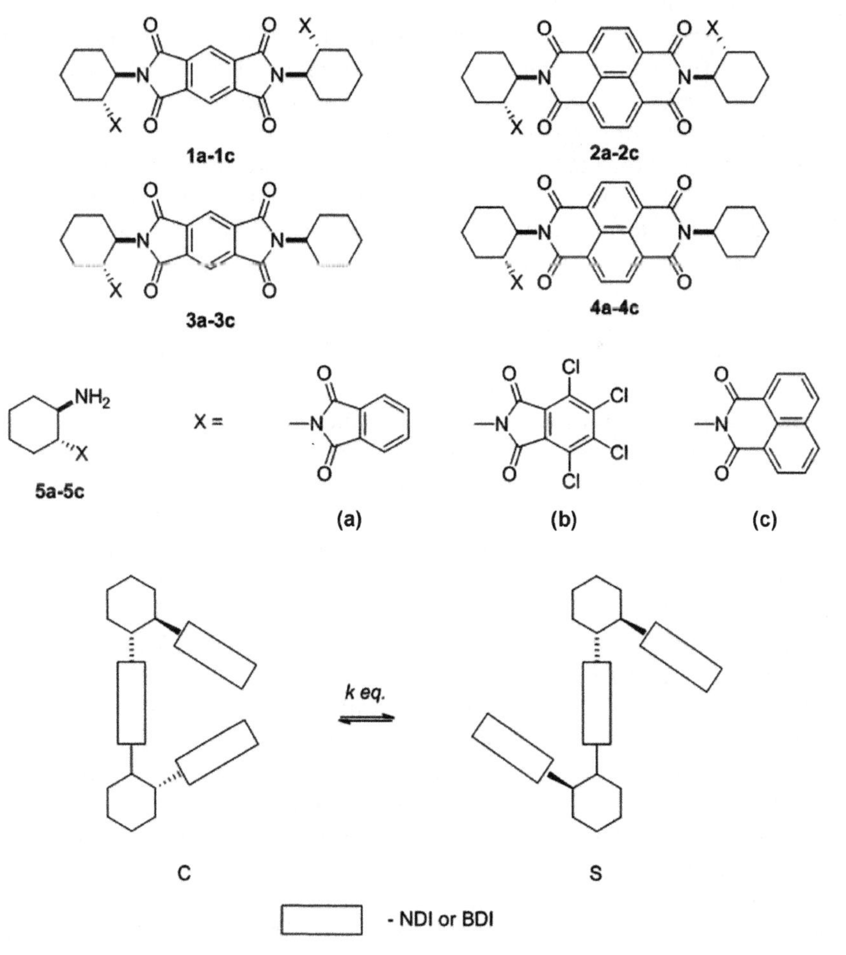

Figure 1.25 The two stable foldamers of imide triads with the (*R*,*R*)-1,2-diaminocyclohexane linkers.

Experimentally, the team were able to show, using ^{1}H NMR studies, that for all NDI-based triads (**1a**, **1b**, **1c**, **2a**, **2c**) except for **2b**, conformational preference was towards the *C* isomer in $CDCl_3$, and that by lowering the temperature from 298 K to 223 K, they were able to drive the conformer ratio towards 100% *C*. In the case of triad **2b**, however, a small preference towards the *S* conformer was exhibited. This was explained by the bulkiness of the naphthalene imide functional group destabilising the *C* conformer due to steric repulsions.

The ability of triads **1a** and **2a** to act as hosts was explored by the crystallisation of each triad with planar aromatic molecules. By the addition of an aromatic guest, π-stacking is observed between the host and guest, which forces the host to adopt an *S* conformation in a host to guest ratio of 1 : 2. This shows that upon inclusion of aromatic molecules, the *anti*-(*S*) conformer is stabilised due to preferential stacking on either side of the naphthalene core.[45]

Stoddart and Wasielewski *et al.* employed 1,2-diaminocyclohexane in a reaction with naphthalenedianhydride to yield three cyclic oligomers: a dimer,[46] a trimer[47] and a tetramer.[48] These molecules (Figure 1.26) not only have interesting geometries—an ellipsoid, a triangle and a square, respectively—but they also possess an interesting electronic communication between the NDI cores. All these species are chiral, the chirality being dictated by the starting diamine. A mixed NDI–BDI triangular system has recently been proposed for applications in energy storage.[49]

1.4 Supramolecular Nanotubes and Receptors

Supramolecular aggregation between host–guest molecules has also been used in the pursuit of molecular switches that are able to turn ON and OFF in response to complexation. Such interactions are key components in data storage, molecular electronics and chemical signalling. Predominantly, the ON state of the supramolecular switch requires an intact host–guest complex; this is usually difficult to achieve due to complex fragility. To maintain the ON state, a constant guest presence is essential and removal of the guest molecule returns the switch to an OFF state.

Shimizu and co-workers have used an atropoisomeric NDI diacid **1** that can act as a stable supramolecular switch (Figure 1.27). They have shown that once turned ON, the diacid **1** can maintain the ON (or OFF) state in the absence of the guest, retaining the memory of the host–guest complex. The diacid exhibits this feature due to the restricted rotation between two conformers, *syn*- and *anti*-. At room temperature, both the *syn*- and *anti*-conformers of the diacid **1** are isolable due to the steric hindrance around the C–N bond restricting rotation.[50]

The formation of the *syn*-diacid is favoured by the introduction of a hydrogen bonding guest, in particular adenine. The team probed the ability of this molecule to act as a molecular switch by heating the NDI diacid with adenine in 1,1,2,2-tetrachloroethane (TCE) at 70 °C for 12 hours followed by

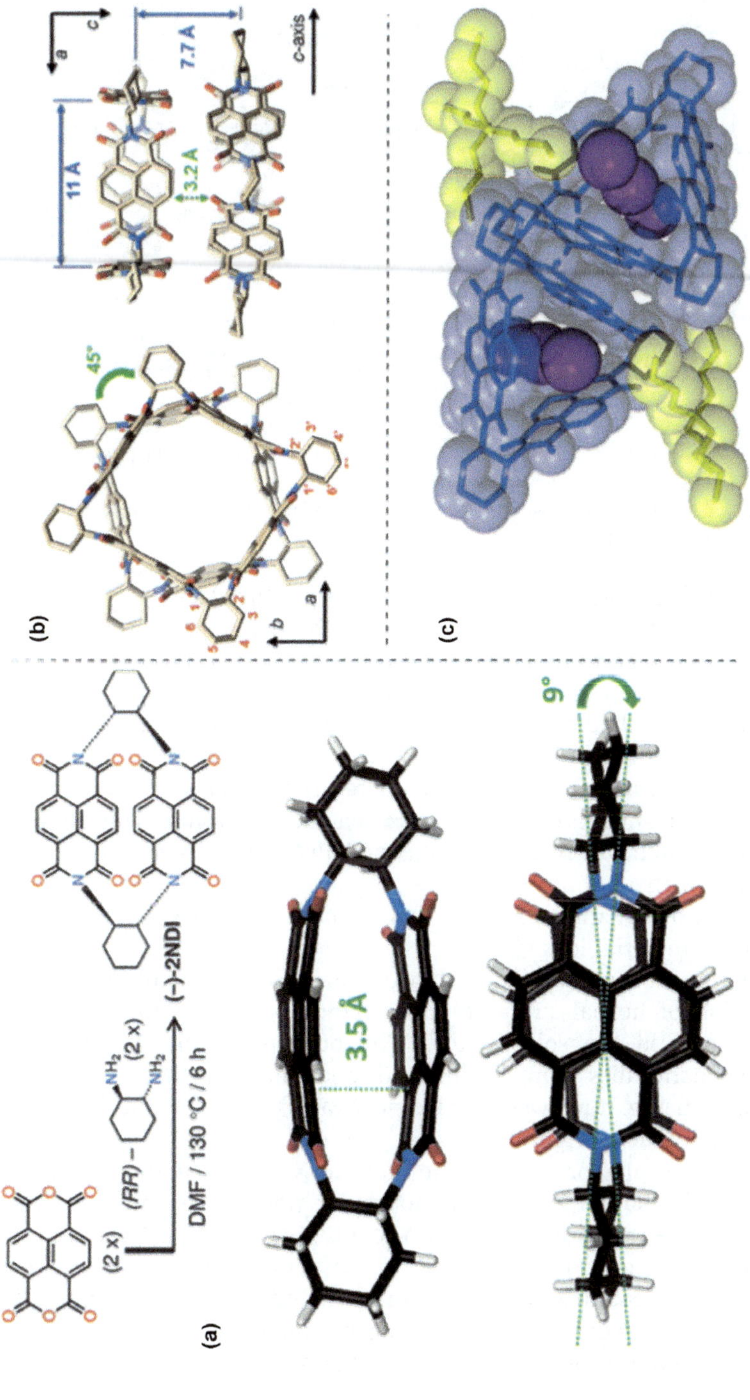

Figure 1.26 (a) Synthesis and X-ray structure of NDI dimer,[46] (b) X-ray structure of molecular square[48] and (c) triangle as the TBA I_3^- complex.[47]

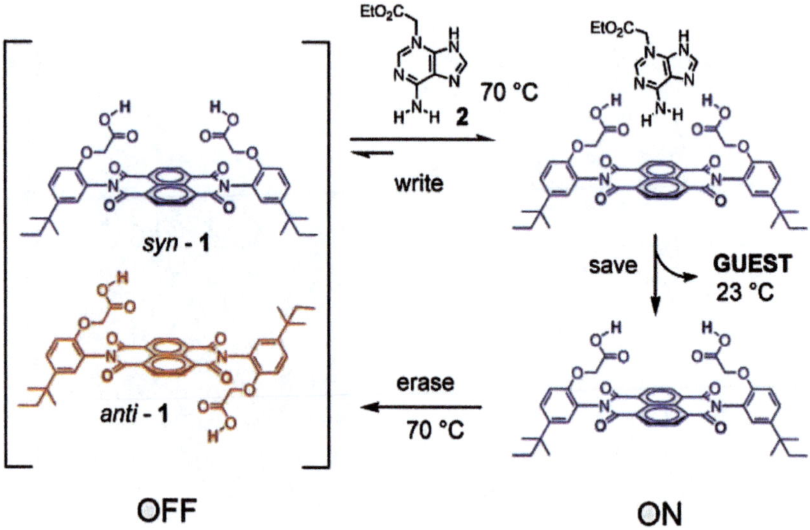

Figure 1.27 A molecular host that retains the conformational memory for binding the guest.[50]

removal of adenine with a 1 M HCl wash. This produced a 93 : 7 *syn–anti* ratio of the NDI diacid, which was identified as the ON state for the molecular switch. Once cooled to room temperature, regardless of the presence of the guest molecule, the *syn–anti* ratio remained unchanged. To return to the OFF state, the *syn*-enriched sample is heated under the same initial conditions to give a conformer ratio of 52 : 48 *syn–anti.*

This conformer mixture at room temperature is remarkably stable despite its solvent environment or the presence of a guest molecule. This allows the state of the switches to be identified using HPLC and ^{1}H NMR. The switch is also able to withstand multiple cycles of guest addition and removal, alternating between ON and OFF states, whilst maintaining switching efficiency and without material loss.[50]

Sanders and co-workers have shown the use of NDI molecules for the construction of helical supramolecular nanotubes in aprotic low polarity solutions and in the solid state.[51] The building blocks for the supramolecular nanotubes were amino-acid-functionalised NDIs, which were synthesised using microwave dielectric heating of NDA and the corresponding amino acid (Figure 1.28).

Proof of nanotube formation came from the X-ray diffraction of L-**1**. In the solid state the supramolecular nanotubular polymer is formed through hydrogen bonding between the amino acid COOH moieties. The amino acid side chains orientate themselves in a *syn* conformation that allows for dimerisation to lead to a helical structure. The structure of this nanotube is reinforced by secondary C–H· · ·O hydrogen bonds between the NDI cores i and $i+3$, which are coplanar in this arrangement.

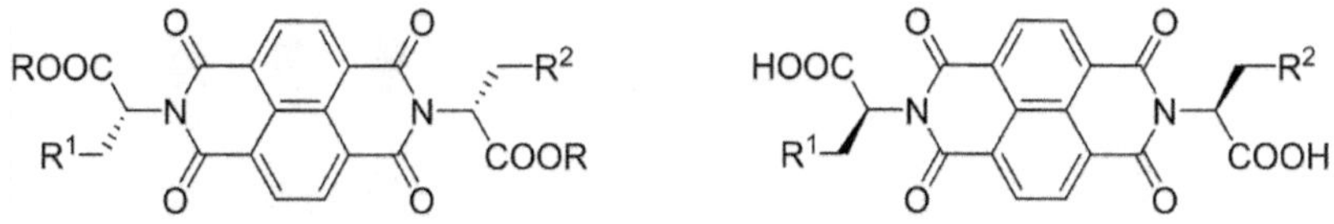

L-**1**: $R^1 = R^2$ = STrt, R = H
L-**2**: $R^1 = R^2$ = OBzl, R = H
L-**3**: $R^1 = R^2 = (CH_2)_3NHBoc$, R = H
4: $R^1 = R^2$ = STrt, R = Me
5: R^1 = STrt, R^2 = OBzl, R = H
6: $R^1 = CH_2CO_2tBu$, $R^2 = CH_2CONHEt$, R = H
7: $R^1 = (CH_2)_3NHAc$, $R^2 = (CH_2)_3NHBoc$, R = H

D-**1**: $R^1 = R^2$ = STrt
D-**2**: $R^1 = R^2$ = OBzl
D-**3**: $R^1 = R^2 = (CH_2)_3NHBoc$

Figure 1.28 Amino acid NDI derivatives used in supramolecular nanotubes. Boc = *tert*-butyloxycarbonyl, Bzl = benzyl, Trt = trityl.[51]

A key feature of these nanotubes is that helical chirality is induced into the supramolecular assembly through the two stereocentres provided by the amino acid side chains. In the case of compound L-**1**, an *M*-helical nanotube is formed due to the *R,R* stereocentres of the amino acid side chains. The opposite enantiomer, the *P*-helical nanotube, was achieved when the stereocentres from the amino acids were *S,S* as seen with compound D-**1**.

Circular dichroism (CD) was utilised to observe the helical assemblies in aprotic solvents such as chloroform. The molar ellipticities of L-**1** and D-**1** in anhydrous $CHCl_3$ were -9.5×10^4 and 9.5×10^4 deg cm^2 dmol^{-1}, respectively, indicating that the different NDI enantiomers led to nanotubes of opposite helical chirality. MeOH destroyed the CD signature of the nanotube (the spectrum becomes identical to a monomeric NDI building block), which confirmed the hydrogen bonded nature of the nanotube.[51] The NDI nanotubes incorporate achiral NDIs provided that they satisfy a geometrical requirement. These mixed chiral–achiral NDI nanotubes have been shown to follow a sergeants-and-soldiers behavior.[52,53]

Sanders and co-workers went on to show that these nanotubes can be receptors of C_{60} by forming a one-dimensional array in which the C_{60} molecules are tightly packed in the tubular cavity as indicated by a series of UV-vis experiments (Figure 1.29).[54] This complexation led to an increased concentration of C_{60} in chloroform up to 16× over the saturation concentration of C_{60} in this solvent. Remarkably, the chirality of the host was transferred to the achiral guest (C_{60}) as indicated by the presence of a circular dichroic signal in the visible region (600–700 nm).

The same group followed on with studies[55] on the complexation of C_{70} with the expectation of a similar interaction due to C_{60} and C_{70} sharing the same small diameter. Combination of a solution of **1** in $CHCl_3$ with a saturated solution of C_{70} in $CHCl_3$ showed again the distinctive colour

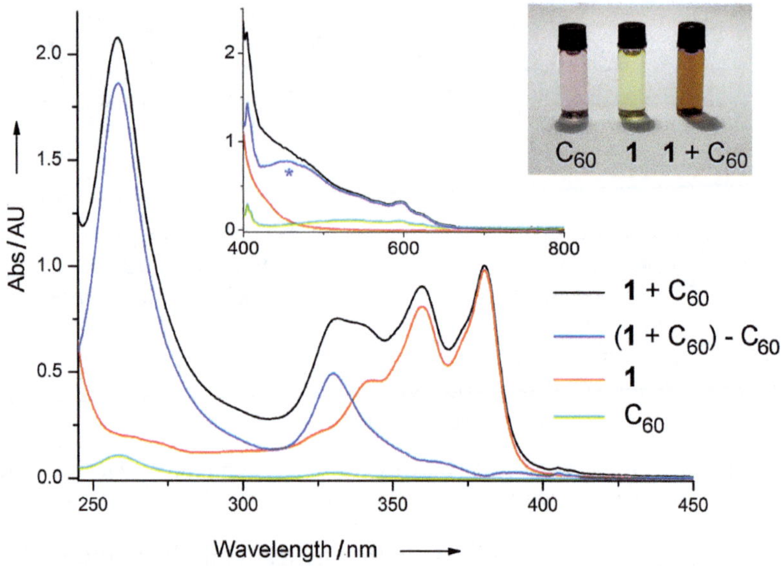

Figure 1.29 UV-vis spectra of **1**, C_{60}, and **1** $+ C_{60}$, and the C_{60} contribution to the **1** $+ C_{60}$ solution (blue trace) in $CHCl_3$. The inset contains the visible region of the spectrum and a photographic comparison of the solutions. The concentration of 1 for the measurement of the visible spectrum was $40\times$ higher than for the UV region.[54]

change of complexation. Unlike its C_{60} counterpart, the complexation of **1** and C_{70} did not show a UV-vis band indicative of C_{70} close contact, implying the formation of discrete NDI $\cdot$ C_{70} complexes. This was confirmed through the comparison of CD spectra of the nanotube and the NDI $\cdot$ C_{70} complex (Figure 1.30). The marked change implied that a different supramolecular architecture is formed, which was subsequently identified using a combination of ^{1}H, ^{13}C and NOESY NMR of the host guest specie and symmetry arguments. The authors proposed a structure with D_{3h} symmetry in which there are six NDI molecules around a C_{70}. In this capsule, there are three NDIs at either pole held together by an unusual hydrogen bonding motif while at the equator, the COOH dimerisation was observed (Figure 1.30(b)).[55]

In a subsequent study, they showed that the supramolecular nanotube–nanotube $\cdot$ C_{60} complex–C_{70} capsule is fully dynamic and interconvertible when using acid–base equilibria to modulate the H-bonding ability of the amino acid derived NDIs. By using sequentially Et_3N and *p*-toluenesulfonic acid, the system can cycle between the nanotube $\cdot$ C_{60} complex and the C_{70} capsule. The whole process was monitored by CD and ^{1}H NMR.[5]

The Sanders group have also shown the uptake ability of these nanotubes by studying the binding affinities for fullerenes and ion pairs.[56] Colorimetric studies were carried out on a fullerene-saturated solution of **1** by the

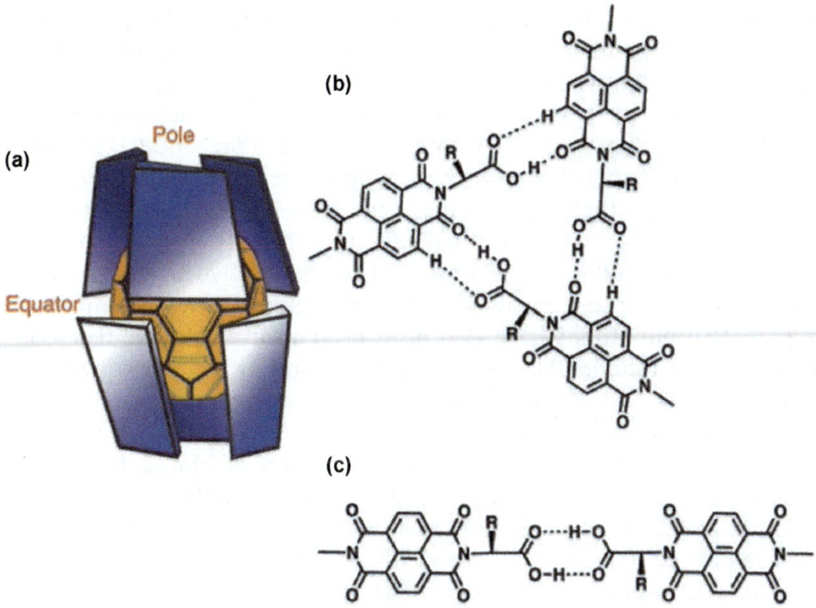

Figure 1.30 (a) Cartoon representation of the C_{70} receptor. (b), (c) Proposed interaction motifs between NDI molecules at the poles (b) and the equator (c) of the C_{70} receptor.[55]

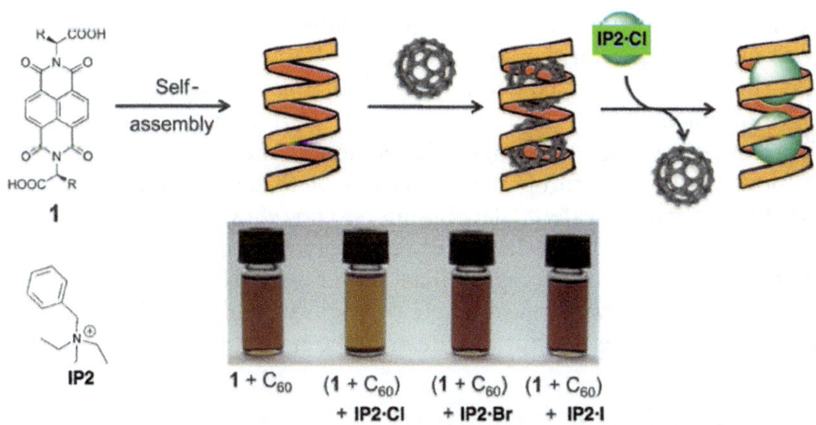

Figure 1.31 Schematic representation of the self-assembled nanotube and the competition experiment with C_{60} and ion pairs. Bottom panel: photographical comparison of a solution of **1** + C_{60} in the presence of different ion pairs.[56]

introduction of ion pairs **IP2 · Cl**, **IP2 · Br**, and **IP2 · I** (Figure 1.31). The addition of one equivalent of these ion pairs to a 10 mM solution of **1** + C_{60} showed that, in the case of **IP2 · Cl**, C_{60} displacement occurred with a cloudy precipitate being formed, whilst **IP2 · Br** and **IP2 · I** showed no visible interaction. This was confirmed through UV-vis spectra where a decrease in intensity of the bands at 593 nm and 600 nm for **1** + C_{60} were seen upon addition of **IP2 · Cl**, whilst they remained largely unchanged when **IP2 · Br** and **IP2 · I** were added. These studies show that C_{60} can be readily displaced from the cavity when a suitable competing guest is present. CD studies also show that the nanotube architecture again remains intact despite the introduction of another guest. By performing such competition experiments, they were able to show that NDI helical nanotubes act as selective receptors for ion pairs. The selectivity is governed by the relative size of the ion pair and is a direct measure of the association strength between nanotube and ion pair. NDI nanotubes have also been used as size selective receptors for aromatic and aliphatic ion pairs and pyrenes and by comparison with other hydrogen bonded receptors, these NDI nanotubes are remarkably easier to synthesise as they self assemble.[57]

The thermodynamics governing the formation of chiral, helical supramolecular structures generally follow a cooperative pathway as the number of interactions between each monomer increases as the first turn of a helix is completed.[58] By treating the hydrogen-bonded equilibria of the NDI nanotubes as a dynamic combinatorial system, Sanders and co-workers were able to show that the formation of NDI-based supramolecular nanotubes follows an isodesmic polymerisation. Utilising variable temperature and concentration-dependent ^{1}H NMR, they were able to elucidate thermodynamic functions governing the self assembling process. During these studies, it was also determined that the C_{60} guest stabilised the supramolecular nanotube thus allowing the quantitative measurement of an associating constant between the nanotube and fullerene.[59]

Recently, Pantoş and co-workers developed covalently-linked oligomeric NDI molecules, which also self-assemble into helical nanotubes. They utilised the methods developed by Sanders and Pantoş to probe the thermodynamics governing the formation of these supramolecular nanotubes formed from covalent NDI oligomers. They show that not only does the covalently-linked NDI isodesmically aggregate into the desired helical nanotube, but that the terephthalate linker used to bind the NDI cores together significantly increases their robustness against H-bonded donors that would otherwise disrupt and destroy the nanotubes.[60]

1.5 Biological Applications

Perylenediimides (PDI) have been used in the metal-mediated construction of supramolecular assemblies. Tian and coworkers have developed an *N,N′*-di(4-pyridyl)-1,6,7,12-tetra(4-*tert*-octylphenoxy)PDI receptor dye, which is axially bound to a Zn^{II} complex.[61] Ruthenium compounds have been found to be promising anticancer agents, in particular arene Ru^{II} complexes, having shown different affinities to biomolecules when compared to their Pt-counterparts. Chi and co-workers have studied supramolecular coordination complexes (SCCs) of arene–Ru complexes as anticancer agents. They

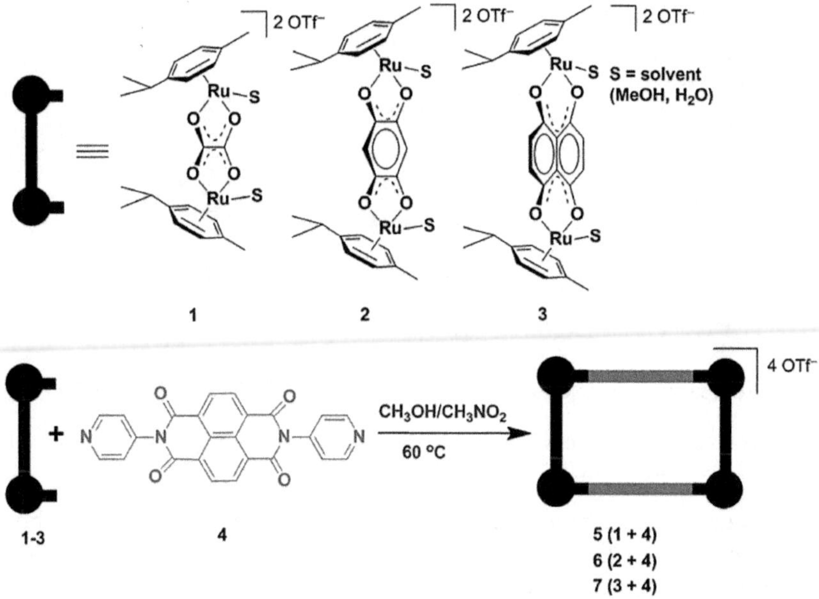

Figure 1.32 Self-assembled arene–ruthenium metallarectangles 5–7.[62]

report an array of cationic Ru [2 + 2] rectangles (**5–7**) with *N,N′*-di-(4-pyridyl)-NDI bridging ligands (Figure 1.32).[62]

Each SCC was screened for its *in vitro* anticancer activity against gastric (AGS) and colon cancer (HCT-15) by exposing the cell lines to increasing concentrations of the compounds over 24 hours. The results show that acceptors **1**, **2** and **4** are inactive (IC_{50} > 200 μM) and acceptor **3** and rectangles **5** and **6** show poor activity (IC_{50} between 130–160 μM). Rectangle **7**, however, showed higher activity (IC_{50} = 14.63 μM) than cisplatin (IC_{50} = 102.81 μM) for AGS and comparable activity (IC_{50} = 16.81 μM) to both cisplatin (IC_{50} = 10.36 μM) and doxorubicin (IC_{50} = 17.30 μM) for HCT-15. From these studies, rectangle **7** was shown to have growth-inhibitory activity against HCT-15 using an *in vivo* hollow fibre model and so is a promising drug for use against MDR cancer cells.[62]

Biological macromolecules, such as the DNA double helix, have driven the academic community towards the development of small molecules that can self-assemble into nanostructures for potential applications in catalysis, sensing, electronics, medical diagnostics and drug delivery. PDI molecules have been previously employed to help visualise drug delivery systems.[63] Qu and co-workers have developed a metal-mediated G-quartet-based nanostructure in dilute solution that uses PDI as a marker for drug release. They show formation of a G-quadruplex using Sr^{2+} templating to stabilise vertical stacking of G-quartets. Sr^{2+} was used as it has been shown to induce G-wire formation.[64] Using this G-wire, a controlled delivery system has been developed probing the drug release process using PDI as it is a

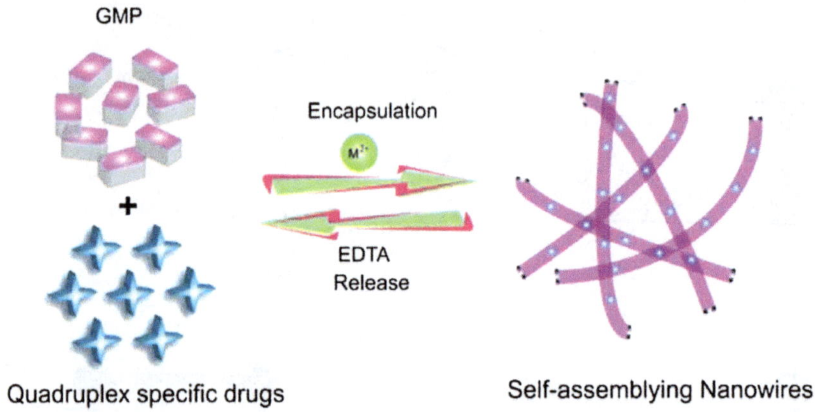

Figure 1.33 Proposed model showing the controlled G-quadruplex specific drug release system.[64]

strong G-quadruplex binder. PDI was present in solution prior to the assembly process so that upon supramolecular assembly, the PDI became trapped within the nanowire structure. It was shown that after several days no ligand release from nanowires was evident. The PDI was only released upon the introduction of Na_2EDTA, which chelates to the Sr^{2+} causing the disassembly of the nanowires and release of PDI (Figure 1.33). By using a combination of SEM and fluorescence techniques, they were able to show that these nanowires are able to encapsulate G-quadruplex specific molecules and be used as a controlled drug release system.[64]

Pascu *et al.* have shown[65] a way to functionalise single-walled carbon nanotubes (SWNTs) *via* supramolecular interactions using an amino acid derived NDI. The NDI forms composites with carbon nanomaterials leading to fluorescent amino acid tagged SWNTs, which are dispersible in common organic solvents ($CHCl_3$, DMSO) as well as in a biocompatible cell medium (Eagle's modified essential medium). The hybrid materials, NDI@SWNT, were characterised as dispersed in organic solvents and aqueous media and in the solid state by high resolution transmission electron microscopy, tapping mode atomic force microscopy, scanning electron microscopy, circular dichroism, Raman and fluorescence spectroscopies (steady-state single and two-photon techniques). DFT calculations were used to identify the geometry and the binding energies of an NDI molecule loaded onto a SWNT strand and the possibility of charge transfer interactions within the hybrid. The NDI@SWNT composite translocates and localises into cells' (*e.g.* FEK-4, HeLa, MCF-7) cytoplasm and partially in the nucleus as indicated by confocal fluorescence imaging and fluorescence lifetime imaging techniques. It was demonstrated that the NDI coating enhances the biocompatibility of SWNTs and mediates their intracellular localisation. Following this study, the same groups described supramolecular assemblies formed between thermally reduced graphene oxide (TRGO) and coronene with a series of D- and L-*p*-halogenated phenylalanine derivatised NDIs where the halogen substituents (X = F, Cl, Br, I) are varied systematically.[66] The formation of charge transfer complexes between the NDI and coronene was demonstrated by UV-vis and ^{1}H NMR studies indicating that the increasing polarisability and general electronegativity of the halide substituent influences the magnitude of the association constants in the ground state. The complexes NDI · coronene and NDI · TRGO have stable photoexcited states within the solution (coronene) and the dispersed phase (TRGO), which puts them forward as prostate cancer imaging agents as demonstrated by confocal fluorescence microscopy. MTT assays carried out in PC-3 cells demonstrate that these complexes improve the cellular viability of such graphene-like materials.[66]

Pascu and Pantoş *et al.* have developed a fluorescent peptide conjugate based on the coupling of cyclo(RGDfK) to a tryptophan-NDI. The cellular behavior regarding the molecular environment and biolocalisation of TrpNDI and TrpNDI–cyclo(RGDfK) in cancer cells was investigated using confocal fluorescence microscopy coupled with fluorescence lifetime imaging mapping, single and two-photon fluorescence excitation.[67]

1.6 Sensors

Conjugated systems have been prominent in the amplification of fluorescent polymers used in the sensing of volatile organic compounds. P-type materials have been the centre of much focus as they are suited to sensing of oxidative reagents, whilst n-type materials, suited for the sensing of reductive reagents, have been reported much less frequently. Zhang and coworkers have developed an n-type material based upon π–stacking aggregation of electron-poor perylene for the probing of organic amines.[68] Organic amines are a byproduct commonly produced by the decomposition of organic compounds and can be used as quality indicators for a variety of food types and even in the identification of certain diseases.

Guo and coworkers have developed highly reversible solid state fluorescence materials by a combination of PDI with cyclodextrin. PDI acts as a highly fluorescent reporter whilst the cyclodextrin serves as a receptor unit (Figure 1.34).[69]

PDI is an n-type material whose strong fluorescence makes it an ideal candidate for optical sensing. In solution, PDI is capable of aggregation through π-stacking and it is well known that the stacking is highly dependent on the solvent, and hence the stronger the π-stacking interaction the greater the aggregation within solution and therefore the greater the fluorescence quenching. The UV-vis spectrum of **1** showed bands between 450–550 nm in a variety of organic solvents (chloroform, acetonitrile, acetone, toluene and methanol) indicating the non-aggregated state of **1**. In water, however, the π-stacking aggregation is evident through low absorbance and broader peaks at the λ_{max}. They show that by using a combination of cyclodextrin with PDI and this linker that the aggregation ability of this

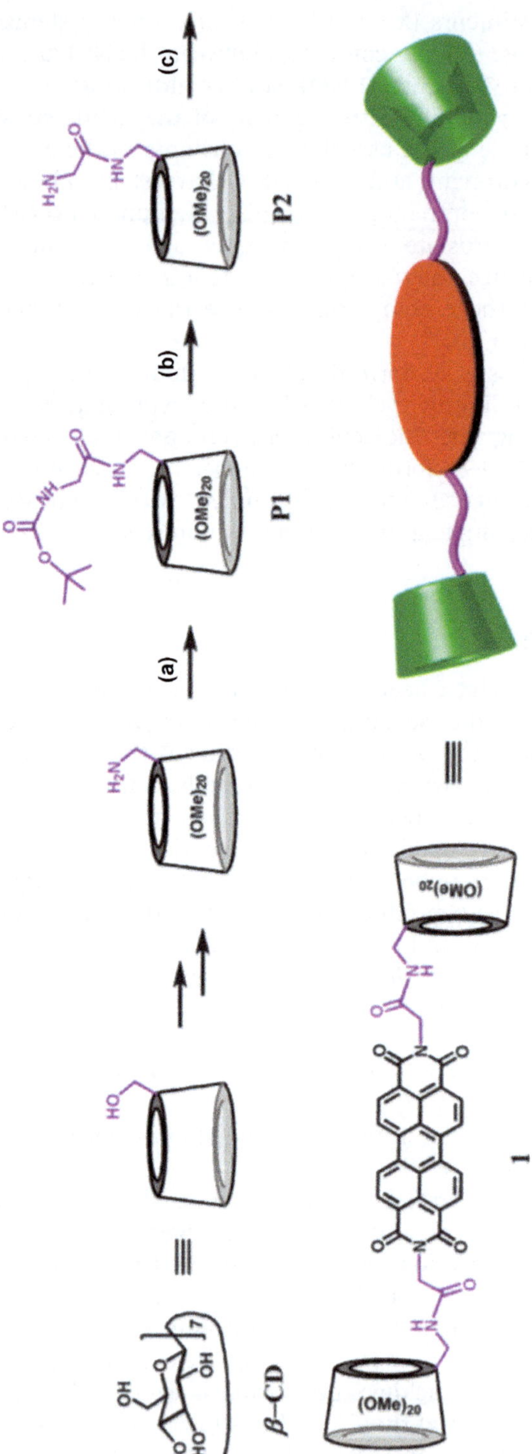

Figure 1.34 Synthetic route and structural illustration of the PDI and cyclodextrin fluorescent sensor.[69]

molecule is 1–2 times stronger (aggregation constant of 2.6×10^6 M^{-1}) than those previously reported.[69]

The ability of **1** to act as a viable amine sensor was probed using solid state fluorescence. Sensor **1** was embedded in poly(vinylidene fluoride) (PVDF) by air drying a sample of PDVF immersed in a 1.0×10^{-4} M aqueous solution of **1**. Prior to amine exposure, the UV-vis spectrum of the substrate was recorded and compared with that of **1** in an organic solvent; this showed that the substrate had both flattened and broadened bands consistent with that of the aggregated PDI structure. A bathochromic emission was also observed at 635 nm, which was attributed to strong π interactions formed as a result of PDI aggregation. The fluorescence of the cyclodextrin PDI derivatives can be quenched either by PET from the analyte to the PDI, or through disruption of the π-stacking within the molecules. They probed the latter interaction, which can occur from the simple inclusion of an organic molecule, and were able to show that no appreciable fluorescence quenching was observed.

The amine vapour detection was performed using a wide range of saturated amine vapours with an exposure time of 10 s and they found that the majority of amines tested, with the exception of triethylamine, show partial quenching with aniline achieving nearly 100% quenching. This shows that by embedding a cyclodextrin–PDI molecule within PVDF, Guo and coworkers were able to develop a practical solid state sensor for amines that can specifically target aniline and avoid background fluorescence quenching through strong PDI binding.[69]

Zhou and co-workers have developed crystalline-branched supramolecular structures (CBSSs) based on pyromellitimide. They show the construction of poly imides from a combination of pyromellitimide and an array of aromatic diamines. Once formed, these polyimides self-assemble into zigzag macromolecular chains with different configurations. These materials have been targeted for use as catalyst carriers and materials with a low dielectric constant and low refractive index.[70]

The surface area and pore volumes of the materials, using Brunauer–Emmett–Teller (BET) and the Langmuir model, were able to show that the largest BET surface area was 352 m^2 g^{-1} and the largest amount of nitrogen adsorbed was 0.73 cm^3 g^{-1}. This shows that through *in situ* self-assembly, polyimides can form macromolecules with large surface areas and a high degree of crystallinity, which could potentially lead to new covalent organic frameworks (COFs) based upon imido bonds.[70]

1.7 Dyads and Triads for Electron Transfer Studies

NDIs and PDIs have been used in supramolecular assemblies used to harvest solar energy.[71] The N-desymmetrised amino acid derived NDIs have found application in an unsymmetrical triad assembly with an aluminium porphyrin and a ruthenium porphyrin being the other members of this supramolecular system (Figure 1.35).[72] A number of control supramolecular dyads

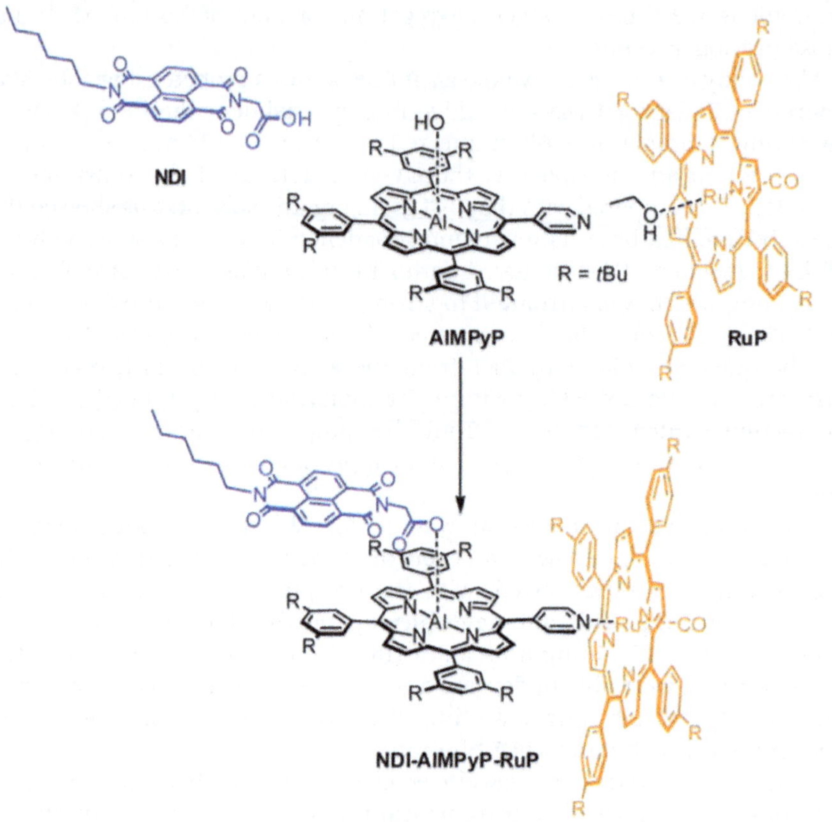

Figure 1.35 Self-assembling of the non-symmetric triad NDI–AlMPyP–RuP.[72]

have been synthesised in order to allow the elucidation of the photophysical properties of these supramolecules.

In related work, NDIs were used in combination with aluminium porphyrins and tetrathiafulvalene to assemble dyads and triads with long-lived charge separated states.[73] A related study used similar N-desymmetrised NDI with phthalocyanines and naphthalocyanines to achieve photoelectron transfer in these supramolecular assemblies.[74]

1.8 Conclusions

Naphthalenediimides and its congeners have a rich supramolecular chemistry, ranging from simple aromatic donor–acceptor studies, to molecules with complex topologies; from biological applications to receptors, chemosensors and pressure devices.[75] The ease of synthesis, the ability to tune the optoelectronic properties of the aromatic cores, the multitude of supramolecular handles preexistent in the NDI and its congeners' core, make these molecules particularly attractive for theoretical and applicative studies. This is an ever-expanding field that will produce new and exciting results for decades to come.

References

1. M. S. Cubberley and B. L. Iverson, *J. Am. Chem. Soc.*, 2001, **123**, 7560–7563.
2. G. M. Prentice, S. I. Pascu, S. V. Filip, K. R. West and G. D. Pantoş, *Chem. Commun.*, 2015, **51**, 8265–8268.
3. R. E. Dawson, A. Hennig, D. P. Weimann, D. Emery, V. Ravikumar, J. Montenegro, T. Takeuchi, S. Gabutti, M. Mayor, J. Mareda, C. A. Schalley and S. Matile, *Nat. Chem.*, 2010, **2**, 533–538.
4. J. Mareda and S. Matile, *Chem. – Eur. J.*, 2009, **15**, 28–37.
5. A. R. Stefankiewicz, E. Tamanini, G. D. Pantoş and J. K. M. Sanders, *Angew. Chem., Int. Ed.*, 2011, **50**, 5725–5728.
6. P. Pengo, G. D. Pantoş, S. Otto and J. K. M. Sanders, *J. Org. Chem.*, 2006, **71**, 7063–7066.
7. K. Tambara, N. Ponnuswamy, G. Hennrich and G. D. Pantoş, *J. Org. Chem.*, 2011, **76**, 3338–3347.
8. S.-L. Suraru and F. Würthner, *Angew. Chem., Int. Ed.*, 2014, **53**, 7428–7448.
9. C. L. Lyall, C. C. Shotton, M. Pérez-Salvia, G. Dan Pantoş and S. E. Lewis, *Chem. Commun.*, 2014, **50**, 13837–13840.
10. M. A. Kobaisi, S. V. Bhosale, K. Latham, A. M. Raynor and S. V. Bhosale, *Chem. Rev.*, 2016, **116**, 11685–11796.
11. N. Sakai, J. Mareda, E. Vauthey and S. Matile, *Chem. Commun.*, 2010, **46**, 4225–4237.
12. A. J. Zych and B. L. Iverson, *J. Am. Chem. Soc.*, 2000, **122**, 8898–8909.
13. G. J. Gabriel, S. Sorey and B. L. Iverson, *J. Am. Chem. Soc.*, 2005, **127**, 2637–2640.
14. Y. Chu, D. W. Hoffman and B. L. Iverson, *J. Am. Chem. Soc.*, 2009, **131**, 3499–3508.
15. D. G. Hamilton, J. E. Davies, L. Prodi and J. K. M. Sanders, *Chem. – Eur. J.*, 1998, **4**, 608–620.
16. D. G. Hamilton and J. K. M. Sanders, *Chem. Commun.*, 1998, 1749–1750.
17. Q. Zhang, D. G. Hamilton, N. Feeder, S. J. Teat, J. M. Goodman and J. K. M. Sanders, *New J. Chem.*, 1999, **23**, 897–903.
18. D. G. Hamilton, L. Prodi, N. Feeder and J. K. M. Sanders, *J. Chem. Soc., Perkin Trans. 1*, 1999, 1057–1066.
19. Y. Tong, D. G. Hamilton, J.-C. Meillon and J. K. M. Sanders, *Org. Lett.*, 1999, **1**, 1343–1346.
20. J. G. Hansen, N. Feeder, D. G. Hamilton, M. J. Gunter, J. Becher and J. K. M. Sanders, *Org. Lett.*, 2000, **2**, 449–452.
21. S. A. Vignon, T. Jarrosson, T. Iijima, H.-R. Tseng, J. K. M. Sanders and J. F. Stoddart, *J. Am. Chem. Soc.*, 2004, **126**, 9884–9885.

22. T. Iijima, S. A. Vignon, H.-R. Tseng, T. Jarrosson, J. K. M. Sanders, F. Marchioni, M. Venturi, E. Apostoli, V. Balzani and J. F. Stoddart, *Chem. – Eur. J.*, 2004, **10**, 6375–6392.
23. A. Coskun, S. Saha, I. Aprahamian and J. F. Stoddart, *Org. Lett.*, 2008, **10**, 3187–3190.
24. C. J. Bruns, J. Li, M. Frasconi, S. T. Schneebeli, J. Iehl, H.-P. Jacquot de Rouville, S. I. Stupp, G. A. Voth and J. F. Stoddart, *Angew. Chem., Int. Ed.*, 2014, 1953–1958.
25. H. M. Colquhoun, D. J. Williams and Z. Zhu, *J. Am. Chem. Soc.*, 2002, **124**, 13346–13347.
26. H. M. Colquhoun, Z. Zhu and D. J. Williams, *Org. Lett.*, 2003, **5**, 4353–4356.
27. H. M. Colquhoun, Z. Zhu, D. J. Williams, M. G. B. Drew, C. J. Cardin, Y. Gan, A. G. Crawford and T. B. Marder, *Chem. – Eur. J.*, 2010, **16**, 907–918.
28. T. Nakagaki, K. Shin-ichiro, A. Harano and T. Shinmyozu, *Tetrahedron*, 2010, **66**, 976–985.
29. F. B. L. Cougnon and J. K. M. Sanders, *Acc. Chem. Res.*, 2012, **45**, 2211–2221.
30. H. Y. Au-Yeung, G. D. Pantoş and J. K. M. Sanders, *Proc. Natl. Acad. Sci. U. S. A.*, 2009, **106**, 10466–10470.
31. H. Y. Au-Yeung, G. D. Pantoş and J. K. M. Sanders, *J. Am. Chem. Soc.*, 2009, **131**, 16030–16032.
32. H. Y. Au-Yeung, G. D. Pantoş and J. K. M. Sanders, *J. Org. Chem.*, 2011, **76**, 1257–1268.
33. F. B. L. Cougnon, H. Y. Au-Yeung, G. D. Pantoş and J. K. M. Sanders, *J. Am. Chem. Soc.*, 2011, **133**, 3198–3207.
34. F. B. L. Cougnon, N. A. Jenkins, G. D. Pantoş and J. K. M. Sanders, *Angew. Chem., Int. Ed.*, 2012, **51**, 1443–1447.
35. F. B. L. Cougnon, N. Ponnuswamy, N. A. Jenkins, G. D. Pantoş and J. K. M. Sanders, *J. Am. Chem. Soc.*, 2012, **134**, 19129–19135.
36. H. Y. Au-Yeung, P. Pengo, G. D. Pantoş, S. Otto and J. K. M. Sanders, *Chem. Commun.*, 2009, 419–421.
37. N. Ponnuswamy, F. B. L. Cougnon, G. D. Pantoş and J. K. M. Sanders, *J. Am. Chem. Soc.*, 2014, **136**, 8243–8251.
38. N. Ponnuswamy, F. B. L. Cougnon, J. M. Clough, G. D. Pantoş and J. K. M. Sanders, *Science*, 2012, **338**, 783–785.
39. W. Wang, L. Wang, B. J. Palmer, G. J. Exarhos and A. D. Q. Li, *J. Am. Chem. Soc.*, 2006, **128**, 11150–11159.
40. J. J. Han, A. D. Shaller, W. Wang and A. D. Q. Li, *J. Am. Chem. Soc.*, 2008, **130**, 6974–6982.
41. A. D. Shaller, W. Wang, H. Gan and A. D. Q. Li, *Angew. Chem., Int. Ed.*, 2008, **47**, 7705–7709.
42. R. Bishop, *Aust. J. Chem.*, 2012, **65**, 1361–1370.
43. K. Kishikawa, C. Iwashima, S. Kohmoto, K. Yamaguchi and M. Yamamoto, *J. Chem. Soc., Perkin Trans. 1*, 2000, 2217–2221.

44. J. Gawronski, M. Kaik, M. Kwit and U. Rychlewska, *Tetrahedron*, 2006, **62**, 7866–7877.
45. J. Gawronski, K. Gawronska, J. Grajewski, K. Kacprzak and U. Rychlewska, *Chem. Commun.*, 2002, 582–583.
46. Y. Wu, M. Frasconi, D. M. Gardner, P. R. McGonigal, S. T. Schneebeli, M. R. Wasielewski and J. F. Stoddart, *Angew. Chem., Int. Ed.*, 2014, **53**, 9476–9481.
47. S. T. Schneebeli, M. Frasconi, Z. Liu, Y. Wu, D. M. Gardner, N. L. Strutt, C. Cheng, R. Carmieli, M. R. Wasielewski and J. F. Stoddart, *Angew. Chem., Int. Ed.*, 2013, **52**, 13100–13104.
48. Y. Wu, S. K. M. Nalluri, R. M. Young, M. D. Krzyaniak, E. A. Margulies, J. F. Stoddart and M. R. Wasielewski, *Angew. Chem., Int. Ed.*, 2015, **54**, 11971–11977.
49. S. K. M. Nalluri, Z. Liu, Y. Wu, K. R. Hermann, A. Samanta, D. J. Kim, M. D. Krzyaniak, M. R. Wasielewski and J. F. Stoddart, *J. Am. Chem. Soc.*, 2016, **138**, 5968–5977.
50. J. M. Lavin and K. D. Shimizu, *Chem. Commun.*, 2007, 228.
51. G. D. Pantoş, P. Pengo and J. K. M. Sanders, *Angew. Chem., Int. Ed.*, 2007, **46**, 194–197.
52. T. W. Anderson, J. K. M. Sanders and G. D. Pantoş, *Org. Biomol. Chem.*, 2010, **8**, 4274.
53. T. W. Anderson, G. D. Pantoş and J. K. M. Sanders, *Org. Biomol. Chem.*, 2011, **9**, 7547.
54. G. D. Pantoş, J. L. Wietor and J. K. M. Sanders, *Angew. Chem., Int. Ed.*, 2007, **46**, 2238–2240.
55. J.-L. Wietor, G. D. Pantoş and J. K. M. Sanders, *Angew. Chem., Int. Ed.*, 2008, **47**, 2689–2692.
56. E. Tamanini, G. D. Pantoş and J. M. Sanders, *Chem. – Eur. J.*, 2010, **16**, 81–84.
57. C. Zazza, A. Palma, A. Amadei, N. Sanna, S. Tatoli and M. Aschi, *Faraday Discuss.*, 2010, **145**, 107.
58. T. F. A. De Greef, M. M. J. Smulders, M. Wolffs, A. P. H. J. Schenning, R. P. Sijbesma and E. W. Meijer, *Chem. Rev.*, 2009, **109**, 5687–5754.
59. N. Ponnuswamy, G. D. Pantoş, M. M. J. Smulders and J. K. M. Sanders, *J. Am. Chem. Soc.*, 2012, **134**, 566–573.
60. K. Tambara, J.-C. Olsen, D. E. Hansen and G. D. Pantoş, *Org. Biomol. Chem.*, 2014, **12**, 607.
61. B. Gao, Y. Li, J. Su and H. Tian, *Supramol. Chem.*, 2007, **19**, 207–210.
62. A. Dubey, J. W. Min, H. J. Koo, H. Kim, T. R. Cook, S. C. Kang, P. J. Stang and K.-W. Chi, *Chem. – Eur. J.*, 2013, **19**, 11622–11628.
63. M. Sun, K. Müllen and M. Yin, *Chem. Soc. Rev.*, 2016, **45**, 1513–1528.
64. D. Hu, J. Ren and X. Qu, *Chem. Sci.*, 2011, **2**, 1356.
65. Z. Hu, G. D. Pantoş, N. Kuganathan, R. L. Arrowsmith, R. M. J. Jacobs, G. Kociok-Koehn, J. O'Byrne, K. Jurkschat, P. Burgos, R. M. Tyrrell, S. W. Botchway, J. K. M. Sanders and S. I. Pascu, *Adv. Funct. Mater.*, 2012, **22**, 503–518.

66. J. A. Tyson, V. Mirabello, D. G. Calatayud, H. Ge, G. Kociok-Köhn, S. W. Botchway, G. Dan Pantoş and S. I. Pascu, *Adv. Funct. Mater.*, 2016, **26**, 5641–5657.
67. Z. Hu, R. L. Arrowsmith, J. A. Tyson, V. Mirabello, H. Ge, I. M. Eggleston, S. W. Botchway, G. D. Pantoş and S. I. Pascu, *Chem. Commun.*, 2015, **51**, 6901–6904.
68. J. Zhang, F. J. M. Hoeben, M. J. Pouderoijen, A. P. H. Schenning, E. W. Meijer, F. C. Schryver and F. De Steven, *Chem. - Eur. J.*, 2006, **12**, 9046–9055.
69. B.-P. Jiang, D.-S. Guo and Y. Liu, *J. Org. Chem.*, 2011, **76**, 6101–6107.
70. S. Zeng, L. Guo, F. Cui, Z. Gao, J. Zhou and J. Shi, *Macromol. Chem. Phys.*, 2010, **211**, 698–705.
71. P. D. Frischmann, K. Mahata and F. Würthner, *Chem. Soc. Rev.*, 2013, **42**, 1847–1870.
72. E. Iengo, G. D. Pantoş, J. K. M. Sanders, M. Orlandi, C. Chiorboli, S. Fracasso and F. Scandola, *Chem. Sci.*, 2011, **2**, 676–685.
73. P. K. Poddutoori, N. Zarrabi, A. G. Moiseev, R. Gumbau-Brisa, S. Vassiliev and A. van der Est, *Chem. - Eur. J.*, 2013, **19**, 3148–3161.
74. M. E. El-Khouly, A. G. Moiseev, A. van der Est and S. Fukuzumi, *ChemPhysChem*, 2012, **13**, 1191–1198.
75. S. Littlejohn, A. Nogaret, G. M. Prentice and G. D. Pantoş, *Adv. Funct. Mater.*, 2013, **23**, 5398–5402.

CHAPTER 2

NDI as a DNA Intercalator

AMY RHODEN SMITH AND BRENT IVERSON*

Department of Chemistry, The University of Texas at Austin, Austin, TX 78712, USA

*Email: iversonb@austin.utexas.edu

2.1 DNA as a Drug Target

The central dogma of molecular biology describes, in its simplest form, the flow of sequence information from DNA to RNA to protein. The DNA double helix stores the genetic blueprint for all living organisms, and transcription to RNA with subsequent translation to protein is a carefully controlled and complex process dictating cellular behavior.[1] As such, the ability to sequence-specifically bind and manipulate DNA has long been a therapeutic target. In recent years, this field has expanded rapidly, most notably with the discovery of the bacterial-inspired programmable CRISPR/Cas9 RNA-guided DNA endonuclease, along with zinc-finger nucleases (ZFNs) and TAL effector nucleases (TALENs).[2,3] In addition to these protein-derived technologies, small synthetic molecules have also shown the ability to target DNA with surprising DNA binding properties. These types of molecules include triplex-forming oligonucleotides (TFOs), peptide nucleic acids (PNAs), minor groove-binding hairpin polyamides, and intercalators. This review will focus on naphthalene diimide (NDI) and similar derivatives and their development as DNA intercalative binding agents.

Monographs in Supramolecular Chemistry No. 21
Naphthalenediimide and its Congeners: From Molecules to Materials
Edited by G. Dan Pantoş

Published by the Royal Society of Chemistry, www.rsc.org

2.2 Overview of DNA Binding Motifs

A number of designed, synthetic molecules have been developed to bind DNA including triplex-forming oligonucleotides (TFOs), peptide nucleic acids (PNAs), minor groove-binding polyamides, and intercalators. Section 2.2 will give a brief overview of these DNA recognition motifs, with intercalation discussed in detail in Section 2.3.

2.2.1 DNA Structure

An overview of DNA structure is necessary to understand the complexity of and difference between the aforementioned DNA binding modes. B-form DNA consists of a right-handed helix with approximately 10 base pair (bp)/turn, 3.4 Å between bp, and 36° rotation/bp.[4] Base pairing is essentially perpendicular to the helical axis, but because the phosphodiester backbones of the complimentary strands are not directly across from one another, there are two differently sized grooves in the helix as pictured in Figure 2.1. The major groove is wide and shallow while the minor groove is much narrower and deeper. Each base pair also presents different recognition elements, in the form of potential hydrogen bond donors and acceptors, in each of the grooves as illustrated in Figure 2.2.[5] It must be noted that while these measurements and figures are suitable for B DNA in its ideal form, DNA is flexible and exhibits a conformationally dynamic behavior in solution.

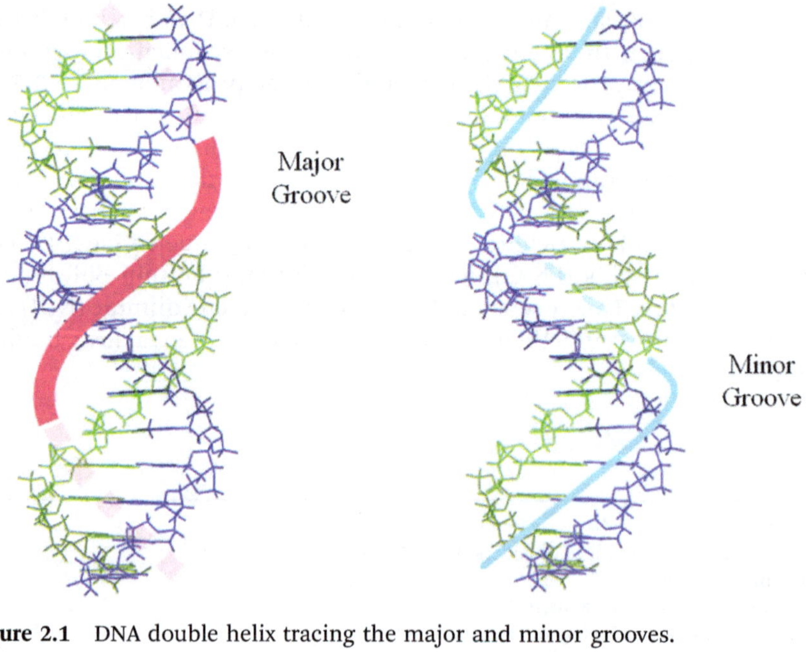

Figure 2.1 DNA double helix tracing the major and minor grooves.

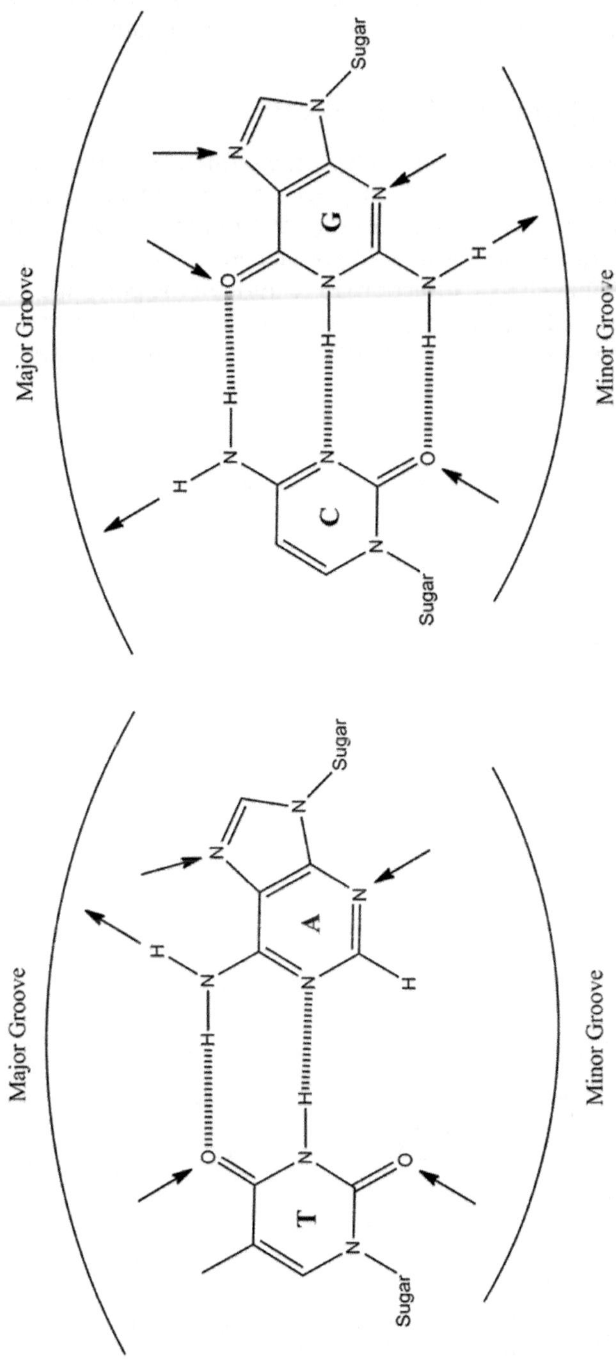

Figure 2.2 DNA base pairs showing the exposed functional groups in the major and minor grooves with arrows showing possible H bond donor and acceptors.

This structural complexity allows it to accommodate binding of small molecules (and proteins) in the major groove, minor groove, in between base pairs, and, as will be discussed further, all three simultaneously.

2.2.2 Triplex-forming Oligonucleotides

Oligonucleotides are able to bind polypurine sequences in the major groove through H-bond patterns termed Hoogsteen base pairing, as opposed to the typical Watson–Crick base pairing in the DNA duplex, to form a DNA triple helix.[6,7] These triple-helix forming oligonucleotides (TFOs) can either bind parallel (in the same 5′ to 3′ direction) in relation to the polypurine strand in the duplex to form Hoogsteen triplets or in the anti-parallel direction to form reverse-Hoogsteen hydrogen bonds (Figure 2.3). Polypyrimidine (TC) third

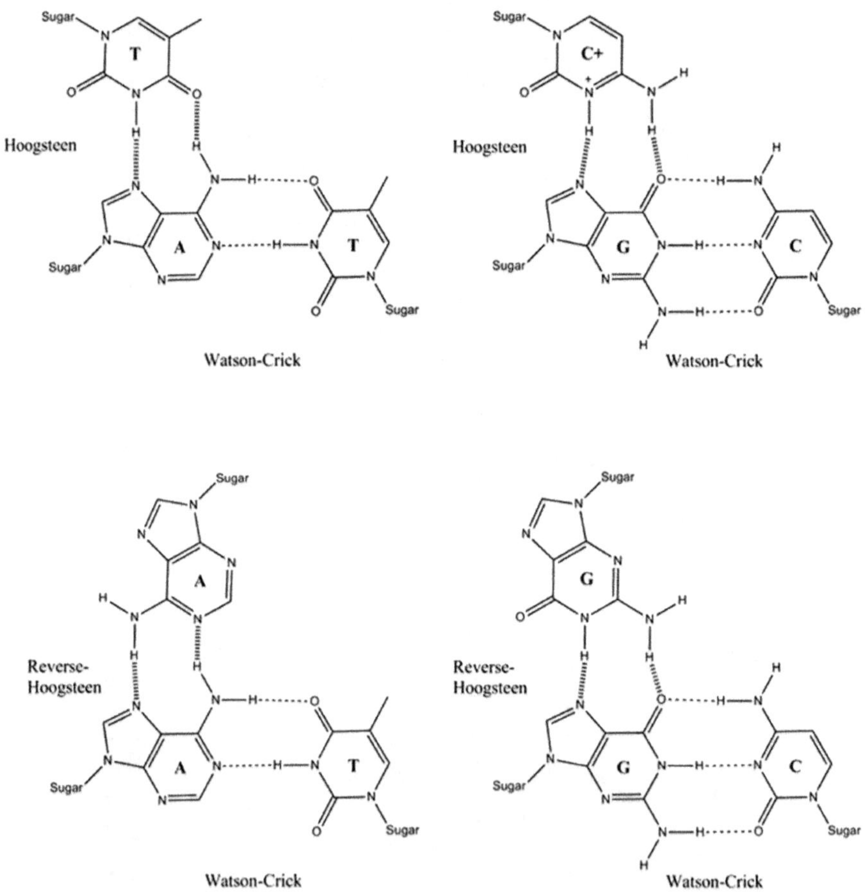

Figure 2.3 Hoogsteen and reverse-Hoogsteen hydrogen bonding motif in triplex formation. Canonical Watson–Crick base pairing is shown by the thin dashed lines while Hoogsteen base pairing is shown by the thick dashed line.

stands will form parallel triplexes in the form C^+:G-C and T:A-T. Since cytosine must be protonated to form a stable triplex, this binding is pH dependent and occurs best at acidic pH. A mixed GT third strand is able to form either parallel or anti-parallel triplexes by forming Hoogsteen or anti-Hoogsteen triplets, respectively, of the nature G:G-C and T:A-T. Finally, a polypurine (GA) third strand binds in an anti-parallel fashion making G:G-C and A:A-T Hoogsteen hydrogen bonds.[8,9]

Limitations for triplex formation include the requirement for a polypurine target, pH dependent formation, and overcoming charge repulsion from the three negatively-charged phosphodiester backbones. A number of nucleobase, sugar, and backbone modifications have been reported to overcome these difficulties, a few of which will be described here (Figure 2.4). One of the greatest advances was the discovery of the nucleobase modification 5-methylcytosine (mC), which ameliorates pH dependency and allows for triplex formation at physiological pH.[10] While B-DNA favors nucleotides with C2′-*endo* sugar geometry, a TFO with C3′-*endo* geometry produces more stable triplexes. Thus, modifications to the nucleotide sugars have focused on stabilizing the triple strand sugars in the C3′-*endo* conformation, and it has been shown that RNA (where C2′-*endo* geometry predominates) also forms very stable triple helices. These analogues, which are also resistant to degradation, include 2′-methoxy ribose and locked nucleic acid (LNA).[11]

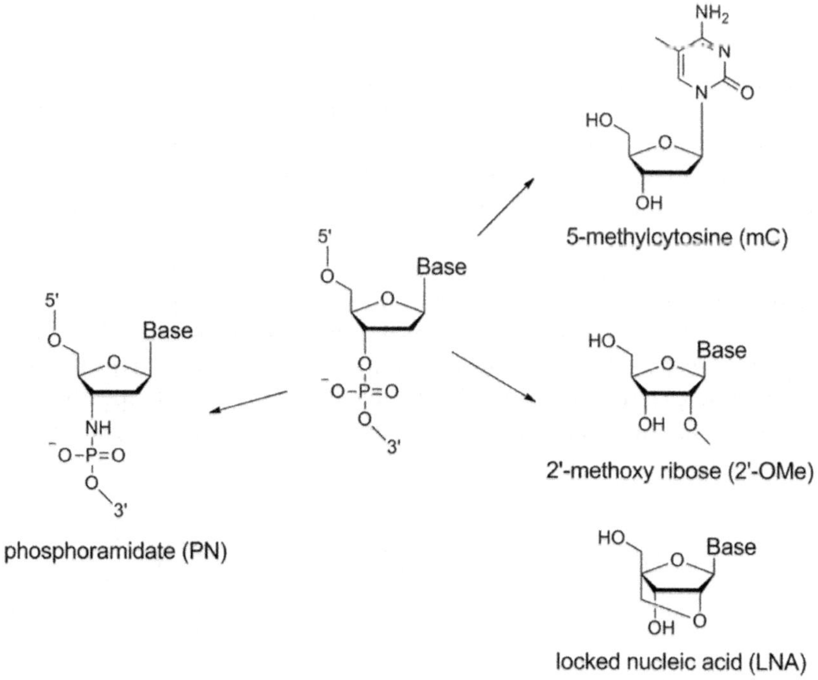

Figure 2.4 Selected nucleobase, sugar, and backbone TFO modifications.

2.2.3 Peptide Nucleic Acid

The peptide nucleic acid (PNA) approach was developed to be a mimic of DNA by combining nucleobase recognition with a neutral and achiral peptide-like backbone composed of 2-aminoethyl-glycine (Figure 2.5). Instead of simply binding as a third strand to a DNA duplex, as originally designed, PNAs most commonly bind in a mode termed triplex invasion to form a $(PNA)_2DNA$ complex, where one PNA strand binds the complementary DNA strand in the duplex by Watson–Crick base pairing, displacing the other DNA strand to make a single-stranded loop, while the second PNA strand binds in the major groove of the PNA–DNA duplex *via* Hoogsteen hydrogen bonds.[12] The combination of this triplex and displaced DNA single strand is often referred to as a P-loop. PNA, being neutral, does not suffer any of the deleterious effects of electrostatic repulsion when binding to DNA, unlike TFOs, so these P-loop structures are far more stable with extrapolated lifetimes into hundreds of days at physiological conditions.[13] In addition, these complexes are very sequence-specific, and even a single mismatch can reduce affinity by 100-fold.[14]

2.2.4 Minor Groove-binding Polyamides

Arguably the most successful synthetic DNA-binding molecule design to date, polyamides bind in the minor groove of the DNA with programmable sequence specificity. These molecules, developed by the Dervan lab, were inspired by the naturally-occurring, polypyrrole antibiotic distamycin, which binds to five contiguous AT base pairs in the minor groove (Figure 2.6).[15] Using analogues of *N*-methylpyrrole, synthetic polyamide structures bind in a 2 : 1 stoichiometry in the minor groove with the two ligands stacked in an anti-parallel head-to-tail fashion. Side-by-side pairings of *N*-methylpyrrole (Py), *N*-methylimidazole (Im), and *N*-methyl-3-hydroxypyrrole (Hp) allow for a recognition code to distinguish all four base pairs, as listed in Table 2.1. Linking the two Py-Im polyamides with γ-aminobutyric acid led to the so-called hairpin polyamides, which prevent slippage of the paired dimers and allow for high sequence specificity and affinity often in the subnanomolar range (Figure 2.7).[16] The curvature of polyamides is greater than that of the minor groove, so the insertion of a flexible β-alanine spacer, specific for A,T base pairs, has allowed for targeting of longer sequences up to 16 bp.[17]

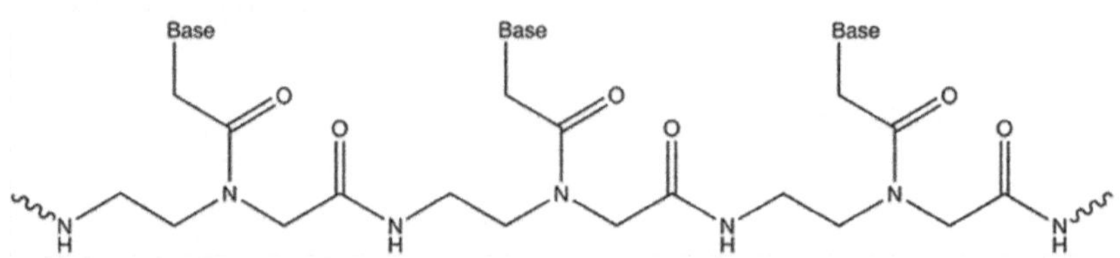

Figure 2.5 General structure of peptide nucleic acid (PNA) with a pseudopeptide 2-aminoethyl-glycine backbone.

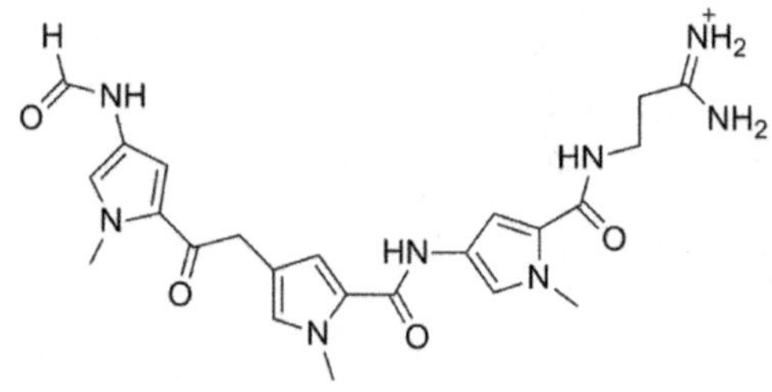

Figure 2.6 Structure and binding site of naturally occurring minor groove-binder distamycin.

Table 2.1 Pairing rules for polyamides. Im = *N*-methylimidazole, Py = *N*-methylpyrrole, Hp = *N*-methyl-3-hydroxypyrrole.

Pair	GC	CG	TA	AT
Im/Py	+	−	−	−
Py/Im	−	+	−	−
Hp/Py	−	−	+	−
Py/Hp	−	−	−	+

While some sequences are still difficult to target, polyamides have led the way in making designed molecules with programmable sequence specificity. More recent research has focused on high throughput screenings of polyamides to aid in redesign strategies and uncover new binding motifs.[18,19] In promising new research, the Lee lab has been exploring the use of minor groove-binding and alkylating *seco*-cyclopropylpyrido[*e*]indolone analogs of CC-1065 to target cancer and especially parasitic diseases. These interesting molecules target extended AT tracts of DNA that are particularly common in the genomes of certain parasites.[20]

2.3 Intercalation

Intercalation, first described by Lerman in 1961, involves the insertion of a typically flat aromatic molecule in between the base pairs of DNA.[21] This type of binding is necessarily accompanied by lengthening and stiffening of the DNA and unwinding of the helix (Figure 2.8). Due to this rearrangement of the DNA helix to accommodate an intercalator, only one intercalator can bind for every two base pairs as described by the so-called "nearest neighbor exclusion" principle.[22,23] In other words, binding of an intercalator to one site inhibits binding of another intercalator at the adjacent site. There are many different classes of organic small molecule intercalators, with common frameworks displayed in Figure 2.9: anthracenes, acridines, anthraquinones, phenanthridines, phenanthrolines, ellipticines, naphthalimides (NI),

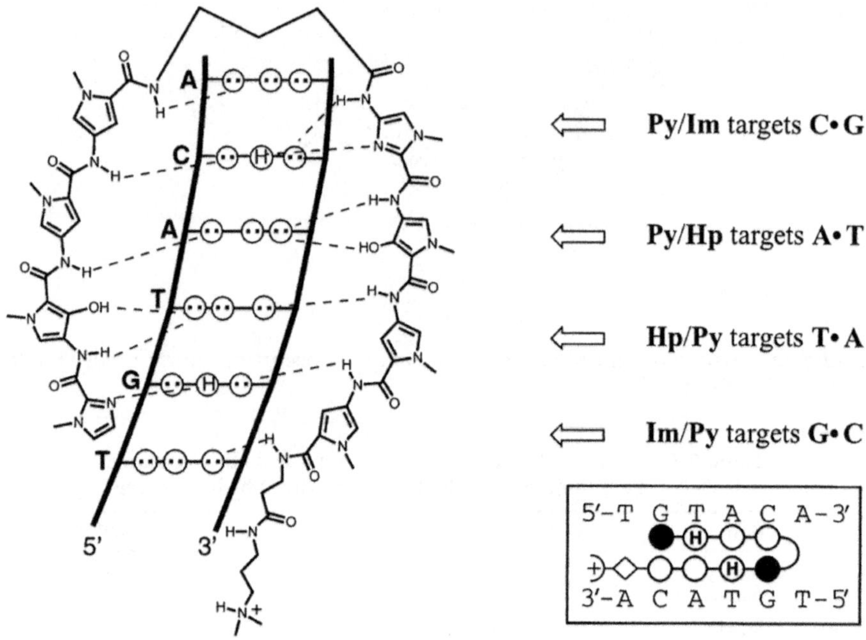

Figure 2.7 Hairpin polyamide motif with ImHpPyPy-γ-ImHpPyPy-β-Dp polyamide with a 5′-TGTACA-3′ binding site. Black circles are Im, open circles are Py, and circles with H are Hp (reprinted from P. B. Dervan, Molecular Recognition of DNA by Small Molecules, *Bioorg. Med. Chem.*, **9**, 2215–2235, Copyright (2001), with permission from Elsevier).

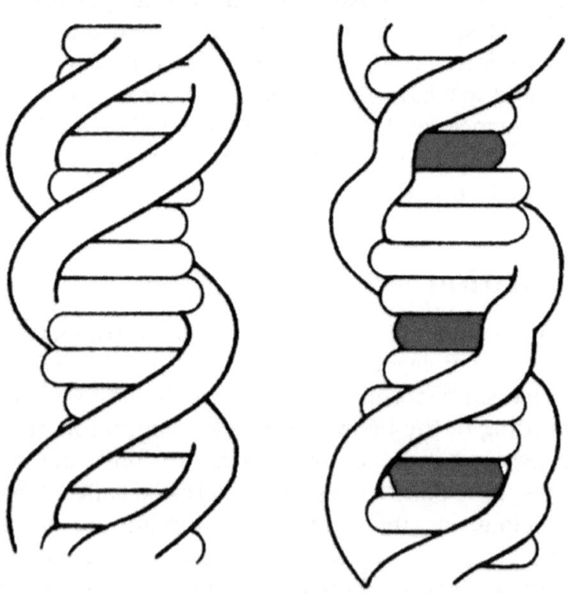

Figure 2.8 Cartoon depicting duplex DNA on the left and DNA with bound intercalator (gray) on the right.

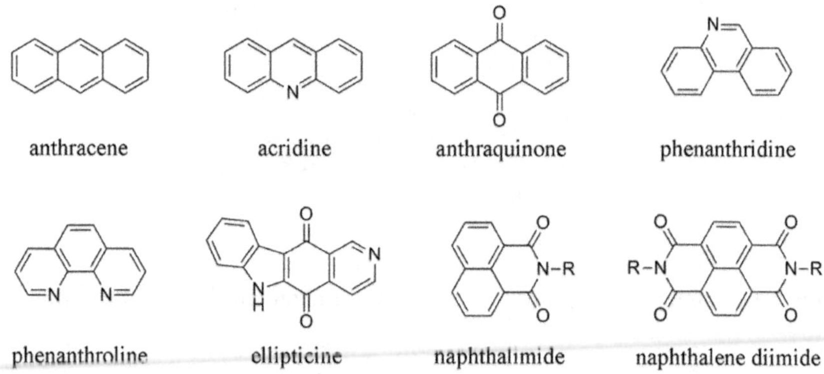

Figure 2.9 Most common organic intercalator frameworks.

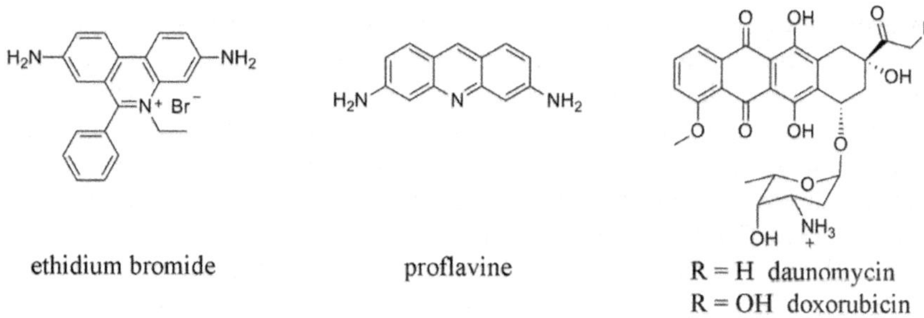

Figure 2.10 Examples of common intercalators.

and naphthalene diimides (NDI).[24] Some of the more well-known intercalating agents include ethidium bromide, a phenanthridine derivative used to image DNA in agarose gel electrophoresis, and proflavine, which was one of the first studied intercalators (Figure 2.10). Daunomycin, also known as daunorubicin, and doxorubicin are members of the anthracycline family of antibiotics (containing an anthraquinone core) and are important drugs for treating a range of cancers.

Metal-containing intercalation complexes, which differ somewhat from the organic intercalators described thus far, also represent a large area of research. Interest in this field was prompted by the discovery of cisplatin, an important anticancer drug that forms covalent adducts with DNA. Most of these DNA interacting molecules contain square planar platinum complexes or octahedral complexes with ruthenium or rhodium in which the aromatic ligands coordinated to the metal centers can intercalate DNA in a reversible manner. Many of these ruthenium and rhodium complexes, such as $Ru(bpy)_2(dppz)^{2+}$ shown in Figure 2.11, exhibit a large increase in luminescence when bound to DNA, which is a phenomenon known as the "light switch" effect and allows for a simple way to monitor intercalative binding (bpy, 2,2′-bipyridine; dppz, dipyrido[3,2-*a*:2′,3′-*c*]phenazine).[25]

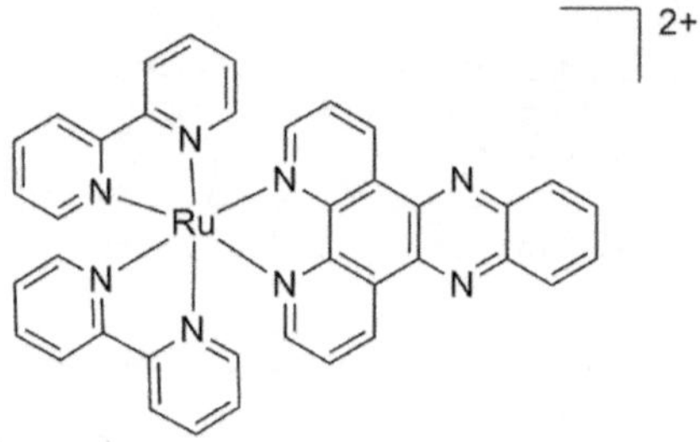

Figure 2.11 Molecular "light switch" rhodium metallointercalator $Ru(bpy)_2(dppz)^{2+}$.

Intercalators commonly bind to DNA in a relatively non-sequence specific fashion, generally only showing a preference for two base-pair steps. Most intercalators are cytotoxic by acting as topoisomerase inhibitors, preventing cellular replication and inducing apoptosis, or by inhibiting transcription. However, many intercalators are not appropriate for use as therapeutics due to uncontrollable side affects arising from their lack of specificity.[26] Threading intercalators are a subclass of intercalators in which one substituent on the intercalator core resides in the minor groove, while simultaneously another resides in the major groove. Polyintercalators, incorporating two or more intercalating units, have demonstrated increased sequence specificity and binding affinity in some cases. The following sections describe studies of threading monointercalators, bisintercalators, and longer polyintercalator derivatives with an emphasis on NDI-based molecules.

2.3.1 Monointercalators

Numerous monointercalators based on the previously mentioned scaffolds (Figure 2.9) have been described. This summary of monointercalators will focus on naphthalimide (NI) and naphthalene diimide (NDI) derivatives. NI intercalators, first introduced by Braña and coworkers, are not typically threading but are able to interact by groove binding with a basic linker, often an amine, extending from the single imide functionality. Series of naphthalimides with various linkers and core substituents have been and continue to be analyzed for their binding abilities and cytotoxic activity against cancer cell lines.[27–31] Often the NI core substituents are small, such as amino or nitro groups, but NI derivatives with longer hydroxyl-alkylamine linkers extending from the core and alkylamine linkers extending from the imide have been reported and likely bind in a threading manner.[32] Two NI derivatives—amonafide and mitonafide (Figure 2.12)—were shown to act as topoisomerase II inhibitors with antitumor properties, and they entered clinical trials but were never approved as drugs due to undesirable side effects.[33,34] A thorough review on naphthalimide intercalators has recently been published by Banerjee *et al.*[33]

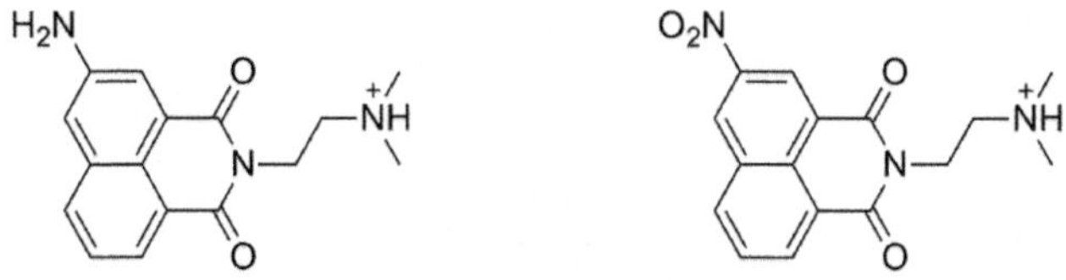

Figure 2.12 Structures of naphthalimides amonafide and mitonafide.

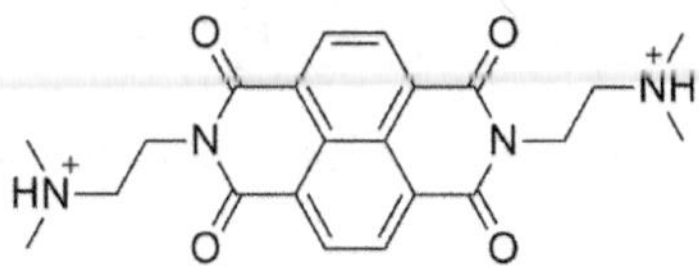

Figure 2.13 Example of a threading NDI intercalator.

NDI was first reported as a threading intercalator by Gabbay *et al.*[35] and described in further detail by Wilson and co-workers.[36,37] The threading topology means that a substituent attached to one imide of NDI will be in the minor groove, while a substituent attached to the other imide will reside in the major groove (Figure 2.13). Compared to analogous non-threading NI intercalators, these intercalators have higher binding affinities, which can be attributed to their having much slower dissociation rates from DNA. These seminal studies also revealed that very bulky substituents, in particular an adamantyl amine, could also thread through the DNA base pairs to allow for intercalation and could be accommodated in the grooves. The identity of the groove-binding linkers for threading NDI monointercalators greatly affects binding affinity and cytotoxic activity. In agreement with studies of NI derivatives, a basic functionality, most often an amine, leads to NDI intercalators with the most desirable properties. Similar studies of NDIs which analyze the effects of altering linker lengths between the imide and amine, changing the steric bulk of the groove-binding substituents, and modifying the basicity of the linker amine functionality continue to be of interest.[27,38–40] NDI intercalators have been shown to prefer binding to GC-rich DNA.[36,41] In addition, NDI derivatives with core substituents (cNDI) have also been evaluated for intercalative ability. As cNDIs often suffer from poor solubility in aqueous solutions, these intercalators contain alkylamine linkers extending from the imides and a PEG linker functionalized with various tertiary amines extending from the core (Figure 2.14). As opposed to typical NDI intercalators, these cNDI intercalators are highly fluorescent in solution, but are quenched once intercalated into DNA.[42]

2.3.1.1 Electrochemical Studies of NDI Intercalators

NDI has been widely studied as an electron acceptor in electron transfer processes, and this property has also been examined alongside its ability to

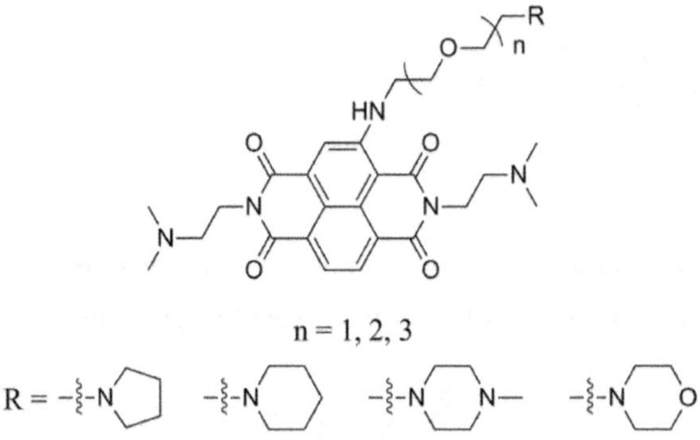

Figure 2.14 PEG-functionalized water soluble NDI derivatives analyzed for DNA binding.

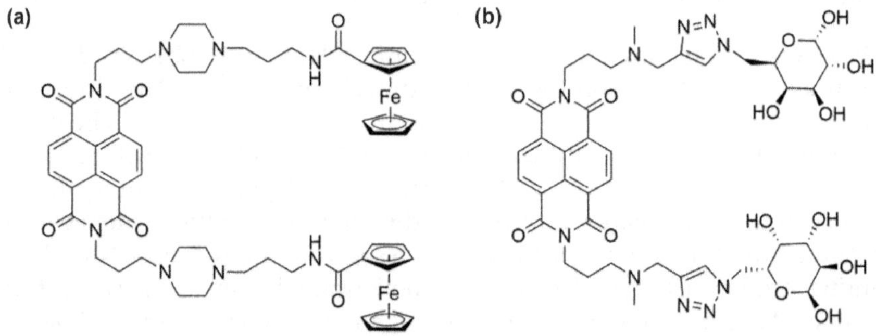

Figure 2.15 (a) Threading NDI intercalator functionalized with ferrocene groups. (b) Threading NDI derivative functionalized with galactose moieties.

intercalate DNA. In electrochemical detection of DNA hybridization, electrochemical response is linked to sequence-selective DNA hybridization, with applications to rapid, sensitive, and cost-effective genome analysis.[43] Takenaka and co-workers have extensively employed NDI modified through the imide linkers with electrochemically active functionalities to develop molecules with reversible redox responses when they bind DNA.[44] In particular, an NDI derivative with ferrocenyl moieties extending from each imide, shown in Figure 2.15, was able to discriminate between dsDNA and ssDNA with a surprisingly low detection limit of 10 zmol for dsDNA.[45] Another threading NDI intercalator containing galactose moieties was shown to give metallization of dsDNA through silver ion reduction for nanowire applications (Figure 2.15).[46] The Takenaka lab and others continue to use modified NDI derivatives toward this purpose.[47–50]

Both NI and NDI derivatives have also shown the ability for photoinduced DNA damage by oxidizing guanine on the 5′ end of GG steps.[51,52] The Barton lab has extensively investigated long range charge transfer through DNA,[53] and has reported that an NDI–TFO conjugate can still mediate photoinduced damage at GG steps at a distance up to 38 bp.[52] In addition, they also observed that NDI was able to repair a thymine dimer in dsDNA from a distance through photooxidation when the NDI unit was covalently attached to one of the oligonucleotide strands.[54] Dixon *et al.* have also described scaffolding intramolecular electron transfer complexes, composed of NDI and a ruthenium center, on DNA by intercalation, and found that intercalation did not significantly affect the electron transfer properties of the NDI–Ru complex.[55]

2.3.1.2 Beyond dsDNA: Targeting Other DNA Structures

NDI monointercalators have also been used for targeting motifs other than simply dsDNA. Dixon and co-workers discovered that NDI derivatives with at least two amine functionalities in one of the threading linkers could bind to and cleave at an abasic site in plasmid DNA.[56] The Neidle lab has reported several crystal structures of tetra-substituted NDIs binding to G-quadruplex DNA. A G-quadruplex is a higher-order DNA secondary structure thought to be important in the G-rich regions of human telomeres. These tetra-substituted NDIs have two linkers extending from the imide groups and two from the NDI core, and positively-charged *N*-methyl piperazine or morpholine groups in the linkers are required for high affinity binding. The NDI derivatives clamp onto the G-quadruplex by binding with the planar core on top of the G-quartet structure with the four linkers in the grooves, stabilizing the structure and inhibiting growth of cancer cells (Figure 2.16).[57,58] Tri-substituted naphthalimides have also been investigated for binding to G-quadruplex DNA.[59]

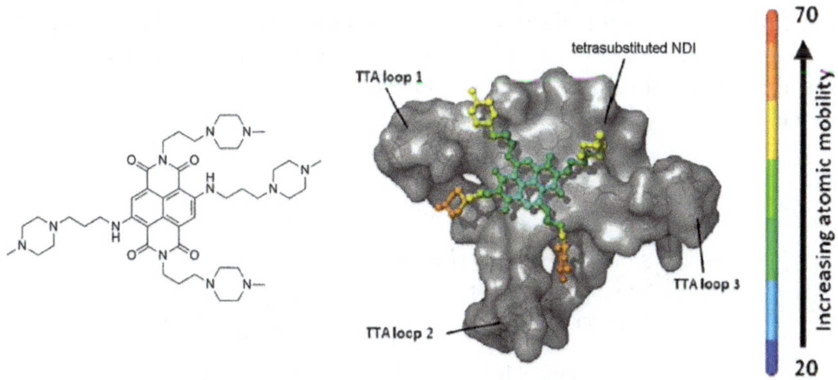

Figure 2.16 Structure of tetra-substituted NDI along with the crystal structure of it bound to the top of G-quadruplex DNA (adapted with permission from G. W. Collie, R. Promontorio, S. M. Hampel, M. Micco, S. Neidle and G. N. Parkinson, Structural Basis for Telomeric G-Quadruplex Targeting by Naphthalene Diimide Ligands, *J. Am. Chem. Soc.*, **134**, 2723–2731, Copyright 2012 American Chemical Society).

2.3.2 Bisintercalators

Bisintercalators contain two intercalating units connected by either a major or minor groove-binding linker. As will be discussed in detail later, the Iverson lab has employed NDI extensively in the design of bisintercalating molecules. The first identified bisintercalator natural product was echinomycin, a pseudosymmetric bicyclic peptide intercalator, whose structure is shown in Figure 2.17.[60] Echinomycin binds preferentially to CpG steps surrounded by AT bases with the peptide linker in the minor groove. Biological function derives from echinomycin's ability to inhibit transcription, with Melillo and co-workers showing that it could block binding of the transcription factor hypoxia-inducible factor-1 (HIF-1) to the hypoxia-responsive element (HRE) sequence, a process important for cancer cell survival and proliferation.[61,62]

Elinafide, also known as LU 79553, was a heavily investigated bisnaphthalimide introduced by the Braña lab (Figure 2.18). This bisintercalator showed antitumor activity through its ability to inhibit topoisomerase II and hence DNA replication.[63] NMR structural studies revealed that elinafide binds preferentially to TpG steps with the alkylamine linker in the major groove.[64] Although elinafide progressed to clinical trials, unwanted side effects hampered its ability to be effectively used in cancer treatment. Structurally similar bisnaphthalimides with core substituents and modified linkers, for example MT02, continue to be of interest in the search for anticancer drugs.[65,66]

While not NDI- or NI-based, another well-studied bisintercalator that bears mentioning is the anthracycline derivative WP631, shown in Figure 2.19, which contains two daunorubicin bisintercalators linked through their amino sugars by *p*-xylene. NMR studies uncovered a binding mode with the linker in the minor groove, intercalation between CpG steps, and 4 bp between intercalators.[67] In the first clear example of an intercalator inhibiting transcription *via* blocking transcription factor binding, WP631 was able to prevent transcription *in vitro* by RNA polymerase II by binding to the GC-rich site for transcription factor activator Sp1.[68,69]

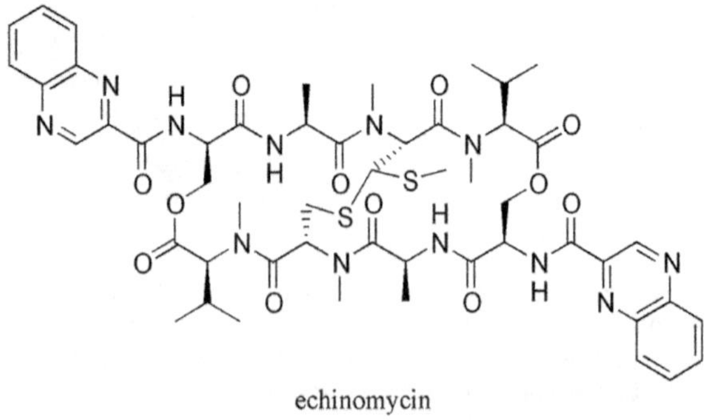

echinomycin

Figure 2.17 Structure of the natural product bisintercalator echinomycin.

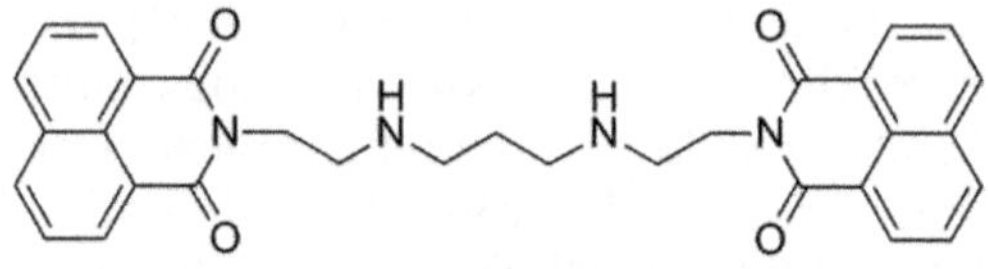

elinafide

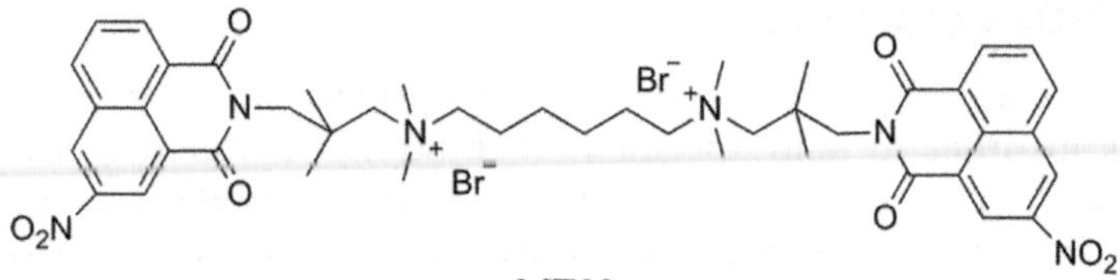

MT02

Figure 2.18 Structures of bisnaphthalimides elinafide and MT02.

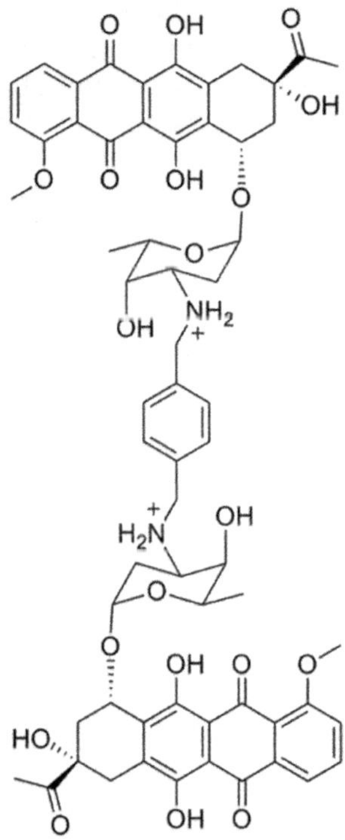

Figure 2.19 Structure of bisanthracycline WP631.

Metallointercalators can also be linked together through attachment of two of their aromatic ligands to generate a bisintercalator complex, as shown by the example from Lincoln, Nordén, and co-workers of the $[Ru(phen)_2dppz]^{2+}$ dimer in Figure 2.20 (phen, 1,10-phenanthroline). This bisintercalator was labeled as a molecular staple for DNA, since the threading binding mode involves the very bulky phen ligands of each ruthenium center residing in one groove, the dppz moieties intercalating, and the aliphatic linker residing in the other groove. Interestingly, binding studies for this bisintercalator indicate that it initially binds as a mono-intercalator, the aliphatic linker disrupts base pairing to pass through the DNA duplex, and finally the second dppz group is able to intercalate.[70,71]

2.3.3 NDI Combilexins

NDI intercalators have often been combined with other DNA recognition moieties in order to lend specificity to binding or impart some kind of functionality. As discussed in Section 2.3.1.1, this combilexin approach has been widely used to take advantage of NDI's electrochemical properties. NDI has been appended to TFOs to create both mono- and bisintercalating molecules.[72,73] The intercalators were found to increase substantially the stability of the triplexes, while still allowing for the TFO to discriminate between cognate and non-cognate sequences. NDI has also been attached to the N-terminus of PNA strands for the targeting of ssDNA. The NDI intercalator was shown through melting studies to stabilize the PNA–DNA duplex, and the intercalator–PNA complex discriminated better between match and

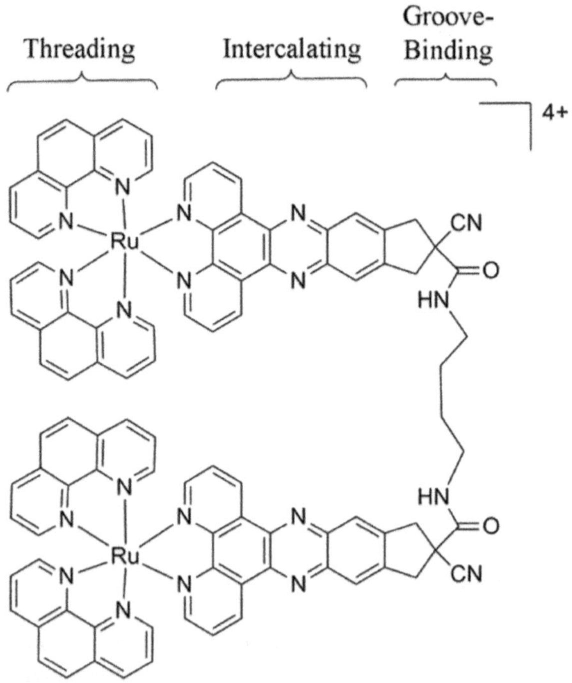

Figure 2.20 Structure of the bisintercalating biruthenium $[Ru(phen)_2dppz]^{2+}$ dimer. The threading, intercalating, and groove binding portions of the molecules are labeled.

mismatch sequences, for the ones tested, compared to unmodified PNA.[74,75] The Iverson lab has employed peptides to link multiple NDI intercalators and develop sequence selectivity, as discussed later on. The Dixon lab developed an NDI–peptide complex containing a metal-chelating Gly-Gly-His motif able to promote single-stranded cleavage in duplex DNA. The intercalator–peptide complex was 100-fold more efficient in nicking plasmid DNA compared to the peptide alone.[76]

2.3.4 Longer Polyintercalator Derivatives

Compared to mono- and bisintercalators, reports of trisintercalators and even longer derivatives are rare. Polyintercalator derivatives containing more than two intercalating units are generally designed to bind DNA in at least two distinct fashions (Figure 2.21).[77] In one approach, all intercalating units are linked on only one side from a linking chain, with the aim of having all intercalators enter from the same DNA groove while the linker follows that groove's curvature. Such intercalating units could therefore be thought of as hanging off the linker in a "pendant" manner, analogous to pendants hanging from a necklace. In a second mode, threading intercalators are linked in a head-to-tail fashion, like beads on a string, so that full polyintercalation requires the linker to thread in and out of the DNA. Note that this binding mode would require the linkers to alter between the major and minor grooves, analogous to the way in which a snake might climb a ladder.

Several of the earliest studies on polyintercalators focused on acridine-based molecules designed using the pendant approach. Roques and co-workers observed only a small increase in binding affinity when going from an acridine dimer to trimer, concluding that the linker chain length and flexibility was crucial in designing high affinity polyintercalators.[78] By using a longer, more flexible aminoalkyl linker, this group generated an acridine trisintercalator (Figure 2.22) specific for polyAT DNA exhibiting an affinity in the range of regulatory proteins, $K_{app} = 10^{14}$ M^{-1}, derived largely from a slow dissociation rate.[79] Nordén and co-workers investigated structurally similar

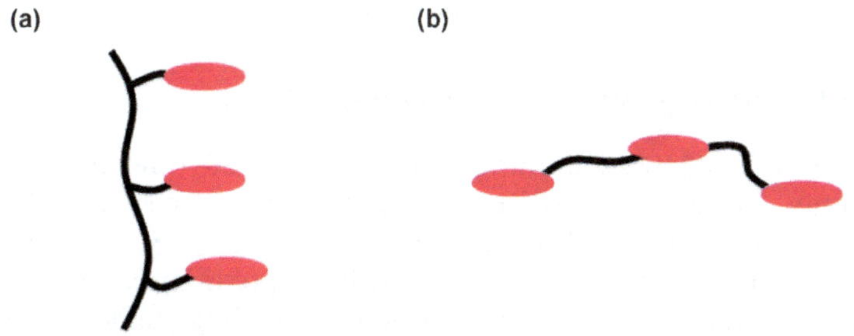

Figure 2.21 Cartoon depicting two different polyintercalation designs with (a) showing a classical "pendant" intercalating mode and (b) showing intercalators connected in a head-to-tail fashion to thread through the DNA.

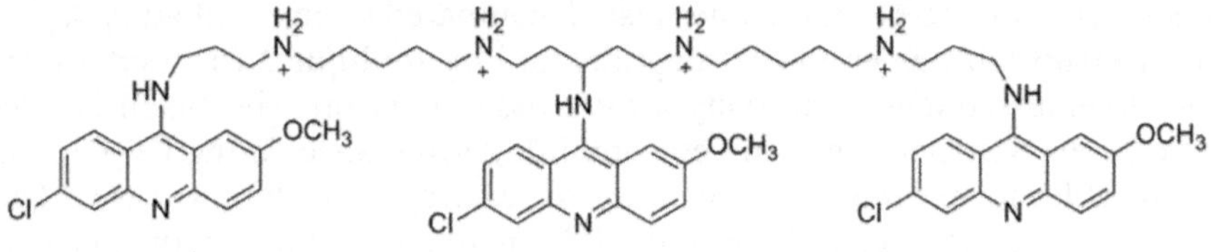

Figure 2.22 Structure of high affinity acridine trisintercalator that binds in the pendant fashion.

acridine-based mono-, bis-, tris-, tetra-, and hexaintercalators. They discovered that while the mono-, bis-, and trisintercalators bound as expected, the tetra- and hexaintercalators only bound with three intercalating units, based on linear and circular dichroism analyses, further establishing the need for linker optimization when designing longer polyintercalators.[80]

For polyintercalators binding in the threading mode, the Zimmerman lab first introduced the idea of a synthetic macrocyclic acridine bisintercalator with the potential to access both grooves simultaneously.[81] A dissertation from that lab by Lamberson presents the idea and synthesis for a threading trisintercalator with a central NDI intercalator with linkers terminating in acridine intercalators, but no binding studies were reported.[82] The Takenaka lab described the synthesis and binding of a polyacridine molecule capable of threading trisintercalation,[83] and we shortly thereafter described the synthesis and binding studies of threading NDI polyintercalators, including the first known tetraintercalator.[84]

More recent studies of polyintercalators include a pendant-type tetraintercalator with a peptide-like backbone and $[Pt(tpy)(py)]^{2+}$ intercalating units (py, pyridine; tpy, terpyridine), although the authors concluded the molecule bound with only two of the $[Pt(tpy)(py)]^{2+}$ units intercalated, again due to non-optimal linker lengths.[85] Far *et al.* describe using molecular modeling to design a pendant trisintercalating porphyrin derivative containing a central intercalating porphyrin with arginyl arms connecting acridine units through flexible linkers on either side of the porphyrin (Figure 2.23), and experimental data supported binding by trisintercalation.[86]

2.4 Threading NDI Polyintercalators

We have developed DNA binding molecules termed threading NDI polyintercalators by linking NDI units together in a head-to-tail fashion (Figure 2.24). These molecules were first reported by Lokey *et al.* and the report includes the first description of a fully bound tetraintercalator, although a structural understanding of the molecules' binding mode was still unavailable.[84] This concept was expanded to include the synthesis and binding studies of the first known octaintercalator, which was concluded to bind with all units intercalated based on unwinding and spectroscopic studies, although only a general sequence preference for GC-rich DNA was noted.[87]

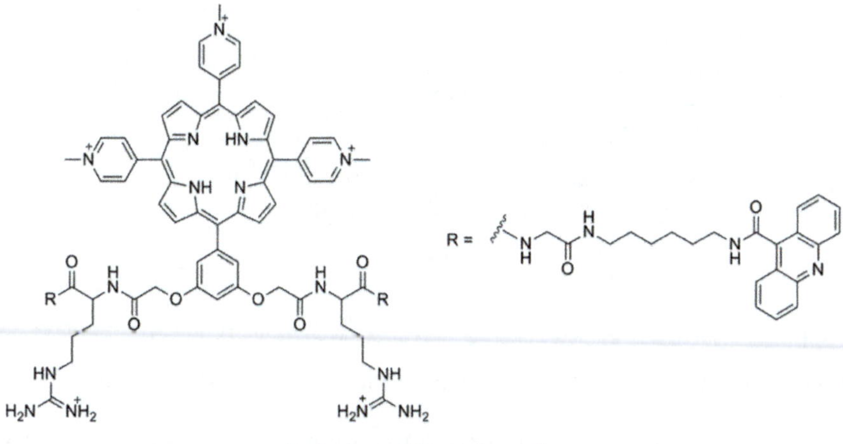

Figure 2.23 Structure of trisintercalating porphyrin–acridine complex.

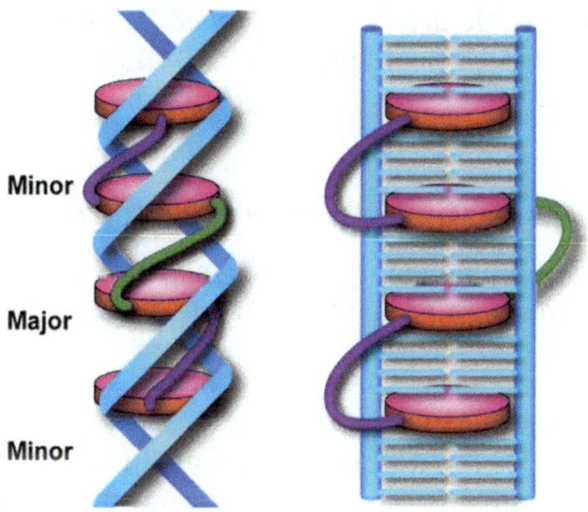

Figure 2.24 Cartoon of threading intercalation.

2.4.1 Synthesis

One of the great advantages of using NDI in the threading polyintercalation scaffold is its ease of synthesis from naphthalene dianhydride and primary amines to create disubstituted diimides. This strategy has been employed by the Iverson lab to synthesize non-symmetric NDI derivatives suitable for use in solid phase peptide synthesis (SPPS).[88,89] Standard Fmoc-based SPPS provides a versatile and robust method for synthesizing NDI polyintercalators, and indeed is likely the reason NDI polyintercalators, most notably containing three or more intercalating units, have been studied in abundance compared to other intercalating small molecules.

A generalized scheme for SPPS-ready NDI monomers is given in Scheme 2.1. Naphthalene dianhydride **1** can be reacted sequentially in a

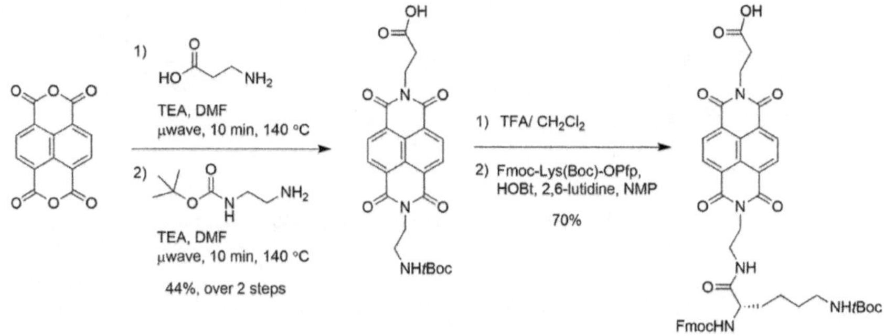

1 2 3

Scheme 2.1 Synthesis of NDI-Lys monomer **3** for use in Fmoc-based SPPS.

microwave reactor under basic conditions with β-alanine followed by mono-Boc-protected ethylenediamine, producing diimide **2**. An Fmoc-protected amine incorporated at this step is unstable, so one solution phase coupling with a lysine amino acid is accomplished by deprotecting the Boc-protected amine with trifluoroacetic acid (TFA), followed by an amide condensation with the activated OPfp ester of Fmoc-Lys(Boc)-OPfp to give the SPPS-ready NDI-Lys monomer **3**.[90,91]

2.4.2 Sequence-specific NDI Bisintercalators

Once the general binding ability of this class of polyintercalators was established, we then modified the approach to these molecules, beginning to consider them from a structural viewpoint with the potential for sequence-specific recognition. A library of bisintercalators that varied the peptide linker between NDI intercalators was synthesized. The library included a wide range of amino acid functionalities and was screened for sequence-specific binding by DNase I footprinting. This analysis resulted in the discovery of two bisintercalators, one having a Gly_3Lys linker (compound **4**) and the other containing a β-Ala_3Lys linker (compound **5**), with different DNA binding preferences for the sequences 5′-GGTACC-3′ and 5′-GATAAG-3′, respectively.[92]

NMR structural analyses of these two bisintercalators with their preferred sequences revealed a series of surprising details. First, the two exhibited entirely different groove-binding topologies, as **4** bound with the linker in the major groove and **5** bound with the linker in the minor groove (Figure 2.25).[93,94] Second, both bisintercalators bound with 4 bp between NDI intercalators rather than adhering to the nearest neighbor exclusion model.

Although no simple story emerged from the structural models, it was possible to hypothesize a few reasons why these linkers direct binding to different sequences. For the major groove-binding Gly_3Lys bisintercalator **4**, hydrogen bonds were observed between the amides adjacent to the NDI

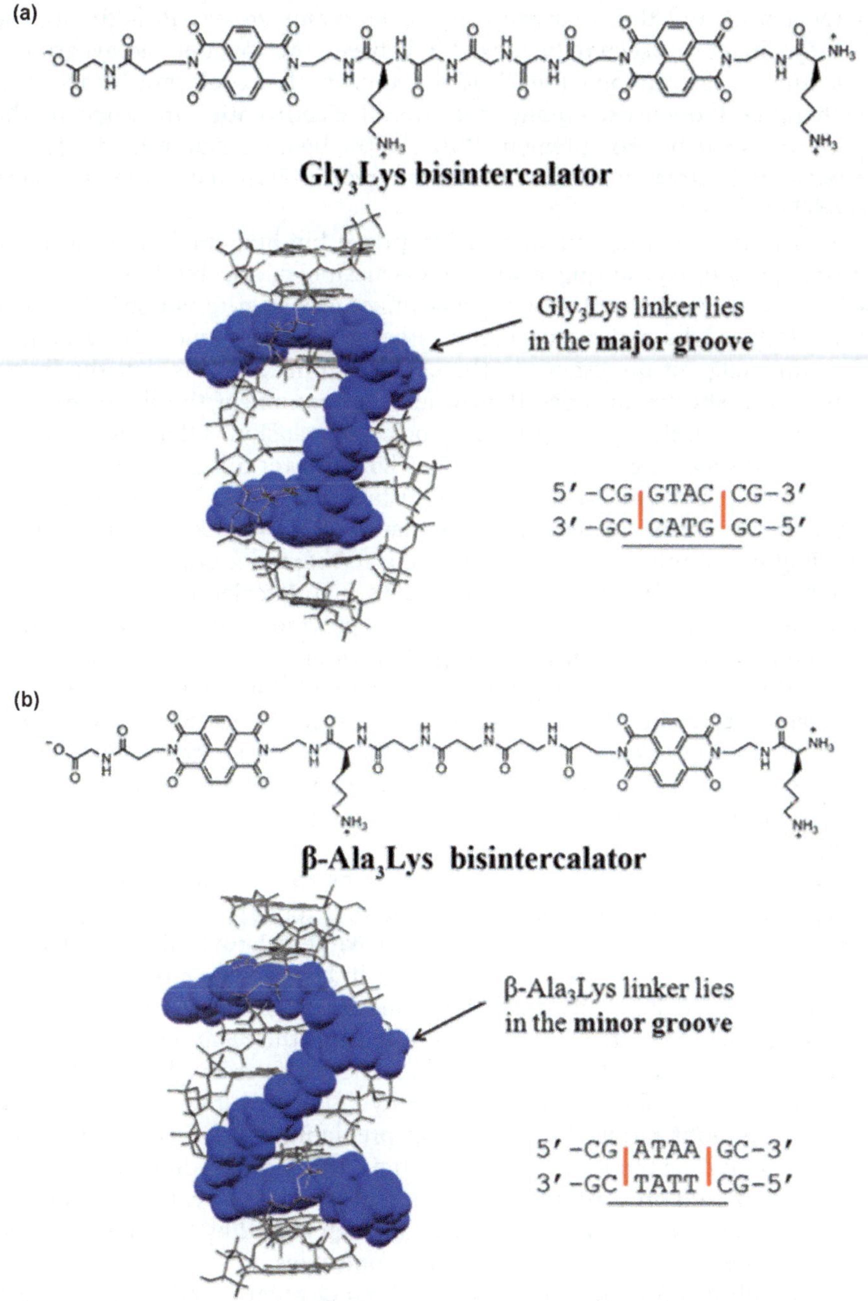

Figure 2.25 The (a) Gly_3Lys bisintercalator **4** and (b) β-Ala_3Lys bisintercalator **5** along with their NMR-determined structures bound to their preferred DNA sequences. The sequences in the NMR structures are shown with the binding sites underlined.

intercalators and the G oxygen (O6) in the major groove. In addition, the Gly_3Lys linker was seen to be tucked in between two thymine methyl groups, resulting in hydrophobic (desolvation) contacts that could not be achieved with other sequences. Finally, the overall electrostatic landscape of the preferred sequence complemented that of the bisintercalator, lending more favorable electrostatic interactions compared to other sequences that were modeled.[93]

It is worth pointing out that major groove binding small molecules are relatively rare, warranting a deeper discussion of the binding mode for Gly_3Lys bisintercalator **4**. The Gly_3Lys linker was seen to not strictly follow the middle of the major groove, but rather it "cut the corner" by extending from one edge of the groove to the other. The net result is that the linker traversed a shorter distance than would be required if the linker were to follow a path in the central groove. A logical conclusion is that overall linker length is critical to groove recognition. The narrower minor groove does not present an opportunity to shorten the distance by cutting across the groove. Therefore, it is reasonable to propose that the Gly_3Lys linker is too short for binding in the minor groove, relegating it to the major groove.

For the minor groove-binding β-Ala_3Lys bisintercalator **5**, the narrower and more hydrophobic minor groove likely provides better accommodation for the β-Ala_3Lys linker. Recall that β-Ala residues have been found to bind A–T sequences in the minor groove in the context of polyamides as well.[16] In the case of threading polyintercalators, the longer length relative to Gly_3Lys allows β-Ala_3Lys to traverse the longer path of the minor groove for binding between 4 bp. Again, hydrogen bonds were seen between the amide hydrogens adjacent to the NDI units and the oxygen (O2) on the thymines in the minor groove.

The discovery and structural analyses of the two bisintercalators with altered sequence specificity and groove-binding topologies laid the groundwork for designing even more complex polyintercalators. Bisintercalator **6** was modeled and synthesized with a rigid linker backbone (Figure 2.26) in an effort to increase binding affinity by minimizing entropy loss upon DNA binding. This bisintercalator was found to bind the *minor groove* of the 5′-GGTACC-3′ sequence, the binding site for the major groove-binding Gly_3Lys bisintercalator **4**, illustrating the versatility of the NDI intercalator scaffold, but also demonstrating the difficulty in predicting *a priori* sequence specificity and binding topology.[95,96] Surprisingly, the expected dramatic increase in affinity based on conformational restriction was not realized with **6**, indicating that a great deal remains to be learned regarding the maximization of binding energy with polyintercalating molecules.

Chu *et al.* also combined the major and minor groove-binding Gly_3Lys and β-Ala_3Lys linkers, respectively, to generate a macrocyclic threading NDI bisintercalator **7**, whose NMR structural analysis represents the first for a pseudocatenane cyclic bisintercalator with duplex DNA (Figure 2.27). This cyclic bisintercalator **7** preferentially bound the 5′-GGTACC-3′ sequence with the linkers in the predicted grooves, with the assumption that the sequence

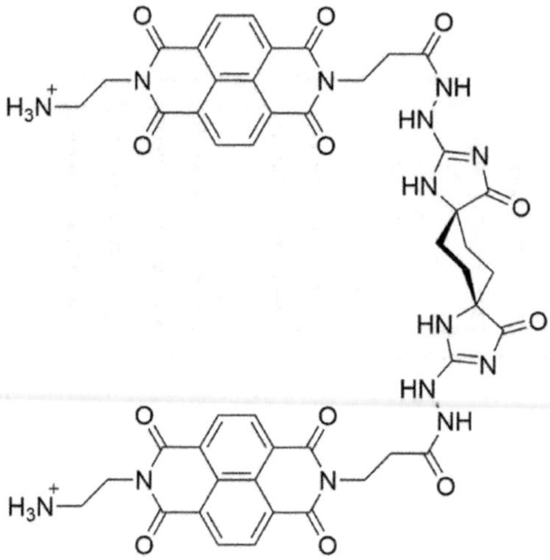

6

Figure 2.26 Structure of rigidified bisintercalator **6**.

specificity was driven by more favorable binding for the Gly_3Lys linker in the major groove of that sequence compared to the binding site preferred by the β-Ala_3Lys linker.[97] Interestingly, it was hypothesized that 7 might be able to scan for its preferred site by initially binding to DNA and sliding up and down along the DNA, disrupting base pairing, until finding its preferred binding sequence. A similar type of "shuffling" or "crawling" association mode has also been proposed for actinomycin D and a trisintercalating acridine polyintercalator.[98,99] No matter the details of the association mechanism, the cyclic bisintercalator 7 was able to associate relatively rapidly, *i.e.* within 1.5 hours, with its preferred binding sequence in the context of a relatively long fragment of double-stranded DNA.

2.4.3 Sequence-specific NDI Tetra- and Hexaintercalators

The modularity of the threading polyintercalator design was exploited in the development of the first tetraintercalating molecule to be structurally characterized when bound to DNA. The lessons learned with the Gly_3Lys and β-Ala_3Lys bisintercalators **4** and **5** were exploited in the design of tetraintercalator **8** (Figure 2.28).[100] In particular, tetraintercalator **8** was designed to bind with linkers residing in the minor, major, and minor DNA grooves, respectively. The minor groove-binding linkers of **8** are composed of the β-Ala_3Lys linkers from **5**. However, the major groove binding portion was chosen to be adipic acid, a linker that matches the overall length and hydrophobic nature of the Gly_3Lys linker of **4**. The adipic acid linker greatly simplified the synthesis of **8**, while imparting overall C_2 symmetry to the molecule. The 14 bp specific binding site for the adipic acid tetraintercalator **8** is a palindromic hybrid sequence of the individual bisintercalator binding

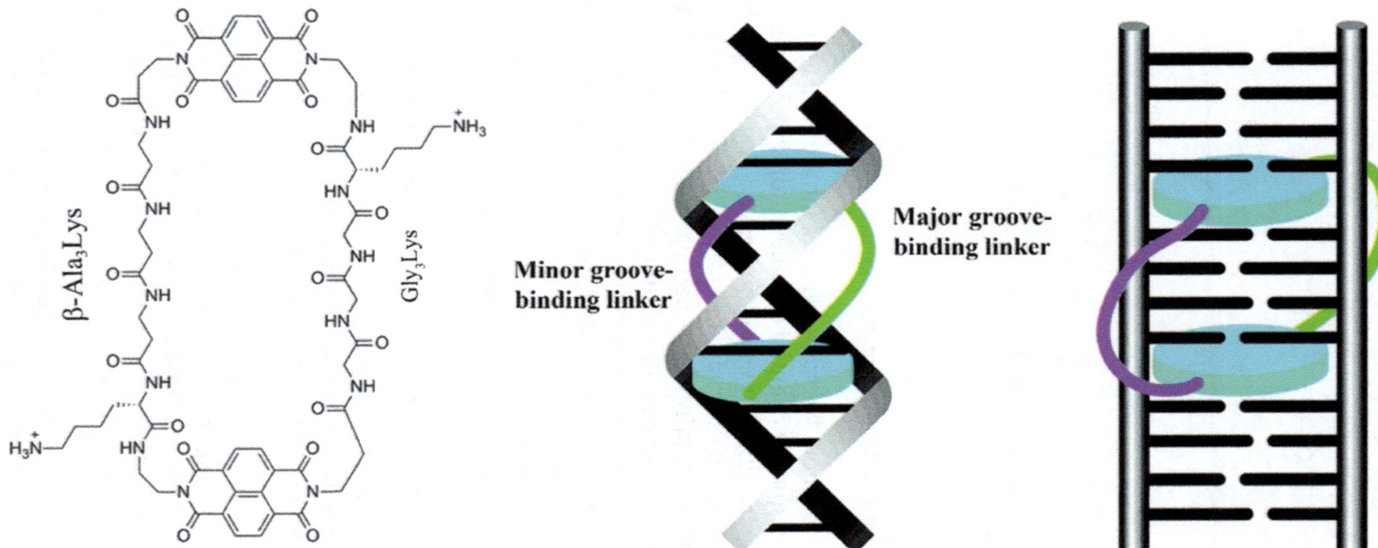

Figure 2.27 Structure of cyclic bisintercalator 7 with groove-binding linkers labeled and cartoon showing 7 forming a pseudocatenane with DNA.
(Adapted from Y. Chu, D. W. Hoffman and B. L. Iverson, A Pseudocatenane Structure Formed between DNA and a Cyclic Bisintercalator, *J. Am. Chem. Soc.*, **131**, 3499–3508, Copyright 2009 American Chemical Society.)

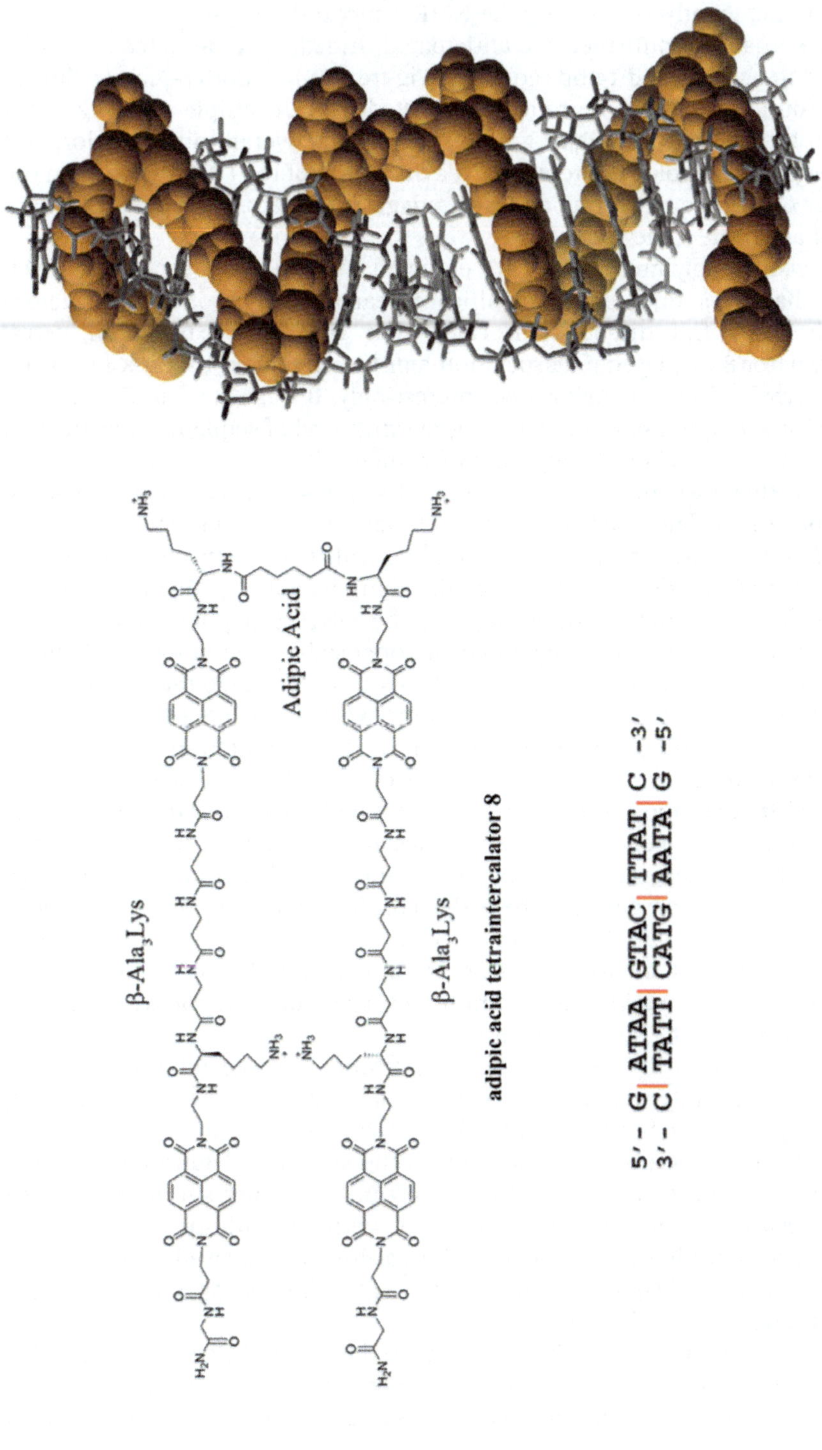

Figure 2.28 Structure of the adipic acid tetraintercalator **8** with its 14 bp binding site along with the NMR-determined structure of the bound complex.

sites, so the complex of the tetraintercalator with the 14 bp site was also C_2 symmetric, greatly simplifying the NMR structural analysis.

NMR analysis confirmed the anticipated threading tetraintercalation of **8** bound to its preferred 14 bp sequence (Figure 2.28).[100] Both β-Ala$_3$Lys linkers were found in the minor groove as expected, and the adipic acid linker was found to reside in the major groove, proving the overall minor, major, and minor groove topology. Interestingly, a distortion of the DNA structure in the center of the sequence was detected, suggesting that the adipic acid linker may have been somewhat short.

Threading polyintercalators are expected to have relatively long dissociation half-lives due to the significant rearrangement of DNA structure required for full dissociation. Consistent with this expectation, tetraintercalator **8** displayed a dissociation half-life of 16 days from its 14 bp site as determined by gel-shift assay. Interestingly, **8** was also found to have a surprisingly rapid association and a gratifying level of sequence specificity in the context of a relatively large DNA fragment.[101]

To further explore the modularity and sequence selectivity of threading NDI polyintercalators, tetraintercalator **9** was designed and synthesized with a different putative major groove binding linker and consequently overall linear N to C amide bond directionality along its entirety (Figure 2.29).[91] In particular, tetraintercalator **9** employs the Gly$_3$Lys peptide linker as the central major groove-binding linker as opposed to the adipic acid linker used in **8**. The critical consequence of this design change relates to overall molecule symmetry, as **9** has one minor groove-binding β-Ala$_3$Lys linker with reversed N to C directionality compared to adipic acid tetraintercalator **8**.

These changes in the linkers were expected to affect the sequence selectivity of **9**, and footprinting analysis revealed that indeed **9** is able to discriminate between a set of hybrid designed sequences with one or two base pair changes, although not in the anticipated manner. Additionally, kinetic studies comparing **8** to **9** did confirm that the adipic acid tetraintercalator **8** displayed a greater than 100-fold preference for its 14 bp site over the preferred site for the linear tetraintercalator **9**, a difference of 3 bp. Tetraintercalator **9** still showed impressive sequence specificity, but not as much as **8** with the sequences tested.

In an effort to increase the binding affinity for NDI tetraintercalators, a systematic study was performed that analyzed the effect of lengthening the major groove-binding linker in the adipic acid tetraintercalator **8** (Figure 2.30).[102] Recall that distortion in the structure determined by NMR indicated that the adipic acid linker may be somewhat shorter than optimum. Astonishingly, tetraintercalator **10** with one additional methylene unit, corresponding to a pimelic acid cross-linker, displayed a dissociation half-life of around 60 days from the same 14 bp site, setting a new record for a synthetic DNA binding molecule.

The pimelic acid cross-linker was used in a new hybrid molecule, **13**, designed to extend sequence-specific threading polyintercalation to the hexaintercalator level. Hexaintercalator **13** is anticipated to bind with linkers

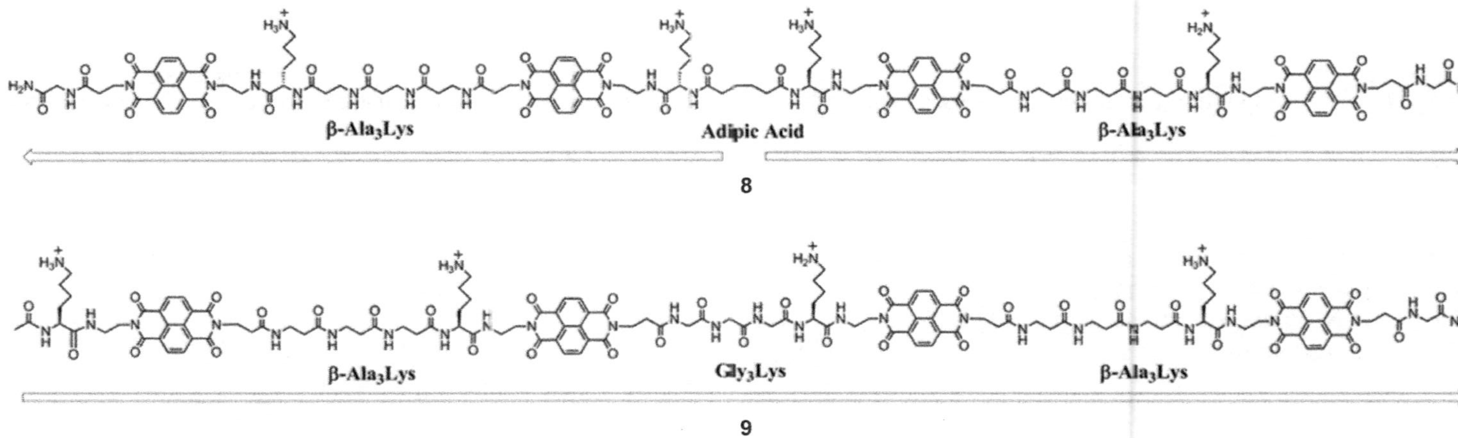

Figure 2.29 Comparison of C_2 symmetric tetraintercalator **8** and linear tetraintercalator **9** with a Gly_3Lys major groove binder and β-Ala_3Lys minor groove binder with reversed N to C directionality compared to **8**. The arrows indicate the N to C directionality of the peptide backbone.

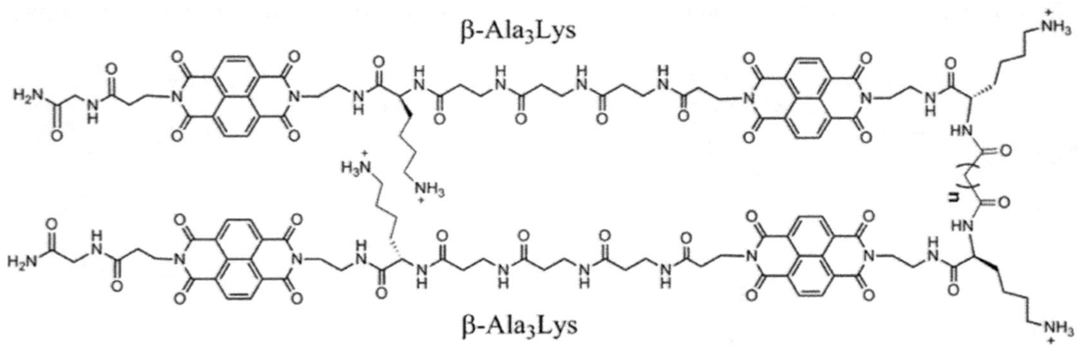

8 n = 4 Adipic Acid 16 days
10 n = 5 Pimelic Acid 57 days
11 n = 6 Suberic Acid 27 days
12 n = 7 Azelaic Acid 18 days

Figure 2.30 Structures of lengthened tetraintercalators **10–12** and corresponding dissociation half-lives for binding to their preferred 14 bp site.

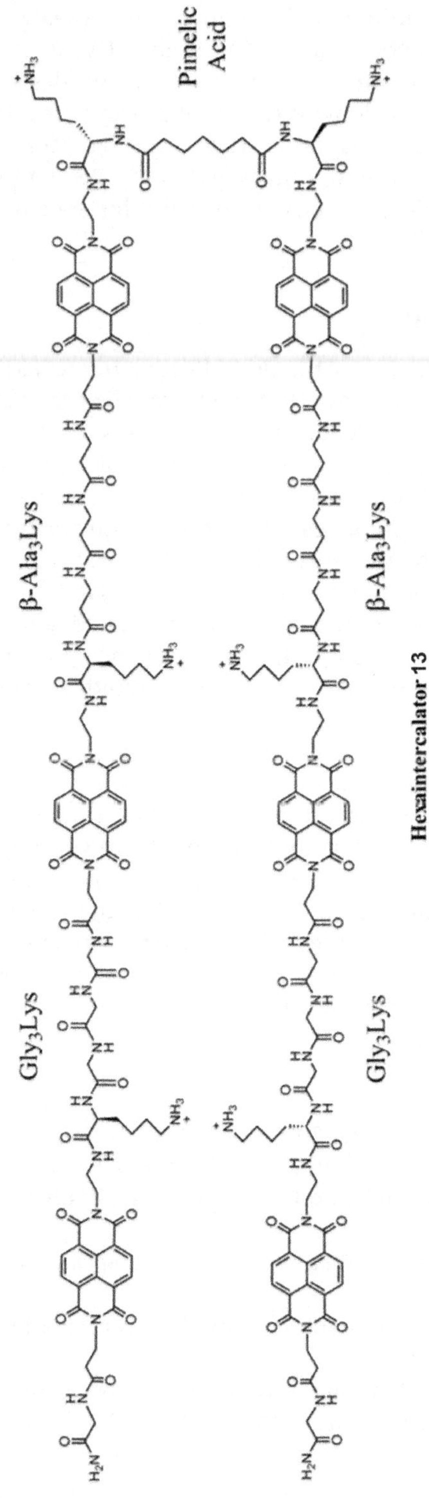

Figure 2.31 Structure of NDI hexaintercalator 13.

residing in the major, minor, major, minor, major DNA grooves, respectively (Figure 2.31). Using DNase I footprinting and UV-vis spectroscopy, it was shown that hexaintercalator **13** fully bound to its designed 22 bp hybrid binding site, which is the longest for a synthetic, non-nucleic acid molecule. Surprisingly, **13** dissociated faster than expected from this sequence, a phenomenon that is still under investigation.[102] For comparison, TFOs and PNAs, which both rely on nucleic acid bases for specificity, have targeted longer sequences.[103,104]

2.5 Conclusion

NDI has proven to be a versatile and effective intercalation platform for a growing family of threading DNA polyintercalators. Future studies will continue to define the scope and limitations of threading NDI polyintercalation as a DNA-binding strategy. New linkers are needed to expand the repertoire of DNA sequences targeted, including new minor groove and major groove specific moieties. On the mechanistic front, it will be important to investigate the surprisingly rapid association rates observed, raising interesting questions about the mechanism(s) by which threading polyintercalators seek out and bind their preferred sequences within long DNA fragments. In this context, the ability of the cyclic bisintercalator **7** to bind its preferred sequence in relatively rapid fashion is particularly intriguing.

NDI has attracted a lot of attention due to its remarkable redox, photophysical and supramolecular properties. It is safe to say that DNA-binding ability is an appropriate addition to the list of reasons why NDI is a fascinating molecule. While early studies of pendant-type polyintercalators with acridine intercalating units were disappointing, our more recent work with threading NDI polyintercalators has revealed the ability to discriminate between similar DNA sequences, bind very long sequences up to 22 bp, and dissociate extremely slowly from preferred sites, all attractive attributes from a DNA-targeted drug design perspective.

References

1. M. Morange, *Resonance*, 2009, **14**, 236–247.
2. J. A. Doudna and E. Charpentier, *Science*, 2014, **346**, 1258096.
3. T. Gaj, C. A. Gersbach and C. F. Barbas III, *Trends Biotechnol.*, 2013, **31**, 397–405.
4. K. Wüthrich, *NMR of Proteins and Nucleic Acids*, John Wiley and Sons, New York, 1986.
5. M. E. Vázquez, A. M. Caamaño and J. L. Mascareñas, *Chem. Soc. Rev.*, 2003, **32**, 338–349.
6. T. Le Doan, L. Perrouault, D. Praseuth, N. Habhoub, J. L. Decout, N. T. Thuong, J. Lhomme and C. Hélène, *Nucleic Acids Res.*, 1987, **15**, 7749–7760.

7. H. E. Moser and P. B. Dervan, *Science*, 1987, **238**, 645–650.
8. M. Duca, P. Vekhoff, K. Oussedik, L. Halby and P. B. Arimondo, *Nucleic Acids Res.*, 2008, **36**, 5123–5138.
9. A. Jain, G. Wang and K. M. Vasquez, *Biochimie*, 2008, **90**, 1117–1130.
10. J. S. Lee, M. L. Woodsworth, L. J. Latimer and A. R. Morgan, *Nucleic Acids Res.*, 1984, **12**, 6603–6614.
11. M. Duca, P. Vekhoff, K. Oussedik, L. Halby and P. B. Arimondo, *Nucleic Acids Res.*, 2008, **36**, 5123–5138.
12. P. E. Nielsen, *Curr. Med. Chem.*, 2001, **8**, 545–550.
13. Y. N. Kosaganov, D. A. Stetsenko, E. N. Lubyako, N. P. Kvitko, Y. S. Lazurkin and P. E. Nielsen, *Biochemistry*, 2000, **39**, 11742–11747.
14. A. Ray and B. Nordén, *FASEB J.*, 2000, **14**, 1041–1060.
15. C. Melander, R. Burnett and J. M. Gottesfeld, *J. Biotechnol.*, 2004, **112**, 195–220.
16. P. B. Dervan, *Bioorg. Med. Chem.*, 2001, **9**, 2215–2235.
17. J. W. Trauger, E. E. Baird and P. B. Dervan, *J. Am. Chem. Soc.*, 1998, **120**, 3534–3535.
18. J. L. Meier, A. S. Yu, I. Korf, D. J. Segal and P. B. Dervan, *J. Am. Chem. Soc.*, 2012, **134**, 17814–17822.
19. C. L. Warren, N. C. S. Kratochvil, K. E. Hauschild, S. Foister, M. L. Brezinski, P. B. Dervan, G. N. Phillips and A. Z. Ansari, *Proc. Natl. Acad. Sci. U.S.A.*, 2006, **103**, 867–872.
20. S. Chavda, B. Babu, S. K. Yanow, A. Jardim, T. W. Spithill, K. Kiakos, J. Kluza, J. A. Hartley and M. Lee, *Bioorg. Med. Chem.*, 2010, **18**, 5016–5024.
21. L. S. Lerman, *J. Mol. Biol.*, 1961, **3**, 18–30.
22. P. J. Bond, R. Langridge, K. W. Jennette and S. J. Lippard, *Proc. Natl. Acad. Sci. U.S.A.*, 1975, **72**, 4825–4829.
23. D. M. Crothers, *Biopolymers*, 1968, **6**, 575–584.
24. N. J. Wheate, C. R. Brodie, J. G. Collins, S. Kemp and J. R. Aldrich-Wright, *Mini-Rev. Med. Chem.*, 2007, 7, 627–648.
25. A. E. Friedman, J. C. Chambron, J. P. Sauvage, N. J. Turro and J. K. Barton, *J. Am. Chem. Soc.*, 1990, **112**, 4960–4962.
26. D. R. Boer, A. Canals and M. Coll, *Dalton Trans.*, 2008, 399–414.
27. V. Tumiatti, A. Milelli, A. Minarini, M. Micco, A. Gasperi Campani, L. Roncuzzi, D. Baiocchi, J. Marinello, G. Capranico, M. Zini, C. Stefanelli and C. Melchiorre, *J. Med. Chem.*, 2009, **52**, 7873–7877.
28. P. Quintana-Espinoza, J. García-Luis, Á. Amesty, P. Martín-Rodríguez, I. Lorenzo-Castrillejo, A. G. Ravelo, L. Fernández-Pérez, F. Machín and A. Estévez-Braun, *Bioorg. Med. Chem.*, 2013, **21**, 6484–6495.
29. Z. Tian, Z. Zhao, F. Zang, Y. Wang and C. Wang, *J. Photochem. Photobiol., B*, 2014, **138**, 202–210.
30. T. Brider, B. Redko, M. Oron-Herman, A. Cohen-Matzlich, G. Gerlitz, G. Gellerman and F. Grynszpan, *Res. Chem. Intermed.*, DOI: 10.1007/s11164-015-2115-1, [Online Early Access]. Published Online: June, 2, 2015, http://link.springer.com/journal/11164.

31. C. A. Johnson, G. A. Hudson, L. K. E. Hardebeck, E. A. Jolley, Y. Ren, M. Lewis and B. M. Znosko, *Bioorg. Med. Chem.*, 2015, **23**, 3586–3591.
32. K. Wang, Y. Wang, X. Yan, H. Chen, G. Ma, P. Zhang, J. Li, X. Li and J. Zhang, *Bioorg. Med. Chem. Lett.*, 2012, **22**, 937–941.
33. S. Banerjee, E. B. Veale, C. M. Phelan, S. A. Murphy, G. M. Tocci, L. J. Gillespie, D. O. Frimannsson, J. M. Kelly and T. Gunnlaugsson, *Chem. Soc. Rev.*, 2013, **42**, 1601–1618.
34. M. F. Braña, M. Cacho, A. Gradillas, B. de Pascual-Teresa and A. Ramos, *Curr. Pharm. Des.*, 2001, **7**, 1745–1780.
35. E. J. Gabbay, R. DeStefano and C. S. Baxter, *Biochem. Biophys. Res. Commun.*, 1973, **51**, 1083–1089.
36. F. A. Tanious, S. F. Yen and W. D. Wilson, *Biochemistry*, 1991, **30**, 1813–1819.
37. S. F. Yen, E. J. Gabbay and W. D. Wilson, *Biochemistry*, 1982, **21**, 2070–2076.
38. R. E. McKnight, A. B. Gleason, J. A. Keyes and S. Sahabi, *Bioorg. Med. Chem. Lett.*, 2007, **17**, 1013–1017.
39. R. E. McKnight, E. Reisenauer, M. V. Pintado, S. R. Polasani and D. W. Dixon, *Bioorg. Med. Chem. Lett.*, 2011, **21**, 4288–4291.
40. A. Milelli, V. Tumiatti, M. Micco, M. Rosini, G. Zuccari, L. Raffaghello, G. Bianchi, V. Pistoia, J. Fernando Díaz, B. Pera, C. Trigili, I. Barasoain, C. Musetti, M. Toniolo, C. Sissi, S. Alcaro, F. Moraca, M. Zini, C. Stefanelli and A. Minarini, *Eur. J. Med. Chem.*, 2012, **57**, 417–428.
41. Z. R. Liu, K. H. Hecker and R. L. Rill, *J. Biomol. Struct. Dyn.*, 1996, **14**, 331–339.
42. H. Wei, M. Lv, X. Duan, S. Li, Y. Yao, K. Wang, P. Zhang, X. Li and H. Chen, *Med. Chem. Res.*, 2014, **23**, 2277–2286.
43. E. G. Hvastkovs and D. A. Buttry, *Analyst*, 2010, **135**, 1817–1829.
44. S. Takenaka, in *Small Molecule DNA and RNA Binders*, ed. M. Demeunynck, C. Bailly, and W. D. Wilson, Wiley-VCH Verlag GmbH & Co., Weinheim, 2004, pp. 224–246.
45. S. Takenaka, K. Yamashita, M. Takagi, Y. Uto and H. Kondo, *Anal. Chem.*, 2000, **72**, 1334–1341.
46. K. Komizo, H. Ikedo, S. Sato and S. Takenaka, *Bioconjugate Chem.*, 2014, **25**, 1547–1555.
47. S. Sato, M. Tsueda and S. Takenaka, *J. Organomet. Chem.*, 2010, **695**, 1858–1862.
48. N. C. Tansil, H. Xie, F. Xie and Z. Gao, *Anal. Chem.*, 2005, **77**, 126–134.
49. M. Haga, M. Ohta, H. Machida, M. Chikira and N. Tonegawa, *Thin Solid Films*, 2006, **499**, 201–206.
50. H. Deng, X. Yang, S. P. X. Yeo and Z. Gao, *Anal. Chem.*, 2014, **86**, 2117–2123.
51. I. Saito, M. Takayama, H. Sugiyama, K. Nakatani, A. Tsuchida and M. Yamamoto, *J. Am. Chem. Soc.*, 1995, **117**, 6406–6407.
52. M. E. Núñez, K. T. Noyes, D. A. Gianolio, L. W. McLaughlin and J. K. Barton, *Biochemistry*, 2000, **39**, 6190–6199.

53. J. C. Genereux and J. K. Barton, *Chem. Rev.*, 2010, **110**, 1642–1662.
54. D. A. Vicic, D. T. Odom, M. E. Núñez, D. A. Gianolio, L. W. McLaughlin and J. K. Barton, *J. Am. Chem. Soc.*, 2000, **122**, 8603–8611.
55. D. W. Dixon, N. B. Thornton, V. Steullet and T. Netzel, *Inorg. Chem.*, 1999, **38**, 5526–5534.
56. V. Steullet, S. Edwards-Bennett and D. W. Dixon, *Bioorg. Med. Chem.*, 1999, **7**, 2531–2540.
57. G. W. Collie, R. Promontorio, S. M. Hampel, M. Micco, S. Neidle and G. N. Parkinson, *J. Am. Chem. Soc.*, 2012, **134**, 2723–2731.
58. M. Micco, G. W. Collie, A. G. Dale, S. A. Ohnmacht, I. Pazitna, M. Gunaratnam, A. P. Reszka and S. Neidle, *J. Med. Chem.*, 2013, **56**, 2959–2974.
59. A. Peduto, B. Pagano, C. Petronzi, A. Massa, V. Esposito, A. Virgilio, F. Paduano, F. Trapasso, F. Fiorito, S. Florio, C. Giancola, A. Galeone and R. Filosa, *Bioorg. Med. Chem.*, 2011, **19**, 6419–6429.
60. M. J. Waring and L. P. G. Wakelin, *Nature*, 1974, **252**, 653–657.
61. D. Kong, E. J. Park, A. G. Stephen, M. Calvani, J. H. Cardellina, A. Monks, R. J. Fisher, R. H. Shoemaker and G. Melillo, *Cancer Res.*, 2005, **65**, 9047–9055.
62. C.-H. Leung, D. S.-H. Chan, V. P.-Y. Ma and D.-L. Ma, *Med. Res. Rev.*, 2013, **33**, 823–846.
63. P. F. Bousquet, M. F. Braña, D. Conlon, K. M. Fitzgerald, D. Perron, C. Cocchiaro, R. Miller, M. Moran, J. George, X. D. Qian, G. Keilhauer and C. A. Romerdahl, *Cancer Res.*, 1995, **55**, 1176–1180.
64. J. Gallego and B. R. Reid, *Biochemistry*, 1999, **38**, 15104–15115.
65. L. González-Bulnes and J. Gallego, *Biopolymers*, 2012, **97**, 974–987.
66. T. M. Menzel, M. Tischer, P. François, J. Nickel, J. Schrenzel, H. Bruhn, A. Albrecht, L. Lehmann, U. Holzgrabe and K. Ohlsen, *Antimicrob. Agents Chemother.*, 2011, **55**, 311–320.
67. H. Robinson, W. Priebe, J. B. Chaires and A. H.-J. Wang, *Biochemistry*, 1997, **36**, 8663–8670.
68. B. Martín, A. Vaquero, W. Priebe and P. José, *Nucleic Acids Res.*, 1999, **27**, 3402–3409.
69. J. Portugal, B. Martín, A. Vaquero, N. Ferrer, S. Villamarin and W. Priebe, *Curr. Med. Chem.*, 2001, **8**, 1–8.
70. B. Önfelt, P. Lincoln and B. Nordén, *J. Am. Chem. Soc.*, 1999, **121**, 10846–10847.
71. M. Bahira, M. J. McCauley, A. A. Almaqwashu, P. Lincoln, F. Westerlund, I. Rouzina and M. C. Williams, *Nucleic Acids Res.*, 2015, **43**, 8856–8867.
72. D. A. Gianolio, J. M. Segismundo and L. W. McLaughlin, *Nucleic Acids Res.*, 2000, **28**, 2128–2134.
73. D. A. Gianolio and L. W. McLaughlin, *Bioorg. Med. Chem.*, 2001, **9**, 2329–2334.
74. A. Mokhir, B. Zohm, A. Fuessl and R. Kraemer, *Bioorg. Med. Chem. Lett.*, 2003, **13**, 2489–2492.

75. A. A. Mokhir and R. Kraemer, *Bioconjugate Chem.*, 2003, **14**, 877–883.
76. V. Steullet and D. W. Dixon, *Bioorg. Med. Chem. Lett.*, 1999, **9**, 2935–2940.
77. S. Takenaka and M. Takagi, *Bull. Chem. Soc. Jpn.*, 1999, **72**, 327–337.
78. B. Gaugain, J. Markovits, J. B. Le Pecq and B. P. Roques, *FEBS Lett.*, 1984, **169**, 123–126.
79. P. Laugâa, J. Markovits, A. Delbarre, J. B. Le Pecq and B. P. Roques, *Biochemistry*, 1985, **24**, 5567–5575.
80. M. Wirth, O. Buchardt, T. Koch, P. E. Nielsen and B. Norden, *J. Am. Chem. Soc.*, 1988, **110**, 932–939.
81. S. C. Zimmerman, C. R. Lamberson, M. Cory and T. A. Fairley, *J. Am. Chem. Soc.*, 1989, **111**, 6805–6809.
82. C. R. Lamberson, Topologically constrained DNA intercalators: Synthesis and DNA binding studies. PhD dissertation, University of Illinois at Urbana-Champaign, Urbana, IL, 1991.
83. S. Takenaka, S. Nishira, K. Tahara, H. Kondo and M. Takagi, *Supramol. Chem.*, 1993, **2**, 41–46.
84. R. S. Lokey, Y. Kwok, V. Guelev, C. J. Pursell, L. H. Hurley and B. L. Iverson, *J. Am. Chem. Soc.*, 1997, **119**, 7202–7210.
85. L. A. Levine, C. M. Morgan, K. Ohr and M. E. Williams, *J. Am. Chem. Soc.*, 2005, **127**, 16764–16765.
86. S. Far, A. Kossanyi, C. Verchère-Béaur, N. Gresh, E. Taillandier and M. Perrée-Fauvet, *Eur. J. Org. Chem.*, 2004, **2004**, 1781–1797.
87. M. M. Murr, M. T. Harting, V. Guelev, J. Ren, J. B. Chaires and B. L. Iverson, *Bioorg. Med. Chem.*, 2001, **9**, 1141–1148.
88. P. Pengo, G. D. Pantoş, S. Otto and J. K. M. Sanders, *J. Org. Chem.*, 2006, **71**, 7063–7066.
89. K. Tambara, N. Ponnuswamy, G. Hennrich and G. D. Pantoş, *J. Org. Chem.*, 2011, **76**, 3338–3347.
90. V. M. Guelev, M. S. Cubberley, M. M. Murr, R. S. Lokey and B. L. Iverson, *Methods Enzymol.*, 2001, **340**, 556–570.
91. A. Rhoden Smith, B. A. Ikkanda, G. G. Holman and B. L. Iverson, *Biochemistry*, 2012, **51**, 4445–4452.
92. V. M. Guelev, M. T. Harting, R. S. Lokey and B. L. Iverson, *Chem. Biol.*, 2000, **7**, 1–8.
93. V. M. Guelev, J. Lee, J. Ward, S. Sorey, D. W. Hoffman and B. L. Iverson, *Chem. Biol.*, 2001a, **8**, 415–425.
94. V. M. Guelev, S. Sorey, D. W. Hoffman and B. L. Iverson, *J. Am. Chem. Soc.*, 2002, **124**, 2864–2865.
95. Y. Chu, V. Lynch and B. L. Iverson, *Tetrahedron*, 2006, **62**, 5536–5548.
96. Y. Chu, S. Sorey, D. W. Hoffman and B. L. Iverson, *J. Am. Chem. Soc.*, 2007, **129**, 1304–1311.
97. Y. Chu, D. W. Hoffman and B. L. Iverson, *J. Am. Chem. Soc.*, 2009, **131**, 3499–3508.
98. D. E. Graves and L. M. Velea, *Curr. Org. Chem.*, 2000, **4**, 915–929.

99. J. B. Hansen, T. Koch, O. Buchardt, P. E. Nielsen, B. Nordén and M. Wirth, *J. Chem. Soc., Chem. Commun.*, 1984, 509–511.
100. J. Lee, V. Guelev, S. Sorey, D. W. Hoffman and B. L. Iverson, *J. Am. Chem. Soc.*, 2004, **126**, 14036–14042.
101. G. G. Holman, M. Zewail-Foote, A. Rhoden Smith, K. A. Johnson and B. L. Iverson, *Nat. Chem.*, 2011, **3**, 875–881.
102. A. Rhoden Smith and B. L. Iverson, *J. Am. Chem. Soc.*, 2013, **135**, 12783–12789.
103. H. Aoki and H. Tao, *Analyst*, 2005, **130**, 1478–1482.
104. K. M. Vasquez, G. Wang, P. A. Havre and P. M. Glazer, *Nucleic Acids Res.*, 1999, **27**, 1176–1181.

CHAPTER 3

NDI Foldamers, Assemblies and Conformational Switching

CAMERON PEEBLES, CHRISTOPHER D. WIGHT AND BRENT L. IVERSON*

The University of Texas at Austin, Department of Chemistry and Biochemistry, Welch Hall 2.204, 105 E. 24th St. STOP A5300, USA
*Email: biverson@mail.utexas.edu

3.1 Introduction

Interactions between aromatic groups play a critical stabilizing role in many naturally occurring systems such as the DNA double helix and proteins. For example, the so-called tryptophan zipper motif stabilizes β-hairpin peptides due to the interactions between four tryptophan residues.[1] In amyloidogenic peptides, it has been suggested that the order and directionality of interactions between aromatic groups play a dominant role in the fibril assembly process.[2,3] On the synthetic side, nanostructures composed of interacting aromatic constituents have shown great potential for optoelectronic devices,[4] n/p-type semiconductors[5,6] and organic electronic devices.[7]

To better exploit interactions between aromatic groups in the biological sciences (incorporation of non-natural, aromatic residues into proteins; designed secondary and tertiary structure, artificial DNAs) and materials sciences (for better charge mobility or optoelectronic uses), a deeper understanding of interactions between aromatic groups is necessary. It is the goal of this chapter to provide a qualitative description of how electrostatics and desolvation dictate aromatic interaction geometries, followed by

Monographs in Supramolecular Chemistry No. 21
Naphthalenediimide and its Congeners: From Molecules to Materials
Edited by G. Dan Pantoş

Published by the Royal Society of Chemistry, www.rsc.org

examples from the literature that illustrate folding, assembly and finally conformational switching behavior.

3.2 Theory Behind Interactions Between Aromatic Groups

3.2.1 "Polar/Pi" Model

A predictive understanding of aromatic stacking geometry was put forth by Hunter and Sanders in the 1990s that was based upon the overall polarization of the aromatic pi electron cloud.[8] This approach, known as the "polar/pi" model, uses an analysis of quadrupole–quadrupole interactions between aromatic molecules to rationalize or predict why some aromatics might exhibit herringbone arrangements or off-set parallel stacking geometries while certain electron-rich and electron-deficient aromatics tend toward alternating face-centered stacking.

The quadrupole moment inherent in aromatic molecules is derived from the amount of pi electron density located both above and below an aromatic ring plane and any counterbalancing charge around the ring periphery. For underivatized aromatics, this amounts to a region of more negative charge both above and below the ring plane (also referred to as the aromatic core), with a region of more positive charge around the periphery. Strongly electron-withdrawing groups on the ring can remove so much electron density from the pi cloud that the quadrupole moment is inverted, having a more positive central region and a more negative region associated with the electron-withdrawing groups. Benzene, the prototypical aromatic molecule, is electron-rich along its central top and bottom surface (core) and electron-deficient around its periphery (see Figure 3.1). Hexafluorobenzene, on the other hand, displays a somewhat inverted quadrupole moment, having a relatively electron-deficient aromatic core and a somewhat electron-rich periphery associated with the fluorine atoms. The calculated quadrupole moments of benzene and hexafluorobenzene are -29.0×10^{-40} C m^2 and 31.7×10^{-40} C m^2, respectively.[9] Based largely on overall quadrupole moments, aromatics are classified according to the nature of their core. For example, benzene would generally be referred to as an electron-rich aromatic, while hexafluorobenzene would be described as electron-deficient.

The "polar/pi" model predicts that when two relatively electron-rich benzene molecules interact, they will maximize complementary quadrupole–quadrupole interactions by stacking in an off-set parallel or an edge-to-face (so-called herringbone) geometry, a prediction that is confirmed experimentally.[10] Note that in the case of electron-rich aromatics such as benzene, a face-centered stacking geometry is predicted to be less favorable due to the repulsion that would occur upon bringing two electron-rich aromatic core surfaces together. On the other hand, the interaction of benzene with the electron deficient hexafluorobenzene is predicted to occur with an alternating face-centered stacked geometry that maximizes the complementarity

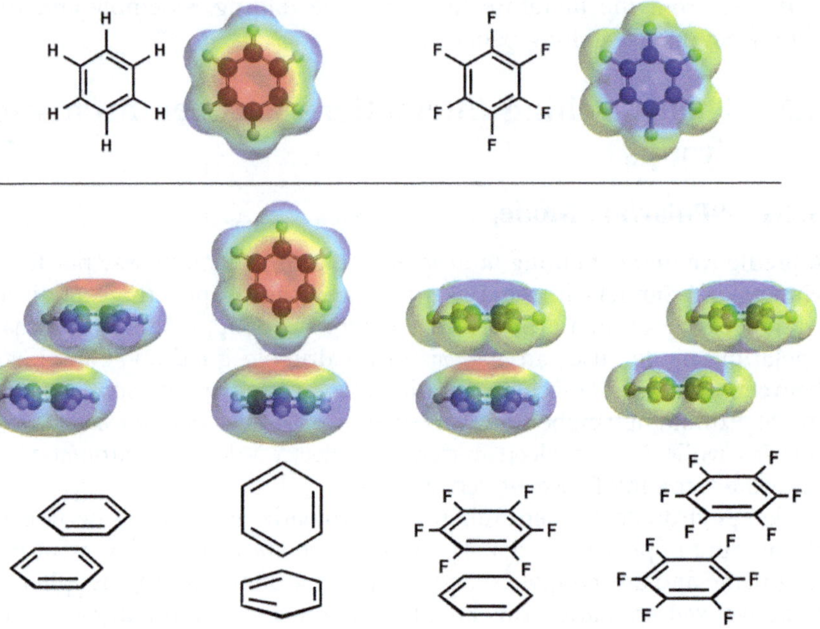

Figure 3.1 Illustration of the "polar/pi" model of aromatic stacking. Top shows electrostatic potentials of electron-rich benzene (top left) and electron-deficient hexafluorobenzene (top right). Bottom shows stacking in the following geometries (from left to right): off-set parallel, edge-to-face, face-centered and off-set parallel. Molecules were generated in Spartan 08 using the DFT B3LYP G-31* method.

of the quadrupole moments between the relatively electron-rich core and electron-deficient periphery of benzene being attracted to the relatively electron-deficient core and the more electron-rich periphery of hexafluorobenzene (see Figure 3.1), a geometry that has been confirmed in the solid state.[11,12]

3.2.2 "Local, Direct Interaction" Model

Some nice early experimental work by Rashkin and Waters used a molecular-rotation system to probe interactions between substituted aromatic rings.[13] In addition to noticing that polarization played a role in the interaction of the rings, they suggested that direct, electrostatic interactions between the substituents on the periphery of both rings should be considered an important factor driving stacking geometry.

Direct complementary interactions between aromatic substituents were examined closely by Wheeler and Houk[14] looking at the interaction energies between substituted face-centered stacked benzene dimers. Importantly, they found that the substituted benzene face-centered stacked dimer interaction energies are not dominated by interactions between the aromatic cores of the substituted benzene rings, as similar trends in interaction energies were replicated in the interaction between an H–X unit and the unsubstituted-benzene ring (see Figure 3.2). Experimental work by Houk *et al.*[15] and further computational work by Wheeler[16] have helped refine this model and painted an alternative picture for predicting the preferred interaction geometries between two or more aromatic units. The model, referred to as the "local, direct interaction" model,[16] theorizes that the local, direct through-space interactions between polarized moieties on the periphery of stacked aromatic units drive the aromatic stacking geometry.

This work, as well as work by Sherill and co-workers,[17–19] Snyder *et al.*,[20] Lee *et al.*,[21] Wheeler *et al.*,[15] and Grimme[22] has helped underscore the importance of electrostatic interactions between the substituent of one aromatic ring and a second aromatic ring.

Dialkoxy naphthalene (DAN) and NDI are the focus of much of the following sections. DAN is a prototypical electron-rich aromatic, and NDI is a prototypical electron-deficient aromatic. The electrostatic potentials for DAN and NDI are dominated by the electron-donating alkoxy and electron-withdrawing imide carbonyl groups, respectively (see Figure 3.3). An important aspect of the DAN and NDI structures is that they are roughly the same size, a parameter that is more important in the solid state.

In the solid state and in solution, mixtures of DAN and NDI are observed to stack in an alternating, face-centered geometry as expected for complementary electron-rich and electron-deficient aromatics (see Figure 3.3). The face-centered stacking geometry and corresponding pi orbital mixing result in a charge-transfer absorbance band and characteristic red-purple color seen when the colorless DAN and NDI are mixed. An alternative terminology refers to electron-rich aromatics such as DAN as a donor (D) and electron-deficient aromatics such as NDI as an acceptor (A) in complexes that exhibit charge-transfer behavior. The charge-transfer absorbance occurs when an electron from the HOMO on the donor is excited to the LUMO of the adjacent acceptor aromatic.

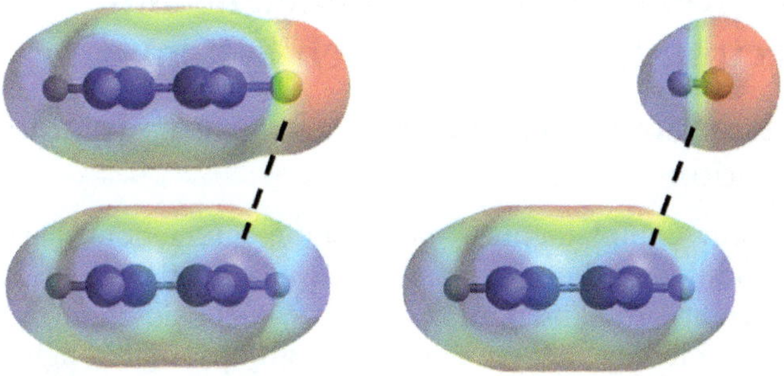

Figure 3.2 Cartoon representation of the "local, direct interaction" model, championed by Wheeler, depicting the interaction of benzene with fluorobenzene (left) and benzene with HF (right). Molecules were generated in Spartan 08 using the DFT B3LYP G-31* method.

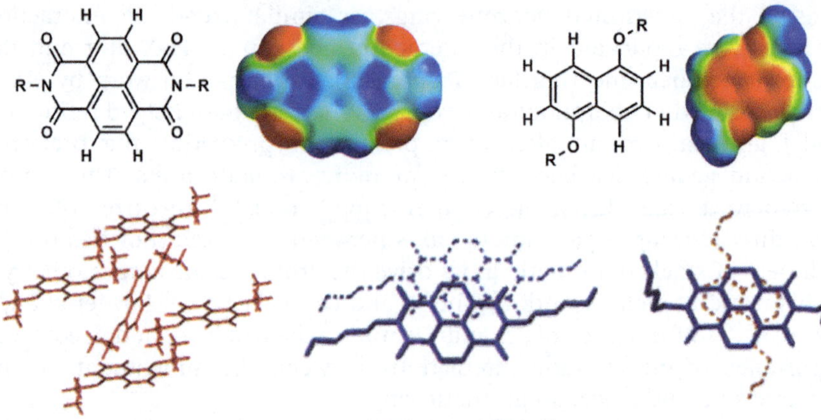

Figure 3.3 Electrostatic potentials of NDI and DAN (top) and crystal structures of DAN, NDI, and 1 : 1 DAN–NDI in the solid state (bottom). Potential maps were generated in Spartan 08 using the DFT B3LYP G-31* method. Adapted with permission from ref. 23 and 24. Copyright (2001 and 2006) American Chemical Society.

DAN self-stacks in a herringbone arrangement, while NDI is observed to stack in an off-set fashion. Examination of the NDI–NDI self-stack reveals close contact between the carbonyl O atom of one NDI and the carbonyl C atom of the adjacent NDI unit, likely providing a qualitative explanation for the observed solid state geometry.

The key insight from this analysis is that the four polarized carbonyls of NDI provide for two distinct stacking geometries with aromatic molecules. With electron-rich aromatics like DAN, alternating face-centered stacking is observed, while an off-set stacking mode is seen when NDI interacts with itself.

3.3 Electron-rich/Electron-deficient and Electron-rich/Electron-rich Systems

3.3.1 Interactions Between Aromatic Groups in Aqueous Solution

Until now, only complementary electrostatic interactions have been discussed as modulating the interactions between relatively electron-rich and electron-deficient aromatic units, but a new consideration arises when discussing interactions between aromatic groups in strongly interacting solvents such as water: solvation/desolvation effects. The key question, then, is how do both solvation/desolvation and complementary electrostatic interactions contribute to the overall geometry and strength of the interaction when two aromatic units are placed in a range of solvents?

Table 3.1 Association constants of mixes of DAN, NDI and 1:1 DAN–NDI in solvents of varying polarity as calculated based on ^{1}H NMR titration data. Adapted with permission from ref. 23. Copyright (2001) American Chemical Society.

	Solvent	K_a (M^{-1}) DAN–DAN	K_a (M^{-1}) NDI–NDI	K_a (M^{-1}) DAN–NDI
1	$CDCl_3$	(1)	(1)	$2 \pm <0.5$
2	Acetone-d_6	$1 \pm <0.5$	$1 \pm <0.5$	$8 \pm <0.5$
3	DMSO-d_6	1 ± 1	$2 \pm <0.5$	$3 \pm <0.5$
4	CD_3CN	1 ± 1	$3 \pm <0.5$	$11 \pm <0.5$
5	CD_3OD	$1 \pm <0.5$	$8 \pm <0.5$	$30 \pm <0.5$
6	3:1 CD_3OD–D_2O	$1 \pm <0.5$	$15 \pm <0.5$	63 ± 2
7	1:1 CD_3OD–D_2O	$2 \pm <0.5$	28 ± 2	254 ± 41
8	1:3 CD_3OD–D_2O	10 ± 2	101 ± 28	952 ± 64
9	D_2O	20 ± 4	245 ± 101	2045 ± 63

We sought to answer this question by determining the association constants of DAN and NDI, individually as well as mixed, in solvents of varying polarities.[23] In particular, Table 3.1 shows the observed interaction energies between DAN–DAN, NDI–NDI and DAN–NDI as a function of solvent polarity. The most obvious trend is that upon increasing solvent polarity, the interaction energies between aromatic units all increased suggesting that solvation/desolvation effects are the dominant energetic parameters in polar solvents. In other words, desolvation of hydrophobic aromatic faces in more polar solvents appears to be the dominant energetic component driving association.

However, solvation/desolvation alone cannot explain all the data. Note the large difference in interaction energies between DAN–NDI and DAN–DAN/NDI–NDI, yet both DAN and NDI have hydrophobic surfaces of roughly the same size. Revisiting the solid-state structures of DAN and NDI in Figure 3.3 provides a reasonable explanation for the observed relative association constants. While the main driving force for association is desolvation of the hydrophobic aromatic faces in highly polar solvents, electrostatics dictate the *geometry* of the interaction and thus the *extent* of hydrophobic surface overlap upon complexation. Whereas the DAN–NDI complementary electrostatic interaction maximizes the burial of hydrophobic surfaces by stacking in a face-centered geometry (thus stronger association), the DAN–DAN interaction occurs in an edge-to-face geometry and buries a minimal amount of hydrophobic surface area, leading to a significantly weaker association. The offset stacking seen with NDI self-assembly results in an intermediate level of buried hydrophobic surface area and correspondingly intermediate association strength.[23,24]

It is worth pointing out that as solvent polarity and presumably the significance of solvation/desolvation effects decrease, the differences between the DAN–DAN, NDI–NDI and DAN–NDI association constants also decrease (Table 3.1). This idea should be kept in mind when considering the design and analysis of supramolecular systems and assemblies involving electron-deficient moieties such as NDI. Greater discrimination between stacking modes (*i.e.* DAN–NDI *vs.* NDI–NDI) is expected for more polar solvents, and conversely, less discrimination would be expected in solvents of low polarity or the solid state.

3.3.2 Survey of Supramolecular Systems

Complementary electrostatic interactions between NDI and relatively electron-rich aromatic moieties such as DAN or pyrene have been exploited extensively in the development of catenanes, hetero-duplex forming oligomers and numerous 1D and 2D nanostructures (see Figure 3.4). By itself, NDI is capable of forming highly ordered structures by way of self-associating in off-set stacking geometries.

The Sanders group has exploited solvaphobic effects in the dynamic combinatorial synthesis of electron-rich and electron-deficient aromatic (D–A) [2]-catenanes.[25,26] Their results, indicated using the donor (D, electron-rich

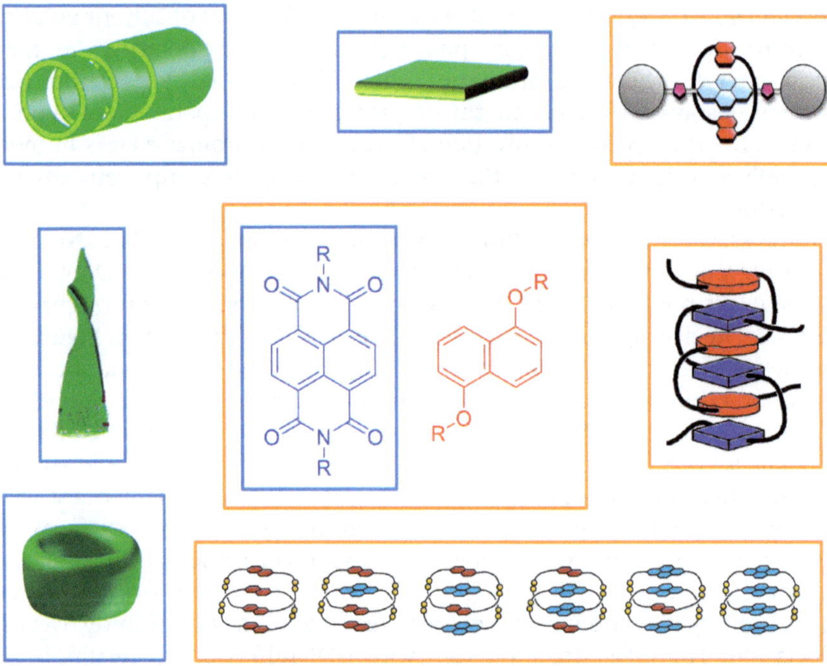

Figure 3.4 Selected cartoon representations of supramolecular structures composed of only NDI (blue boxes) or a mixture of DAN and NDI (orange boxes). Shown clockwise from top left are nanotubes, nanosheets, nanospheres, oligomeric heteroduplexes, [2]-catenanes, [2]-rotaxanes and nanofibrils.
Adapted with permission from ref. 25, 26 and 27. Copyright (2011, 2009 and 2002) American Chemical Society; and from ref. 28 © 2015 WILEY-VCH Verlag GmbH & Co. KGaA, Weinheim.

aromatic) and acceptor (A, electron-deficient aromatic) terminology, have shed light on the formation of DAAD, DADD and ADAA stacked [2]-catenanes, which has challenged the conventional thinking of stacking preference based on a primary consideration of electrostatic complementarity.

Based on the successful development of folding systems, we have used a design based on DAN and NDI oligomers to self-assemble heteroduplexes in water.[27] NMR studies showed that a face-centered DAN–NDI alternate stacking geometry was present in the heteroduplex. Interestingly, as the number of aromatic units in the individual DAN and NDI oligomers was increased from 1 to 4, the resulting heteroduplex binding affinity increased proportionately. Perhaps most interesting of all, there was a surprising temperature independence of the association constant observed, indicating a high degree of enthalpy–entropy compensation.

In 2015, Ramaiah *et al.* reported the synthesis and properties of amphiphilic naphthalene di- and monoimide chromophores that self-assemble in aqueous media to form vesicles capable of transporting the hydrophobic dye Nile red.[28] Investigators found that the hydrophobic cargo was released following a DNA-mediated disassembly of the vesicles, due to the significant intercalating action of the two naphthalene imide conjugates used ($K_{DNA} = 5–8 \times 10^4\ M^{-1}$).

Ghosh *et al.* reported the spontaneous self-assembly of NDI-based bolaamphiphiles in aqueous solution into 2D vesicles that could then be transformed into 1D fibrils upon introducing insoluble, relatively electron-rich pyrene monomers that would insert themselves between two NDI monomeric building blocks.[29] Using similar self-assembling NDI monomeric scaffolds, they also reported the assembly of 1D fibers[30,31] and 2D vesicles[32] derived from NDI–NDI interactions.

Using NDI-appended dipeptides, Parquette *et al.* have shown the formation of 1D nanofibrils facilitated by intermolecular NDI–NDI off-set parallel stacking coupled with favorable charged electrostatic interactions between monomers.[33,34] Intermolecular NDI–NDI off-set parallel stacking was also observed in lysine-based NDI amphiphilic monomers that self-assembled in aqueous solutions to form 1D nanotubes.[6,35]

Govindaraju *et al.* took a different approach to forming highly ordered nanostructures by directly appending either a single, or a strand of, amino acids to the imide of a monomeric NDI unit. A combination of hydrophobic interactions and favorable NDI–NDI off-set parallel stacking interactions led to the formation of 1D chiral assemblies,[36] conductive 2D nanosheets and nanocups[37] and nanospheres, nanobelts and fractals.[5]

Stoddart *et al.* have also employed NDI in the synthesis of switchable donor–acceptor [2]-rotaxanes. In one instance, "click" chemistry was used to afford a neutral [2]-rotaxane composed of a rigid NDI strand interlocked with 1,5-dinaphthol[38]crown-10 (DN38C10).[38] "Shuttling" of the [2]-rotaxane, that is the electrochemically-driven switching interaction between the DN38C10 unit and the NDI and triazole rings, was proposed based on CV and spectroelectrochemistry analysis. In another instance, DN38C10 was shown to "shuttle" between electron-deficient NDI and 4,4′-bipyridinium aromatic units even with a sterically bulky tetraarylmethane moiety between the units.[39]

As illustrated, there are many highly ordered systems that form supramolecular assemblies composed of either alternating stacking between NDI and an electron-rich aromatic such as DAN, or NDI–NDI self-assembly (see Figure 3.4). There is now a growing list of interesting systems containing NDI that are capable of conformational switching between different stacking geometries. On the following pages, several of these systems are described.

3.4 Systems That Undergo Stacking Geometry Switching

3.4.1 Synthetic Ion Channel

The conformational change of static, monofunctional synthetic ion channels into more complex multifunctional channels that can open and close in response to external stimuli is a maturing area of supramolecular chemistry. In 2005, Matile *et al.* introduced a ligand-gated synthetic ion channel incorporating NDI units that open in response to intercalating guest molecules (see Figure 3.5).[40] The design uses four linear "stave-like" monomers, incorporating *p*-octiphenyl rigid-rod scaffolds appended with NDI units, which assemble into cylindrical, barrel-stave pores. The pore is closed initially due to the conformational twisting of the *p*-octiphenyl scaffold to accommodate NDI–NDI interactions between aromatic groups, favorable hydrogen-bonding and side chain steric crowding. In the closed position, one can envision J-type NDI-aggregates, but slightly mismatched, with the NDI faces off-set in a twisted, parallel fashion to one another.

Addition of DAN monomer into the solution leads to the intercalation of DAN in between NDI units, eliminating the twist and opening the ion channel, as corroborated using the exciton chirality method in CD studies.[41] The open ion channel has the characteristic red-purple color due to a strong charge transfer absorbance derived from alternating DAN–NDI stacking. This "ligand-gating" provides a dramatic example of how the two stacking geometries of NDI, off-set stacking with itself, and face-centered stacking with DAN, can be exploited to modulate supramolecular system properties.

This design was later incorporated into a vesicular membrane to act as an electron-transport pathway and an ion channel.[42] Photoexcitation of the system transports an electron from an electron-source outside the membrane to the embedded pi-helix structure and to a final acceptor in the membrane. Addition of DAN monomer opens the pi-helix structure into a synthetic ion channel which then facilitates hydroxide ion transport through the membrane.

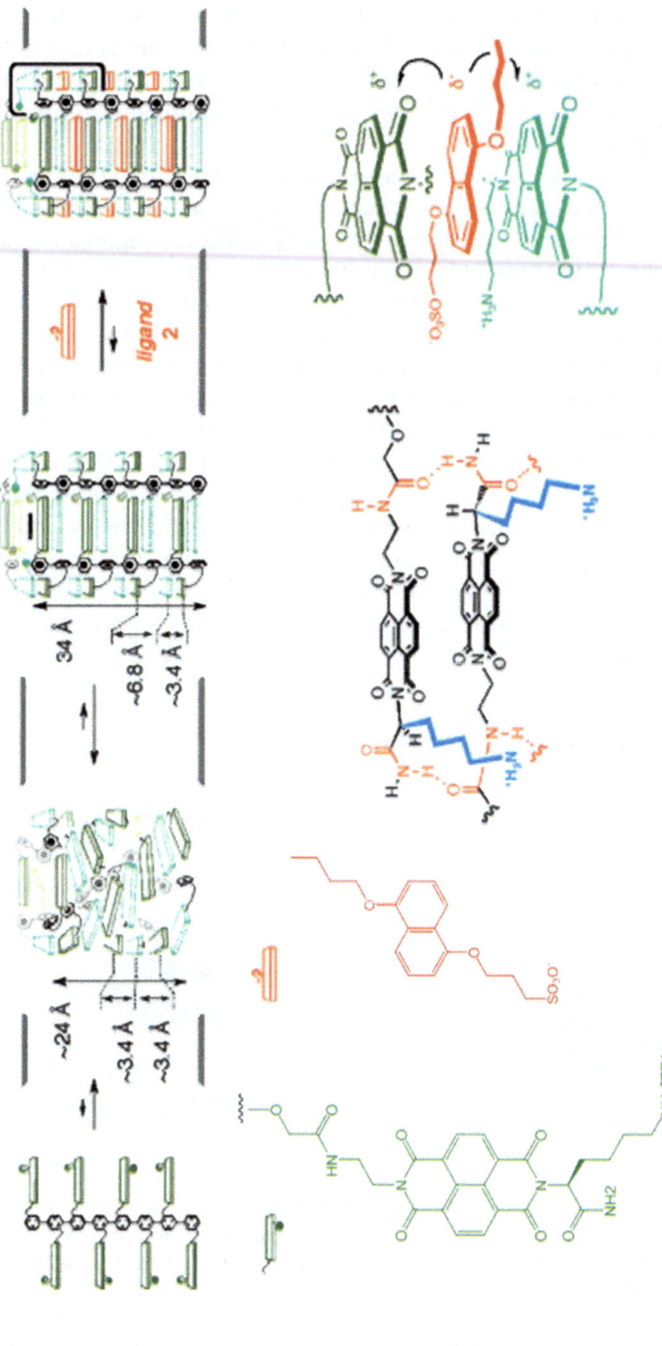

Figure 3.5 The assembly and later opening of the pi-helix structure into the pi-barrel pore upon intercalation of DAN monomers is shown at the top. Structures of the NDI (green) and DAN (red) aromatic units are shown on the bottom left while modeled NDI–NDI, off-set parallel displaced interactions in the pi-helix structure and DAN–NDI face-centered interactions in the pi-barrel pore structure are shown on the bottom right.
Adapted from ref. 39 with permission from The Royal Society of Chemistry.

3.4.2 Self-sorting DAN and NDI Monomers

In 2010, Molla, Das and Ghosh used a trade-off between hydrogen-bonding and interactions between DAN and NDI monomers to facilitate both the alternate and segregated stacking of these monomers in solution (see Figure 3.6).[43] To this extent, DAN and NDI monomers were synthesized with complementary and non-complementary hydrogen-bonding motifs and then mixed in a non-polar solvent.

The absence of a charge-transfer absorbance when the DAN and NDI monomers containing *non-complementary* hydrogen-bonding motifs were mixed together suggested the monomers were forming segregated, self-assembled stacks. This segregated assembly was theorized to be due to the complementary hydrogen-bonding between like monomers (DAN–DAN and NDI–NDI) overriding the preference for alternating stacking (DAN–NDI). Consistent with this hypothesis, mixing of DAN and NDI monomers with *complementary* hydrogen-bonding units resulted in a strong charge-transfer absorbance, indicative of alternating DAN–NDI stacking.

Further refinement of the DAN and NDI monomer units allowed for a time-dependent reorganization of the DAN and NDI monomers from an alternating, stacked DAN–NDI assembly to segregated stacked assemblies, both of which were gels.[7,44] These authors proposed that interactions between aromatic groups in the kinetically stable alternate stacked assembly would reorganize so that stronger hydrogen-bonding interactions could occur in the thermodynamically favored segregated stacked assemblies.

3.4.3 Synthetic Amyloid Fibrils

In 1999, we synthesized an amphiphilic DAN–NDI foldamer that was observed to undergo an irreversible change to form a hydrogel upon heating.[45] While the unheated foldamer solution was red-purple due to the familiar DAN–NDI charge-transfer absorbance of the pleated foldamer, after heating,

Figure 3.6 The transition from a DAN–NDI face-centered, alternating assembly to segregated DAN–DAN and NDI–NDI self-sorted assemblies is schematically shown. NDI monomers are represented as blue squares while DAN monomers are represented as red ovals. Non-complementary hydrogen bonds are represented by the interaction of different colored spheres (yellow and green), while complementary hydrogen bonds are represented by interactions of the same colored spheres.

the hydrogel was a pale yellow in color. Initial characterization of the hydrogel indicated a high degree of aggregation. Later, CD spectroscopy was used to determine that the aggregated state of the hydrogel was highly ordered.[46]

Recent work has shed considerable light upon the structural details of the aggregated state responsible for hydrogel formation.[47] A peak in the UV trace of the aggregated sample, known to represent the NDI short-axis component of the foldamer, intercepted the null of the bisignate CD signal taken of the same sample indicative of NDI off-set, twisted self-assembly. Further, visualization of the foldamer aggregates by TEM and AFM revealed fibrils of uniform width with regular helicities. This morphology is typically associated with a fibril twist and is also found in some protein amyloid fibril aggregates.

Taken together, these studies revealed that the conformational switching process seen with the amphiphilic DAN–NDI foldamer upon heating is the result of an irreversible change to fibrillar aggregates that tangle to form a hydrogel. A detailed model of the aggregation process can be proposed based on all available data: (1) upon heating, the kinetically stable, pleated DAN–NDI secondary structure of the foldamer undergoes unfolding, akin to opening an accordion, to yield a linearly-shaped intermediate (see Figure 3.7(a)); (2) reorganization of the unfolded aggregate occurs by forming new intermolecular interactions with other unfolded aggregates facilitated by desolvation and complementary electrostatics between NDI–NDI off-set parallel stacking (see Figure 3.7(b)). The amphiphilicity of the foldamer imparted by the leucine and aspartate side chains results in distinct hydrophobic and hydrophilic faces of the growing aggregate structure; (3) dimerization of two aggregates sequesters the hydrophobic faces on the interior juxtaposing the hydrophilic face on the exterior of the fibril. It is occlusion of the large hydrophobic faces of the assembled aggregates that ultimately drives assembly and results in the process being essentially irreversible once the aggregates have grown large enough.

The bottom line here is that the unique behavior of the amphiphilic DAN–NDI foldamer is a consequence of the NDI moieties switching between DAN–NDI alternating stacking in the folded state to NDI off-set self-stacking in the aggregate. Interestingly, there are many striking similarities between this synthetic system behavior and natural protein amyloid formation including the requirement for overall amphiphilicity, the irreversible nature of the process and formation of twisted fibrils of similar dimensions. The synthetic DAN–NDI system can therefore be thought of as an amyloid analogue that uniquely "bridges-the-gap" between abiotic foldamers and natural proteins that can transition between a kinetically stable folded state and irreversible fibril formation.

3.4.4 Mesophases Exhibiting Switching Behavior

We also used the complementarity of DAN and NDI monomers to generate highly colored mesophases.[24] Melting of mixtures of DAN and NDI

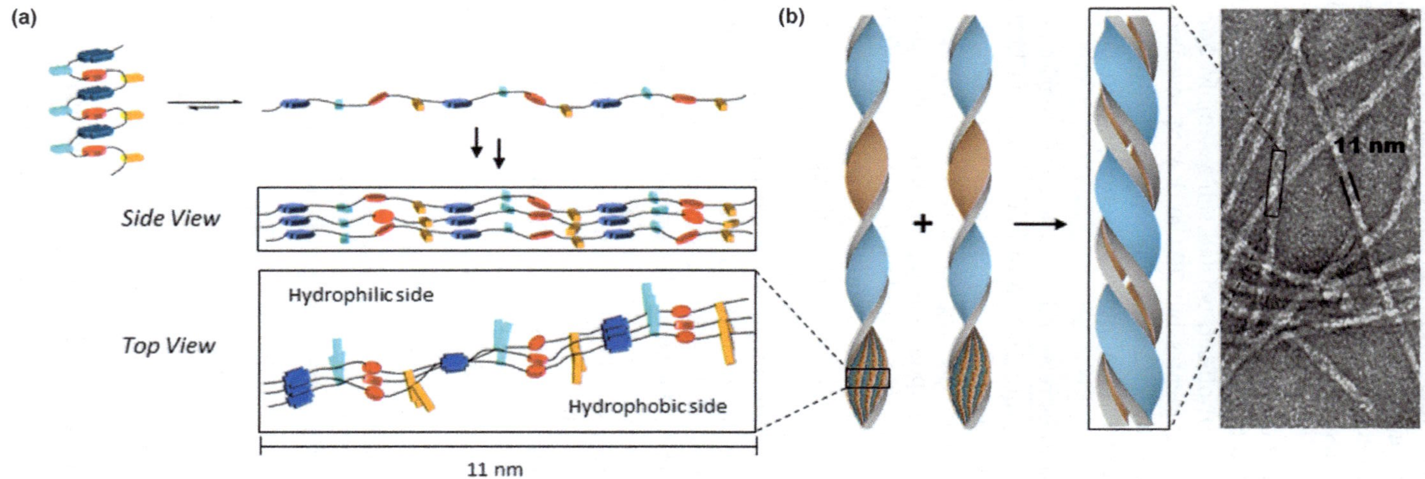

Figure 3.7 Proposed aggregation pathway of the amphiphilic foldamer. (a) Heating of the initial foldamer solution results in complete unfolding of the pleated structure to a new ribbon structure facilitated by NDI–NDI off-set parallel interactions. The ribbon is amphiphilic and has a hydrophobic side and a hydrophilic side; (b) the amphiphilic ribbon dimerizes to sequester the hydrophobic side chains on the interior of the dimer while exposing the hydrophilic side chains on the exterior of the dimer (blue squares represent NDI, red squares represent DAN, teal squares represent aspartic acid side chains, orange squares represent leucine side chains).
Adapted from ref. 47 with permission. Copyright © 2013 WILEY-VCH Verlag GmbH & Co. KGaA, Weinheim.

derivatives yielded a red-purple color indicative of the characteristic aromatic donor–acceptor charge transfer absorbance. This color, along with UV-Vis, single crystal and variable temperature XRD data, provided strong evidence that an alternating, face-centered stacking geometry of DAN and NDI molecules was occurring in the columnar mesophase. The same red-purple color was observed in the crystalline phase for many of these mixtures suggesting that little reorganization occurred during the mesophase-to-crystalline phase transition.

In some cases, however, melting of specific DAN and NDI monomers gave a red-purple mesophase, but upon cooling produced a dramatic thermochromic change to give a yellow-colored crystalline solid (see Figure 3.8). An interesting feature of the mixtures exhibiting thermochromic behavior is that the transition occurs without any detectable change to the macroscopic morphology of the solid (see Figure 3.8). Later work, which sought to further explore the thermochromic behavior, focused on side chain interactions

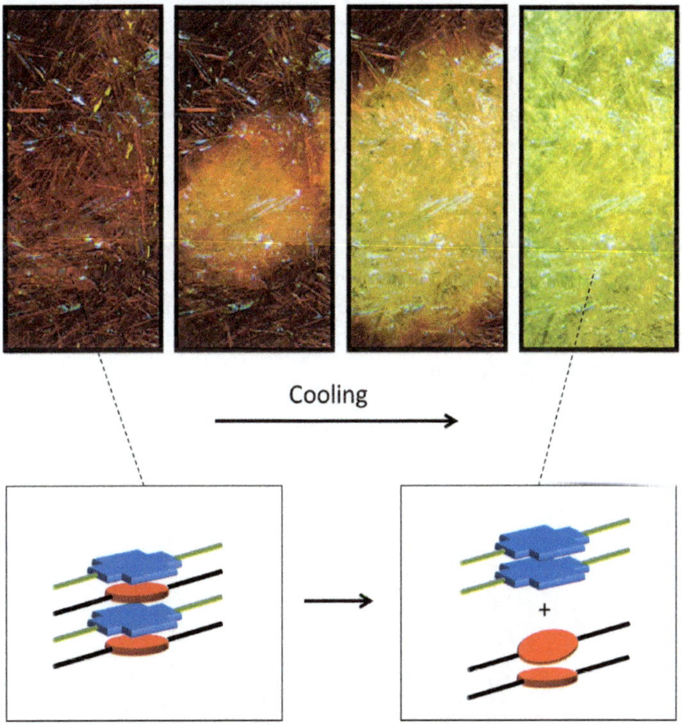

Figure 3.8 Optical microscope images showing the transition from mesophase (red) to crystalline phase (yellow). Notice that the macromolecular morphology remains intact through the transition. The bottom scheme shows the self-sorting behavior of the DAN (red) and NDI (blue) monomers as a function of side-chain incompatibilities (black *vs.* green wedges).
Adapted with permission from ref. 48. Copyright (2010) American Chemical Society.

between the DAN and NDI monomers.[48] These results confirm that manipulating side chain interactions in a systematic way between DAN and NDI monomers can lead to highly tunable mesophases as well as controllable thermochromism.

In analogy to the amphiphilic foldamer described in the previous section, detailed characterization of the mesophase and crystalline phase revealed that the thermochromic transition in these mixtures once again involves a shift from alternating DAN–NDI stacking geometry (mesophase) to NDI–NDI off-set self-stacking (crystalline phase).

Recently, our group investigated the photophysical and stimuli-responsive properties of several conjugated monoalkoxynaphthalene-naphthalimide (MAN–NI) DA dyads.[49] Several of these dyads showed a difference in solid-state color between relatively faster (yellow) and slower (yellow-orange or orange) evaporation times from solution. In the solid-state, one of these dyads was found to change color in response to pressure (orange-to-yellow), heating (orange to yellow), and vapor fuming (yellow to orange), with repeatable cycles of color changing.

Based on single crystal data, XRD patterns, and modeling, the stimuli-responsive behavior of this dyad was determined to result from a 180° molecular rotation relative to the long axis of the molecule. When ground or heated, the orange crystalline form, possessing a head-to-tail (NI–MAN) stacking geometry, interconverted to a yellow soft crystalline mesophase with a head-to-head (NI–NI) stacking geometry that persisted at room temperature in the absence of solvent vapor. When subject to solvent vapor, the yellow mesophase returned to the orange crystalline state. Figure 3.9 shows the conversion of the orange crystalline form to the yellow soft crystalline mesophase, while maintaining macroscopic morphology, when heated at 110 °C for 30 seconds.

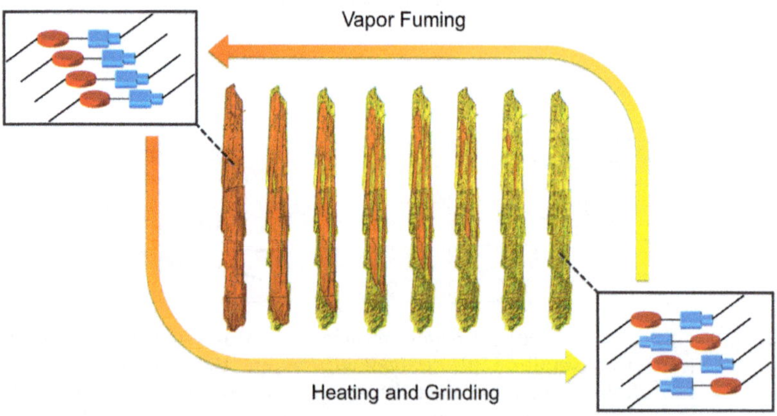

Figure 3.9 Reversible interconversion of orange crystalline NI–MAN derivative (head-to-tail stacking geometry) with the yellow soft-crystalline mesophase (head-to-head stacking geometry) by heating or grinding (orange to yellow) or vapor fuming (yellow to orange).

At this point, the authors feel it necessary to include a couple more examples of molecular switching systems, even if they do not fit into the scope of this review, due to their importance in the field of supramolecular chemistry. In 2014, Kumar, George and co-workers reported a stimuli-responsive helical assembly composed of functionalized NDI units that respond to various adenosine phosphates (APs) with differential chiroptical signaling.[50] NDI cores terminally appended with zinc coordinated dipicolylethylenediamine motifs formed supramolecular helical assemblies that adopt opposite handedness in the presence of ATP when compared to ADP or AMP. Experimental CD studies indicate that the NDI assemblies adopt left-handed helicity in the presence of AMP and ADP, while right-handed helicity is imparted upon the binding of ATP. Molecular dynamics and molecular mechanics simulations indicate that differences in non-covalent interactions between APs and the functionalized NDI units result in the two helical conformations. Rapid helicity switching following enzymatic AP hydrolysis allows for an unprecedented way to probe phosphate interconversion kinetics *in situ*.

While Stoddart's NDI-containing [2]-rotaxanes were briefly mentioned, his work with paraquat-containing molecular switches demonstrates the progress that has been made towards implementing these devices in molecular electronics.[51] Sauvage *et al.* have also made important contributions to molecular machines using transition metal complexes as elements of rotaxanes and catenanes.[52]

3.5 Outlook

NDI has proven to be an important supramolecular building block in large part because it can adopt two different but entirely predictable stacking geometries: alternating face-centered stacks with complementary electron-rich aromatic moieties such as DAN and an off-set self-stacking mode. The alternating face-centered stacking is favored by strongly interacting polar solvents due to dominating solvation/desolvation effects, while NDI off-set self-stacking can predominate in cases in which no electron-rich aromatics are present or in the absence of a polar solvent, including the solid state. These insights allow a systematic manipulation of stacking geometry and therefore properties of assemblies. Taking advantage of these two distinct states has generated several interesting systems capable of conformation switching ranging from artificial ion channels to synthetic amyloid analogues and even mesophases capable of dramatic thermochromism. Moving forward, we can now expect a number of new and exciting applications of NDI and its analogues, limited only by our imaginations.

References

1. A. G. Cochran, N. J. Skelton and M. A. Starovasnik, *Proc. Natl. Acad. Sci. U.S.A*, 2001, **98**, 5578–5583.
2. E. Gazit, *FASEB J.*, 2002, **16**, 77–83.

3. E. Gazit, *Prion*, 2007, **1**, 32–35.
4. Y. Chen, Y. Feng, J. Gao and M. Bouvet, *J. Colloid Interface Sci.*, 2012, **368**, 387–394.
5. M. B. Avinash and T. Govindaraju, *Nanoscale*, 2011, **3**, 2536–2543.
6. H. Shao, M. Gao, S. H. Kim, C. P. Jaroniec and J. R. Parquette, *Chem. – Eur. J.*, 2011, **17**, 12882–12885.
7. A. Das, M. R. Molla, B. Maity, D. Koley and S. Ghosh, *Chem. – Eur. J.*, 2012, **18**, 9849–9859.
8. C. A. Hunter and J. K. M. Sanders, *J. Am. Chem. Soc.*, 1990, **112**, 5525–5534.
9. A. Hori, The importance of pi-interactions in crystal engineering: frontiers in crystal engineering, in *Arene-Perfluoroarene Interactions in Coordination Architectures*, ed. E. R. T. Tiekink and J. Zukerman-Schpector, John Wiley & Sons, Ltd. Published by same, 2012, ch. 7, pp. 163–185.
10. E. G. Cox, D. W. Cruickshank and J. A. Smith, *Proc. R. Soc. London, Ser. A*, 1958, **247**, 1–21.
11. C. R. Patrick and G. S. Prosser, *Nature*, 1960, **187**, 1021.
12. J. H. Williams, J. K. Cockcroft and A. N. Fitch, *Angew. Chem., Int. Ed. Engl.*, 1992, **31**, 1655–1657.
13. M. J. Rashkin and M. L. Waters, *J. Am. Chem. Soc.*, 2002, **124**, 1860–1861.
14. S. E. Wheeler and K. N. Houk, *J. Am. Chem. Soc.*, 2008, **130**, 10854–11085.
15. S. E. Wheeler, A. J. McNeil, P. Müller, T. M. Swager and K. N. Houk, *J. Am. Chem. Soc.*, 2010, **132**, 3304–3311.
16. S. E. Wheeler, *J. Am. Chem. Soc.*, 2011, **133**, 10262–10274.
17. A. L. Ringer, M. O. Sinnokrot, R. P. Lively and C. D. Sherrill, *Chem. – Eur. J.*, 2006, **12**, 3821–3828.
18. M. O. Sinnokrot and C. D. Sherrill, *J. Phys. Chem. A*, 2003, **107**, 8377–8379.
19. S. A. Arnstein and C. D. Sherrill, *Phys. Chem. Chem. Phys.*, 2008, **10**, 2646–2655.
20. S. E. Snyder, B.-S. Huang, Y. W. Chu, H.-S. Lin and J. R. Carey, *Chem. – Eur. J.*, 2012, **18**, 12663–12671.
21. E. C. Lee, D. Kim, P. Jurečka, P. Tarakeshwar, P. Hobza and K. S. Kim, *J. Phys. Chem. A*, 2007, **111**, 3446–3457.
22. S. Grimme, *Angew. Chem., Int. Ed.*, 2008, **47**, 3430–3434.
23. M. S. Cubberley and B. L. Iverson, *J. Am. Chem. Soc.*, 2001, **123**, 7560–7563.
24. J. J. Reczek, K. R. Villazor, V. Lynch, T. M. Swager and B. L. Iverson, *J. Am. Chem. Soc.*, 2006, **128**, 7995–8002.
25. F. B. Cougnon, H. Y. Au-Yeung, G. D. Pantoş and J. K. Sanders, *J. Am. Chem. Soc.*, 2011, **133**, 3198–3207.
26. H. Y. Au-Yeung, G. D. Pantoş and J. K. Sanders, *Proc. Natl. Acad. Sci.*, 2009, **106**, 10466–10470.
27. G. J. Gabriel and B. L. Iverson, *J. Am. Chem. Soc.*, 2002, **124**, 15174–15175.
28. B. H. Shankar, D. T. Jayaram and D. Ramaiah, *Chem. – Eur. J.*, 2015, **21**, 17657–17663.

29. M. R. Molla and S. Ghosh, *Chem. – Eur. J.*, 2012, **18**, 9860–9869.
30. H. Kar, M. R. Molla and S. Ghosh, *Chem. Commun.*, 2013, **49**, 4220–4222.
31. A. Das, M. R. Molla and S. Ghosh, *J. Chem. Sci.*, 2011, **123**, 963–973.
32. A. Das and S. Ghosh, *Macromolecules*, 2013, **46**, 3939–3949.
33. H. Shao and J. R. Parquette, *Chem. Commun.*, 2010, **46**, 4285–4287.
34. H. Shao, T. Nguyen, N. C. Romano, D. A. Modarelli and J. R. Parquette, *J. Am. Chem. Soc.*, 2009, **131**, 16374–16376.
35. H. Shao, J. Seifret, N. C. Romano, M. Gao, J. J. Helmus, C. P. Jaroniec, D. A. Modarelli and J. R. Parquette, *Angew. Chem., Int. Ed.*, 2010, **49**, 7688–7691.
36. M. Pandeeswar, M. B. Avinash and T. Govindaraju, *Chem. – Eur. J.*, 2012, **18**, 4818–4822.
37. M. B. Avinash and T. Govindaraju, *Adv. Funct. Mater.*, 2011, **21**, 3875–3882.
38. H.-P. J. de Rouville, J. Iehl, C. J. Bruns, P. L. McGrier, M. Frasconi, A. A. Sarjeant and J. F. Stoddart, *Org. Lett.*, 2012, **14**, 5188–5191.
39. S. K. Dey, A. Coskun, A. C. Fahrenbach, G. Barin, A. N. Basuray, A. Trabolsi, Y. Y. Botros and J. F. Stoddart, *Chem. Sci.*, 2011, **2**, 1046–1053.
40. P. Talukdar, G. Bollot, J. Mareda, N. Sakai and S. Matile, *J. Am. Chem. Soc.*, 2005, **127**, 6528–6529.
41. N. Sakai, P. Talukdar and S. Matile, *Chirality*, 2006, **18**, 91–94.
42. S. Bhosale, A. L. Sisson, P. Talukdar, A. Fürstenberg, N. Banerji, E. Vauthey, G. Bollot, J. Mareda, C. Röger, F. Würthner, N. Sakai and S. Matile, *Science*, 2006, **313**, 84–86.
43. M. R. Molla, A. Das and S. Ghosh, *Chem. – Eur. J.*, 2010, **16**, 10084–10093.
44. A. Das, M. R. Molla, A. Banerjee, A. Paul and S. Ghosh, *Chem. – Eur. J.*, 2011, **17**, 6061–6066.
45. J. Q. Nguyen and B. L. Iverson, *J. Am. Chem. Soc.*, 1999, **121**, 2639–2640.
46. V. J. Bradford and B. L. Iverson, *J. Am. Chem. Soc.*, 2008, **130**, 1517–1524.
47. C. Peebles, R. Piland and B. L. Iverson, *Chem. – Eur. J.*, 2013, **19**, 11598–11602.
48. P. M. Alvey, J. J. Reczek, V. Lynch and B. L. Iverson, *J. Org. Chem.*, 2010, **75**, 7682–7690.
49. C. Peebles, C. D. Wight and B. L. Iverson, *J. Mater. Chem. C*, 2015, **3**, 12156–12163.
50. M. Kumar, P. Brocorens, C. Tonnelé, D. Beljonne, M. Surin and S. J. George, *Nat. Commun.*, 2014, **5**, 5793.
51. A. H. Flood, J. F. Stoddart, D. W. Steuerman and J. R. Heath, *Science*, 2004, **306**, 2055–2056.
52. B. Champin, P. Mobian and J.-P. Sauvage, *Chem. Soc. Rev.*, 2007, **36**, 358–366.

CHAPTER 4

Naphthalenediimide in Modular Columnar Liquid Crystals: Key Component of Donor–Acceptor Columnar Liquid Crystals

JOSEPH J. RECZEK

Denison University, 500 West Loop, Granville, OH 43023, USA
Email: reczekj@denison.edu

4.1 Liquid Crystals: Brief Overview

Liquid crystalline materials have become staple components of technological devices in the modern world.[1,2] As implied by their name, liquid crystals (LC) are materials that have stable states on their phase diagrams between ordered crystalline solids (Cr) and completely isotropic liquids (IL). More generally referred to as mesophases, these states of matter are typically less rigid than solids, and less fluid than liquids. However, the degree of molecular order in liquid crystalline materials can vary widely between these two extremes, and there are some who would reserve the term liquid crystal for mesophase systems having actual physical properties of both a liquid and a crystal, rather than simply being defined by intermediate molecular order. Liquid crystalline phases can be considered dynamic supramolecular systems, usually governed by weak intermolecular interactions. Frequently,

Monographs in Supramolecular Chemistry No. 21
Naphthalenediimide and its Congeners: From Molecules to Materials
Edited by G. Dan Pantoş

Published by the Royal Society of Chemistry, www.rsc.org

these phases exhibit properties quite distinct from those common to either liquids or solids; they have the ability to self-align, organize, and sometimes self-heal. Importantly, some liquid crystalline systems also exhibit extreme structural sensitivity to external stimuli such as temperature, pressure, or electromagnetic fields, as well as chemical dopants and patterned surfaces. The combination of self-organization and extreme sensitivity to stimuli allows for many of the current, and pending, applications of these fascinating materials.[3]

A wide array of compositional and structural diversity is represented in the variety of liquid crystalline mesophases. Broadly, they can be classified as one of two types of systems: those formed primarily through the addition of solvent or those formed through temperature change, termed lyotropic and thermotropic respectively. Lyotropic systems are usually comprised of amphiphilic molecules that assemble in solution to form micelles; this occurs when a high enough solution concentration has been reached, termed the critical micelle concentration. Upon concentration and/or cooling, the micelles can organize further, such as orienting into cylinders that pack into hexagonal lattices or line up as lamellar sheets. In thermotropic systems, as a material is heated from the crystalline phase, molecules gain the ability to move in translational or rotational fashion, while being restricted in one or several degrees of freedom. This maintains some types of long-range molecular ordering, but molecules are also dynamically moving in some fashion. The majority of systems that exhibit thermotropic liquid crystalline phases are composed of molecules consisting of a rigid section with flexible groups attached, most commonly alkyl chains. Thermotropic mesogens are further subdivided by shape into molecules that have a rod- or blade-like core structure, termed calamitic, and molecules with a flat, disk-like structure, termed discotic (Figure 4.1).

4.1.1 Calamitic Liquid Crystals

The classification of thermotropic liquid crystalline mesogens into these two categories by molecular shape is largely based on the different types of ordered structures commonly adopted in their respective mesophases.

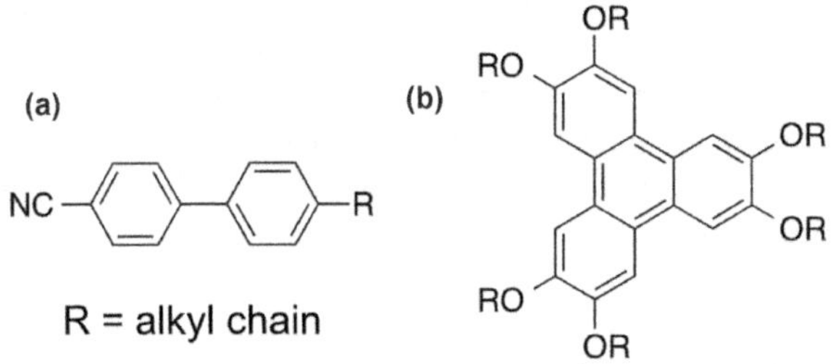

Figure 4.1 Examples of mesogen structures. (a) Calamitic 4-cyano-4′-alkylbiphenyl, the pentyl derivative (5CB) is commonly used in LC displays. (b) Hexa-alkoxy triphenylene (HAT), a common discotic mesogen.

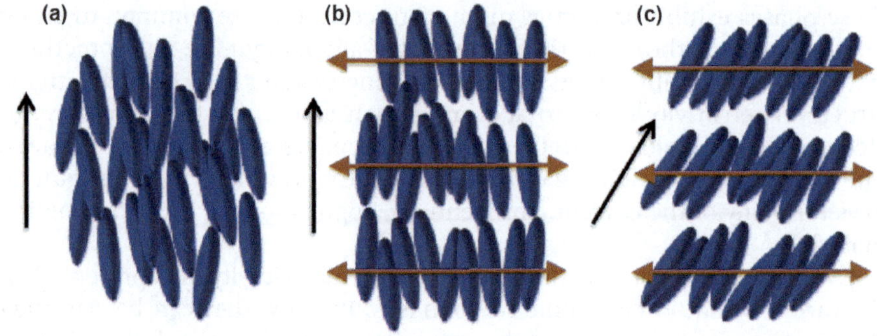

Figure 4.2 Schematics of calamitic phases: (a) nematic, (b) smectic A, (c) smectic C.

Calamitic molecules usually form nematic and smectic phases, while discotics align into a nematic discotic or columnar phase. The nematic phase (N) of calamitic systems is one of the least ordered mesophases (Figure 4.2(a)), where molecules have completely free translational motion, and free rotation about the long axis of the molecule, but are oriented in the same direction locally along the long axes. Calamitic nematic molecules are often designed with a significant molecular dipole along the long axis, such as 5CB (Figure 4.1(a)), allowing them to be easily orientated in an electromagnetic field. This is a critical aspect of LC display technology. As a nematic phase is cooled, molecules may take on some positional order with respect to each other by aligning into sheets or planes, changing from a nematic to a smectic phase. There are several varieties of smectic phases relating to the molecular order present within a sheet of calamitic molecules, and on the tilt of the molecules with respect to the normal plane packing. Molecules in the smectic A phase (SmA) are aligned in rows and oriented orthogonal to the packing planes (Figure 4.2(b)). Smectic C (SmC) phases also have mesogens packed into rows, but they exhibit a tilt with respect to the normal plane (Figure 4.2(c)). Neither the SmA nor SmC phase has any additional positional or rotational order of molecules within the plane; additional smectic phases (B–F) are named to represent more ordered packing, and phases with very little or no translational freedom are sometimes referred to as crystal smectic or plastic phases. Finally, both nematic and smectic phases can exhibit a propagating 'twist' with respect to the molecular orientation throughout the mesophase. Often this is induced by chirality in the mesogens and designated with an asterisk (N* and Sm*), and these chiral or 'twisted' phases have been of particular importance in recent technology applications.

4.1.2 Discotic Liquid Crystals

Although theorized for some time, the liquid crystalline phases of discotic mesogens were discovered relatively recently in 1977.[4] While relatively rare, discotic liquid crystalline systems may also exhibit nematic phases in which molecules take on a particular plane of orientation, but have no positional order (N_D), or only very localized aggregation in small stacks (N_{Col}) (Figure 4.3(a),(b)). Far more common in the mesophases of disc-like molecules are columnar liquid crystal phases (CLC), which exhibit face-oriented stacking of molecules into extended columnar structures. Similar to the smectic phases of calamitic molecules, discotics can be characterized by the order and orientation of molecules within columns, as well as by the two-dimensional inter-columnar packing structure, normal to the columnar direction. Three inter-column packing geometries are generally observed: hexagonal (Col_h), rectangular (Col_r), and oblique (Col_{ob}) (Figure 4.3(c),(d),(e)). Of these three, hexagonal is by far the most common as most discotics typically have three or more flexible alkyl chains around a rigid central core, effectively giving mesogens a 'circular' area that packs in a hexagonal lattice. In addition, the plane of individual molecules within a column may be oriented orthogonally or tilted with respect to the column direction; molecules in a Col_h phase will sometimes tilt out of the normal plane upon cooling, transitioning to Col_r or Col_{ob} phases.

The relative order of intermolecular stacking of molecules within a column can also vary. If mesogens are very evenly spaced, leading to extended positional order along a single column, the phase is termed 'ordered' (Figure 4.4(a)). The Col phase is considered 'disordered' if there is some fluidity between disks within the column (Figure 4.4(b)). In either case, intermolecular spacing within columns is usually in the range of 3.4–3.9 Å, although larger distances are sometimes observed. It is important to

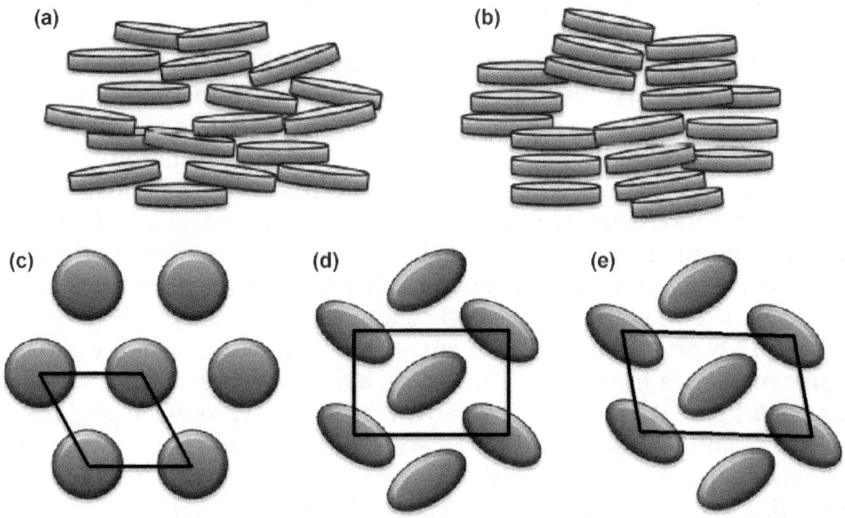

Figure 4.3 Representations of discotic mesophases. Edge-on view of a N_D (a) and N_{Col} phase (b). Top view of inter-columnar packing in Col_h (c), Col_r (d), and Col_{ob} (e) phases.

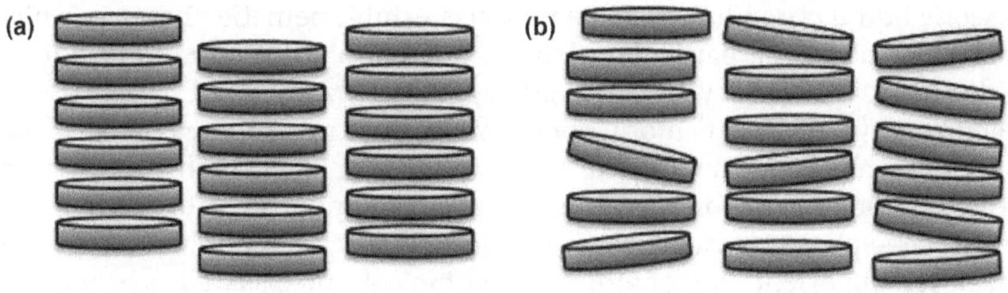

Figure 4.4 Representation of evenly-spaced stacking in a columnar ordered phase (a), and the less ordered stacking within a column in a columnar disordered phase (b).

consider that, while there may be some positional order with respect to the intermolecular spacing of discotics within a column, there is not necessarily any positional relationship of molecules in between columns in the Col_h, Col_r, and Col_{ob} phases. Additionally, molecules in these phases are still dynamic, and are generally free to rotate within and exchange in and out of a column, allowing for some fluidity and self-repair of defects. In some cases, three-dimensional positional order is present in the columnar mesophase, but molecules retain some rotational or other mobility, and the phase is considered columnar plastic (Col_p).

4.1.2.1 Conduction in Columnar Liquid Crystals

One of the most exciting aspects of columnar liquid crystalline materials is their potential to serve as organic semiconducting materials.[5–7] In the case of discotics with aromatic cores, packing into a columnar structure results in a slight overlapping of the π orbitals on neighboring mesogens. If holes or electrons are injected into these systems, charge has the ability to travel between molecules, along the length of a column. The appended alkyl chains that occupy space between columns serve as insulation, inhibiting lateral movement of charge between columns. In this way, columnar phases can resemble molecular wires capable of one-dimensional charge transport. The mechanism of charge-transport along a column is not fully understood or agreed upon in these systems, and likely varies in different systems from completely delocalized banded conduction, to localized polar-to-polar hopping of charge, to a molecular charge-hopping mechanism. Regardless, charge mobilities in some CLC materials have been measured at values as high as 0.1 $cm^2\ V^{-1}\ s^{-1}$, which exceeds that of many processed conjugated polymers, and is comparable practically to many solid-state crystalline systems. Importantly, CLC systems have advantages over polymer and crystalline systems, including relatively high solubility, simple processing, self-assembly of directionality, and the dynamic ability to 'self-heal' structural defects, which are often detrimental to efficient conduction in crystalline systems. These considerations make CLCs of great interest as potential organic semiconducting materials. Overall, there is still much to learn about the phases, properties, and structure–property relationships in these materials, and their further study and development is of great interest for fundamental knowledge and practical applications in organic electronics.

4.1.3 Liquid Crystal Characterization

Several methods are often employed in determining the phase type(s) of liquid crystalline materials. A commonly used technique for initial characterization of liquid crystalline phases is polarizing optical microscopy (POM) of thin films.[3] Due to their particular shapes and restriction of molecular orientation, mesophases are anisotropic, and in polarized light will often show optical textures that are characteristic to the birefringence of different phase types. While often not completely diagnostic, particularly for columnar phases, analysis and comparison of POM images can give excellent information about general mesophase structure. More precise distinction of structure in LC phases can often be achieved with diffraction studies, most commonly X-ray. The relationship of reflections, or the lack thereof, can give excellent insight into the molecular packing and three-dimensional order of phases. However, one should use some caution in assigning absolute structure with diffraction studies as some signals can be extremely weak or misleading, again this is particularly true of columnar phases.

The precise temperatures and change in enthalpies for phase transitions are commonly characterized in liquid crystalline systems using differential scanning calorimetry (DSC).[3] Transition temperatures are of course critical for practical considerations of LC materials as they must be stable over useful temperature ranges. Phase transition enthalpies can give insight into the relative amount of order between two neighboring phases; a larger enthalpy change is typically observed with greater difference in molecular degrees of freedom, although this is not always tightly related due to the high complexity of competing intermolecular interactions in some phase changes. Finally, charge mobilities, of particular interest in the design of CLC materials, are often measured using a time-of-flight (TOF) technique in which the material is sandwiched between transparent electrodes and photo-excited with a laser pulse at one electrode and current measured at the opposite electrode.[5,8] An alternative technique for measuring conduction in CLCs is pulse-radiolysis or flash-photolysis time-resolved microwave conductivity (PR or FP-TRMC).[5,9] This technique also uses photo-excitation of the CLC sample, but relies on a simultaneous change in microwave absorption of the sample, the decay of which relates to the lifetime of the charge-carrier states. While less common, PR-TRMC has several advantages for use with CLC samples, including that there is no need for electrode contacts, and there is no dependence on alignment of columns within a test sample.

4.2 Columnar Liquid Crystals and Aromatic Donor–Acceptor Interactions

4.2.1 Stability in Columnar Liquid Crystals

The structure, stability, and especially the electronic properties observed in CLC materials are thought to be strongly related to the degree of π–π overlap between neighboring molecules within a column.[5] For example, in recent work, two interesting tri-indole discotics were synthesized with different side chains, affecting the intermolecular stacking distance within the columns of the Col_h phase (Figure 4.5(a)). The observed stacking distances were 3.9 and 4.4 Å, and the correlated charge mobilities were measured to be greater in the more closely packed system by over 10 000 times.[10] Generally, an increase in conductivity is also observed with increasing order in the columnar phase, which can correspond to better π overlap. To maximize the semiconducting properties of CLCs, recent trends have moved towards attempts at increasing the amount of π orbital overlap between stacked molecules in columnar materials. Increasing intermolecular π overlap is often attempted by synthesizing larger aromatic cores, increasing ordered stacking and the overall area of potential π overlap. Evidence suggests there is an exponential correlation between charge mobility and aromatic core size.[11] Another approach to increasing π–π overlap is incorporation of second and third row elements into the mesogen core.[9,12] The more diffuse orbitals of the larger atoms allow for better intermolecular π–π overlap over a given separation distance, and this may be why hexathio-triphenylene (HTT) discotics typically exhibit higher charge mobility than the corresponding HAT derivatives (Figure 4.5(b)). While these strategies can be effective, increasing aromatic core size often negatively influences solubility and

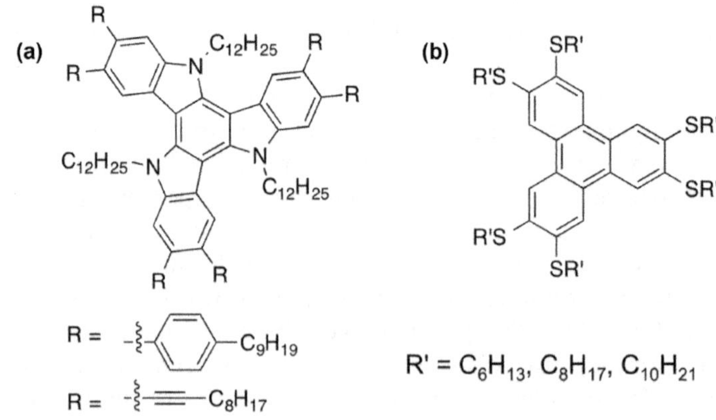

Figure 4.5 (a) The tri-indole CLC mesogen has an intermolecular stacking distance of 3.9 and 4.4 Å, with a corresponding 10 000× decrease in conductivity. (b) Replacing the oxygens with sulfur atoms, these HTT derivatives yield higher conductivity values than the corresponding HATs.

processing of CLCs, and adding second and third row heteroatoms can significantly decrease molecular stability, especially to oxidative degradation. Additionally, both approaches introduce significantly more synthetic challenges towards generating practical organic electronic materials.

4.2.2 Aromatic Donor–Acceptor Interactions

An alternative strategy towards increasing π–π overlap and structural order within the columns of a CLC material is to use non-covalent interactions to reinforce or enhance aromatic stacking. There are several approaches to this including the use of tangential hydrogen-bonding to align and better hold the discotics in place,[13,14] or steric interactions to stabilize stacking by locking mesogens in place.[9,15] However, these are limited to specific structures and not easily generalizable. Alternatively, stacking in columnar liquid crystals can be stabilized inherently by using two components having aromatic surfaces that are attracted electrostatically.

Aromatic donor–acceptor interactions (ADAI) occur between two aromatic molecules with complementary quadrupole moments, one having a relatively electron-rich aromatic surface and the other relatively electron-poor.[16] As theorized by Hunter and Sanders, this electrostatic complementarity allows for stable face-centered stacking between the two aromatics, and when in close contact the electrostatic attraction can provide stable association.[17] It should be noted here that recent work by Wheeler and others suggests this stable association may also be due to interactions between the concentration of charge on aromatic substituents.[18,19] Regardless, this interaction can be strongly enhanced by the solvophobic effect, and it is an important non-covalent force in many biological systems.[20–22] In polar solvents the aromatic surfaces are enthalpically driven together, and ADAIs have been used as design elements for numerous self-organizing and supramolecular systems in solution. In the absence of solvent, ADAIs can also stabilize face-centered stacking in a columnar mesophase, the electrostatic attraction helping to maximize π orbital overlap. In contrast, the face-centered self-stacking of aromatics in most discotic materials is electrostatically unfavorable. For both electron-rich and electron-poor moieties, face-centered stacking requires an associated energetic cost due to the repulsive electrostatic interaction (Figure 4.6). It is of note that the quadrupole charge distribution generally becomes more diffuse as the size of the aromatic core increases, which would decrease the electrostatic repulsion of face-centered stacking and may contribute to the greater CLC stability and π overlap observed with larger core sizes.

4.2.2.1 Charge Transfer Absorption

One of the most striking characteristics of ADAIs is the broad UV/Vis absorbance often observed in these systems. The close face-centered stacking of complementary aromatic components (~3.4 Å) leads to close

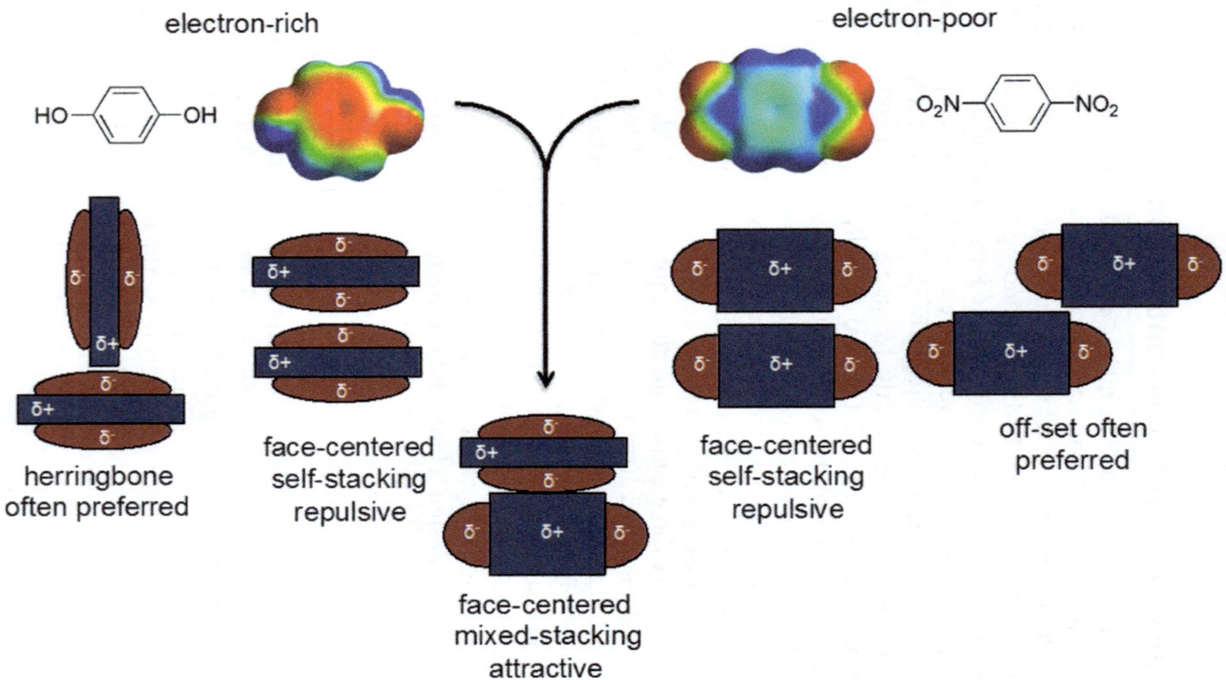

Figure 4.6 Electrostatic potential maps of a representative electron-rich and electron-poor aromatic. (Spartan, Wavefunction, Inc.) The cartoon scheme shows a side view of the unfavorable face-centered self-stacking and favorable co-stacking of these moieties. Relatively favored self-stacking geometries are also shown.

interaction of the pi orbitals. This allows for a charge-transfer (CT) absorbance in which an electron can be excited directly from the HOMO of the electron-rich component to the LUMO of an adjacent electron-poor component. The energy gap (E_G) of the $HOMO_{donor}$–$LUMO_{acceptor}$ transition is often (but not always) red-shifted from either molecular absorbance, which can result in very colorful materials that are quite distinct from the component molecules. For this reason, ADAIs are sometimes also referred to as charge-transfer interactions.

4.2.3 Aromatic Donor–Acceptor-Based Mesophases

While the effect of aromatic donor–acceptor interactions has been observed in crystalline and mesophase systems for over fifty years,[23] Ringsdorf and coworkers were the first to implement them as an intentional design element for generating discotic materials in 1989.[24] Polymers were appended with relatively electron-rich aromatics, including HAT derivatives, and small amounts of the electron-deficient aromatic 2,4,7-trinitro-9-fluorenone (TNF) were added. Even at very low loading, adding TNF to the HAT polymers induced a new Col_h phase, not present in the pure HAT polymer. This phenomenon was soon shown to apply to small molecule systems of monomeric HAT and TNF (Figure 4.7(a)).[25]

Though TNF and a few other electron-poor aromatics were shown to enhance and/or induce mesophase behavior of several discotic molecules, the electron-poor components were initially poorly matched in terms of geometry and symmetry to the electron-rich components. These early examples of donor–acceptor LC mixtures, while empirically interesting, were difficult to characterize structurally. There continues to be a debate about the structural nature of mesophase-stabilizing effects in these systems, either being the result of the electron-poor component being fully integrated within the columns in an alternating fashion *via* aromatic donor–acceptor interactions, or organized in between columns of the electron-rich moiety.[26,27]

Systems in which the two components are geometrically matched are more likely to yield mixed CLC phases for which there is strong evidence for an alternating face-centered stacking geometry of donors and acceptors. In work by Grubbs and coworkers, a C_3 symmetric perfluorotriphenylene (PFTP) electron acceptor was used with HAT derivatives, the two components matching relatively well the area and symmetry of the two components (Figure 4.7(b)).[28] The resulting mesophase of a 1 : 1 PFTP–HAT mixture is stable over a temperature range greater than either component by up to 70 °C. Importantly, an alternating face-centered columnar structure of the mixture is assigned based on a single crystal structure of a 1 : 1 mixture of HAT with PFTP cooled from the mesophase. This confirms the alternating complementary structure of these mixtures, which can be referred to as charge-transfer liquid crystals, or donor–acceptor columnar liquid crystals (DACLCs).

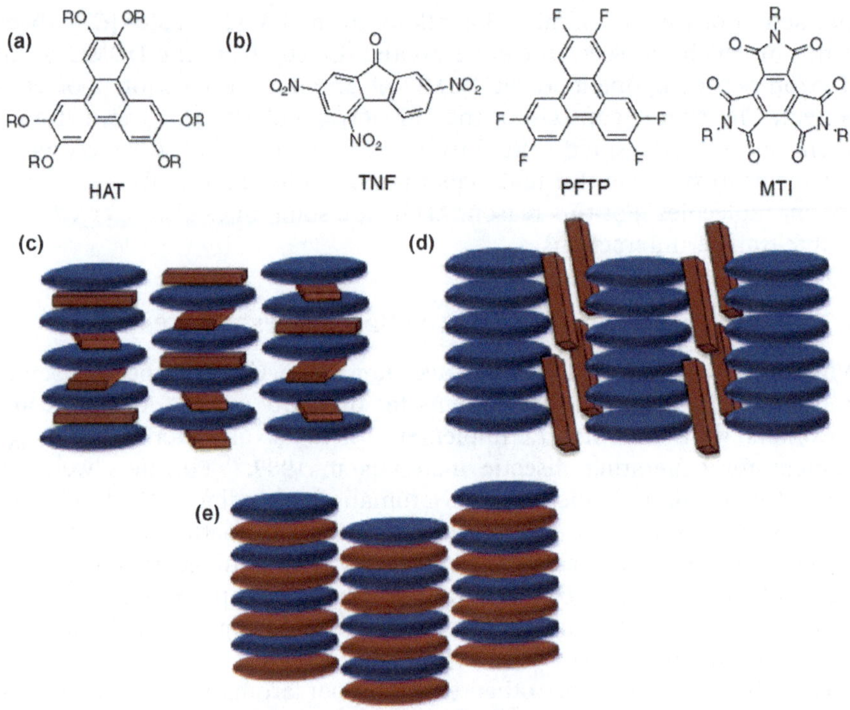

Figure 4.7 (a) Core structure of the HAT discotic, a common electron-rich component. (b) Three electron-poor aromatic derivatives, TNF, PFTP, and MTI. Representation of alternating (c) and inter-columnar stacking (d) of a non-geometrically matched aromatic donor–acceptor system. (e) Representation of alternating donor–acceptor stacking with geometrically matched components.

Another structurally similar aromatic donor–acceptor pair that exhibits a Col_h phase is that of HAT and the electron-poor mellitic triimide (MTI). Park and coworkers were also able to confirm the presence of face-centered alternating stacking with a single crystal structure cooled from the mesophase.[29] Using MTI as the acceptor allowed for both components to be derivatized with peripheral alkyl chains. By independently varying the length of alkyl chains on either the HAT or triimide component, a wide range of mesophase transition temperatures could be obtained through mixing and matching of components, although no predictable trends could be identified.

4.3 Naphthalenediimide-Based Donor–Acceptor Columnar Liquid Crystals

There have been several other aromatic donor–acceptor pairings in columnar liquid crystalline materials over the past decade, but the remainder of this chapter will be devoted to those involving NDI as the acceptor component.[30,31] As mentioned above, NDI has been used extensively as an electron-poor component in aromatic donor–acceptor assemblies. In particular, it has been paired with 1,5-dialkoxy naphthalenes (DAN) in the design of a variety of self-assembled architectures. The two moieties have excellent electrostatic complementarity on the aromatic surface and similar to the MTI and HAT pairing, NDI and DAN are easily derivatized with a variety of alkyl substituents. The two aromatic cores are also well matched in size and symmetry, making them ideal for study in the bulk state as components of highly modular alternating aromatic donor–acceptor columnar liquid crystals.

4.3.1 Independent Naphthalenediimide

Before focusing on the diverse DACLCs in which molecular NDI is a key component, pure NDI LC materials will be discussed. Many NDI derivatives have shown excellent potential as n-type semiconducting materials in the solid state.[32] However, the relatively small size of the NDI means that its electronic properties are highly susceptible to defects in the solid state; the stability and usefulness of NDI towards a number of organic electronic applications could be enhanced if stable columnar mesophases of these semiconducting compounds could be engineered. This has been shown to be possible for the related class of perylenediimides (PDI).[33] In fact, increased charge mobility has recently been observed in the LC *versus* Cr phase of some PDI derivatives, as well as improved device performance of LC PDIs.[34,35] Developments in related PDI materials have been thoroughly covered in a recent review.[36]

As pure substances, the majority of *N*,*N*′-dialkyl NDI derivatives either do not exhibit thermotropic mesophases, or go through a narrow-range transition at relatively high temperature (Figure 4.8, Table 4.1, ref. 37–41).

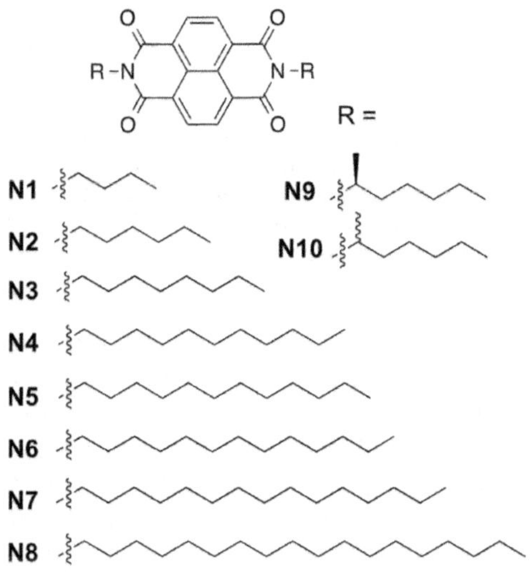

Figure 4.8 Representative *N*,*N*′-dialkyl NDI derivatives.

Table 4.1 Phase transitions (DSC) of several *N,N'*-dialkyl NDI derivatives.

NDI	T_1 $(\Delta H)^a$ (°C) (kcal mol^{-1})		T_2 $(\Delta H)^a$
N1	226 (23)	→	160 (3)
N2	204 (22)	→	150 (15)
N3	181 (32)		
N4	162 (17)	→	144 (5)
N5	160 (27)	→	145 (8)
N6	162	→	143
N7	156	→	139
N8	149	→	138
N9[b]	156 (23)		
N10[c]	117 (24)		

[a]Values reported when available for exothermic transition on cooling.
[b]Enantiomerically pure.
[c]Diastereotopic mixture of *R,R*; *R,S*; *S,S* derivatives.

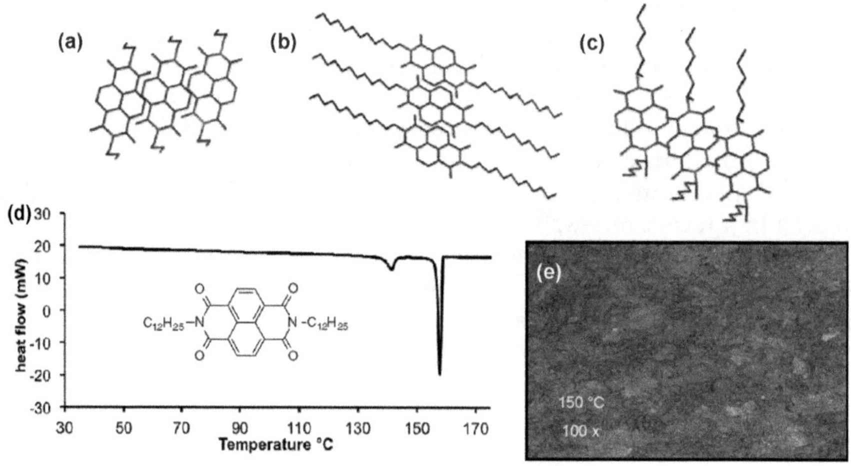

Figure 4.9 Single crystal X-ray diffraction of several *N,N'*-dialkyl NDI derivatives; chains shown are C_4H_9 (a), $C_{12}H_{25}$ (b), $CH(CH_3)C_5H_{11}$(R) (c). DSC trace on cooling (d) and POM image of the mesophase (e) for the $C_{12}H_{25}$ *N,N'*-dialkyl NDI. Reprinted with permission from P. M. Alvey *et al.*, *J. Org. Chem.*, 2010, 75, 7682. Copyright 2010 American Chemical Society.

Perhaps as a result of this, the mesophases of NDI derivatives have not been highly characterized, although the crystalline phases and ordered thin films have undergone significant study with respect to structure and charge mobility.[37,38] As may be expected from the above discussion of quadrupoles in electron-poor aromatics, *N,N'*-dialkyl NDI derivatives typically stack in a slipped-face geometry, where molecules are off-set on the short axis (Figure 4.9(a)–(c)).[38] The IL–LC transition on cooling is significantly more exothermic compared to the LC–Cr transition for almost all NDIs.

This implies a highly ordered mesophase, and with the mosaic pattern observed for the mesophase texture under POM, these are likely calamitic-plastic mesophases of some type.[39,40] Recent work with some block-structure *N*,*N'*-azomethine-NDI derivatives has shown evidence of SmA and SmC phases, although multiple transitions are observed and it is unclear what the role of the side chains is in inducing mesophase behavior.[41]

Overall, pure molecular NDI derivatives are not yet practical as liquid crystalline materials. In contrast with the larger PDI core, the side chains needed to induce a broad thermotropic mesophase in NDI often dominate the phase structure, dampening the functional properties. Still, it has been suggested that NDIs that exhibit a mesophase should have enhanced potential as n-type semiconductors, as annealing past the crystallization point has been shown to allow for optimal core alignment of molecular packing, critically enhancing charge mobility.[37]

4.3.2 Naphthalenediimide Donor–Acceptor Columnar Liquid Crystals

As noted above, ADAIs have been used to achieve many elegant solution phase assemblies and one prominent component pairing is that of electron-poor NDI with electron-rich 1,5-DAN first developed in the aedamer systems of Iverson and coworkers.[42] These two components are well matched in terms of electrostatics, size, and geometry, and as such are ideally suited as components of DACLCs (Figure 4.10). Importantly, X-ray crystallography done on co-crystals of this pairing confirms their ability to stack in a completely alternating face-centered geometry, with a relatively close interfacial stacking of 3.38 Å.[39,40,42]

Even more so than with the MTI–HAT system, it is straightforward to modify both NDI and DAN with a variety of alkyl substituents. Several 1,5-DAN derivatives have been studied as 1:1 molar mixtures with many of the NDI compounds above (Figure 4.11).[40] In general, the DAN–NDI mixtures

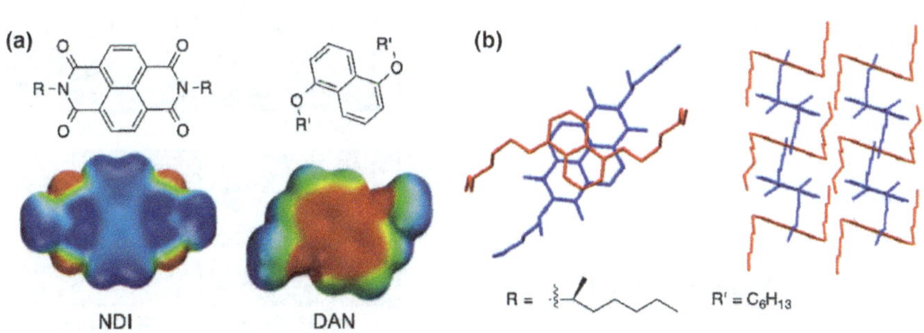

Figure 4.10 (a) Electrostatic potential maps of NDI and DAN. (b) Co-crystal structure of DAN and NDI showing the top and side view of the alternating face-centered stacking. Hydrogens omitted for clarity.
Reprinted with permission from P. M. Alvey *et al.*, *J. Org. Chem.*, 2010, 75, 7682. Copyright 2010 American Chemical Society.

behave as a single new material, and all combinations exhibit one or two thermotropic mesophases. This can be clearly seen in the DSC traces of the mixtures, which do not resemble those of the individual components, and often show a low-energy IL–LC phase transition with significantly larger LC–C transitions, typical of LC materials (Figure 4.12(a)). POM images of the donor–acceptor mesophases typically show either dendritic branching or rectangular mosaic textures, strongly suggestive of Col_r and Col_p phases respectively (Figure 4.12(b)). Several of these mixtures have been studied

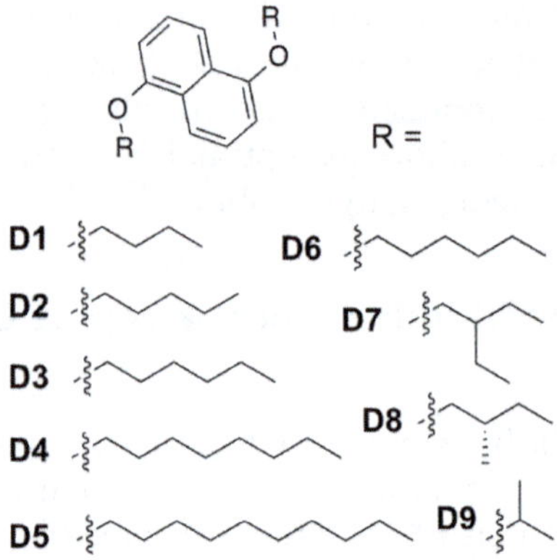

Figure 4.11 Sample variety of 1,5-dialkoxy naphthalenes mixed with NDI derivatives.

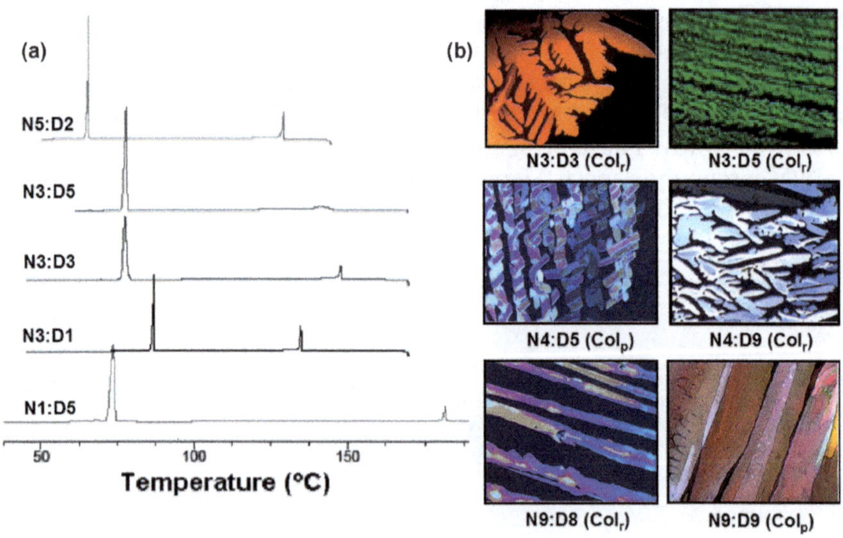

Figure 4.12 Representative DSC scans (a) and POM images (b) of several 1:1 NDI–DAN materials.
Reprinted with permission from J. J. Reczek *et al.*, *J. Am. Chem. Soc.*, 2006, **128**, 7995. Copyright 2006 American Chemical Society.

Table 4.2 Phase transition temperatures of NDI–DAN mixtures.

DAN	D1	D2	D3	D6	D7	D9
NDI	(T_{c1}, T_{c2}, $^{a}T_{c3}$) (°C)					
N3	137, 86, 81	139, 64	148, 78	144, 41	—	—
N4	—	—	—	—	136, 26	140, 79
N5	134, 85, 73	129, 66	131, 77	131, 120	—	132, 125
N9	116, 103, 83	121, 99, 65	129, 122, 72	122, 20	123, 27	129, 81
N10	98, 89, 63	95, 61	111, 77	—	—	106, 81

[a]All temperatures recorded by DSC on cooling.

using powder X-ray diffraction techniques, confirming the Col_r packing geometry for these systems.[40] In contrast to the pure NDI components discussed above, the DAN–NDI mixtures often exhibit mesophases over broad temperature ranges, and some at relatively mild temperatures (Table 4.2). For example, the LC–Cr transitions for mixtures **N4–D7**, **N9–D6**, and **N9–D7** are all around room temperature, and each has a mesophase range spanning over 100 degrees. Finally, all of the NDI–DAN combinations result in a DACLC material deep red in color due to the CT absorbance, a striking contrast with the components that are generally white or light yellow.

A rare feature of NDI–DAN DACLCs is the relatively predictable temperature of their phase changes. The upper clearing point temperature of mixtures is observed to track fairly well with the melting point of the NDI component in the mixture, as the lower crystallization transition tracks with that of the DAN component used. This may be correlated to the C_2 symmetry of NDI and DAN components, an element that distinguishes this system from the vast majority of CLC discotics. Taking into account several of the NDI–DAN co-crystal structures suggests that the alkyl chains of the respective components are oriented on orthogonal axes, perhaps to optimize columnar stacking of the DAN and NDI cores (Figure 4.10(b)). This may relate to the empirical phenomena of mesophase temperature transitions in these materials being somewhat independently and predictably adjusted, through the simple mixing and matching of different components. In theory, the ability to rationally adjust components to independently change mesophase properties could yield critical insight into structure–property relationships in CLC design.

4.3.3 Thermochromic Phase Properties of NDI–DAN Materials

In initial studies with DAN–NDI liquid crystals, it was observed that the CT absorbance was strongly present in the liquid crystalline phase of all mixtures, but in some cases the red CT color would persist upon crystallization from the mesophase, and in others the color would dissipate (Figure 4.13).[40] Analysis of the powder X-ray diffraction patterns of the crystalline phases of

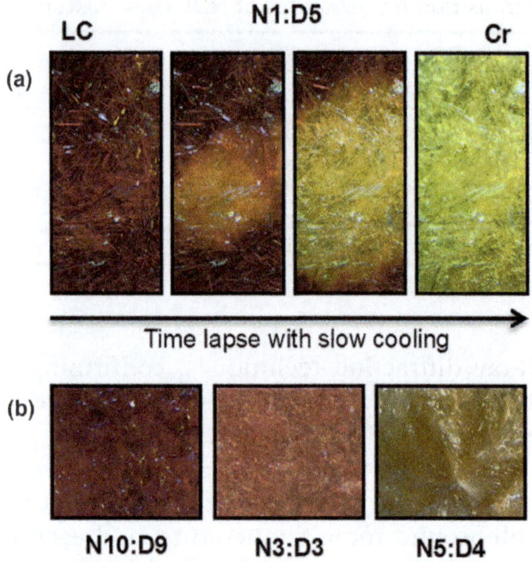

Figure 4.13 (a) Thermochromic phase change upon cooling of N1–D5. (b) Crystalline phase of different NDI–DAN materials showing retention (N10–D9), partial retention (N3–D3), and loss (N5–D4) of CT absorption. All samples are the same deep red color in the mesophase.
Reprinted with permission from P. M. Alvey *et al.*, *J. Org. Chem.*, 2010, **75**, 7682. Copyright 2010 American Chemical Society.

several mixtures showed that in materials that lost the CT absorbance, the diffraction pattern changed significantly from that of the mesophase, and actually began to resemble a sum of patterns seen with the independent components. Mixtures that maintained their color while crystalline exhibited a diffraction pattern unique from either crystalline component. It was reasoned that upon crystallization, the NDI–DAN systems exist in a delicate balance between the electrostatic interactions allowing for stable face-centered stacking, and crystal packing forces of the individual components. If the latter were greater, then the components would micro-separate upon crystallization, breaking the close π–π contact between components, and therefore no longer exhibiting the red CT absorbance. A systematic study of the effect of different DAN and NDI alkyl substituents on this phase-chromic phenomenon revealed subtle trends that reinforce the delicate balance between forces causing either retention or loss of donor–acceptor stacking in the crystalline phase.[39] For straight-chain derivatives, no trends were associated with the DAN components. However, a slight preference for maintaining D–A stacking in the crystalline phase was observed for NDI chains of moderate length (**N3**) over both short (**N1**) and long (**N12**) chains. Adding a methyl branch next to the aromatic core had an interesting effect. A core adjacent methyl branch on the NDI component (**N9**, **N10**) led to retention of the CT color. Examination of the co-crystal structure suggests that this methyl group may serve to lock the DAN component in place, creating a kinetic barrier to separation. In contrast, core-adjacent methyl branching on the DAN component drives component separation upon crystallization.

While there is still more work to be done on specific factors that determine the dramatic phase-chromic behavior, there is clearly a delicate balance of forces that can be influenced towards retention or loss of donor–acceptor stacking upon crystallization. This has the potential to be exploited in the design of solid-state systems, and perhaps as a 'switch' for columnar assembly and associated properties.

4.3.4 NDI with Other Electron-Rich Naphthalene Components

The general ability of NDI to serve as the electron-poor component in a variety of DACLC materials has been explored using alternative naphthalene derivatives as donor components.[43] NDI compound **N3** was mixed with a series of electron-rich naphthalenes, differing from the traditional 1,5-DAN component with respect to the position and heteroatom identity of the substituents (Figure 4.14). Importantly, even as substitution on the core-structure of the electron-rich naphthalenes is altered, all derivatives exhibit mesophase behavior by DSC when mixed with **N3** (Table 4.3), and POM textures again support the presence of Col_r and Col_p phases. Similar to the 1,5-DAN mixtures above, clearing temperatures matched the crystallization temperature of **N3** relatively closely (180 °C), although they ranged from 123 °C to 167 °C. This supports a strong general correlation between the NDI component crystallization temperature, and the corresponding clearing point of DACLC mixtures with naphthalene-core donors. The same cannot be said for the crystallization point of these materials, which varied widely

Table 4.3 Phase transitions for **N3** with donor series. Reprinted with permission from K. R. Leight *et al.*, *Chem. Mater.*, 2012, **24**, 3318. Copyright 2012 American Chemical Society.

Donor–**N3**	$T_1/\Delta H$ (°C)/(J g^{-1})		$T_2/\Delta H$ (°C)/(J g^{-1})		$T_3/\Delta H$ (°C)/(J g^{-1})
D10	150/10.0		→		80/51.6
D11	159/15.4		→		75/30.9
D12	123/33.9	→	99/20.2	→	22/9.4
D13	143/6.5	→	124/3.2	→	39/51.9
D14	162/10.7		→		17/40.2
D15	128/8.4		→		34/19.8
D16	167/22.1		→		25/32.5
D17	155/3.1	→	143/11.4	→	69/55.0
D18	157/2.8	→	147/8.9	→	21/44.8
D19	154/27.9	→	40/1.2	→	1/6.1
D20	146/12.4	→	100/5.0	→	17/38.8
D21	145/21.7		→		41/49.4

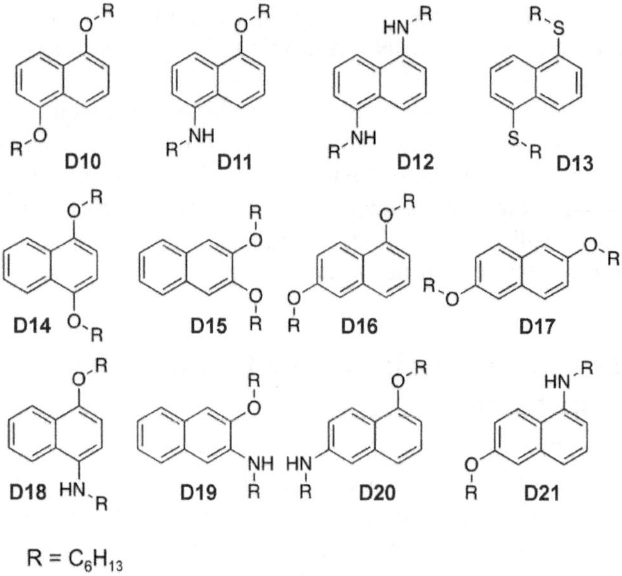

Figure 4.14 Series of electron-rich naphthalenes combined with **N3**.

and did not track with the respective crystallization temperatures of the donor components. However, the variety and range of crystallization temperatures for the mesophases of these DACLCs, all with hexyl alkyl chains, is impressive, and significantly expands the ability to independently tune phase transition temperatures by mixing different components.

4.3.5 Predictable Absorption in Modular NDI DACLC Materials

Typically, tuning the absorption characteristics of organic molecules requires a synthesis-heavy and somewhat empirical approach. Similar to the ability to independently tune phase-change temperatures, the CT-based absorption in DACLCs offers a creative design strategy for simple manipulation of optical properties. As absorption in these materials depends on the energy gap between the $HOMO_{donor}$ and $LUMO_{acceptor}$, simply mixing components with different relative orbital energy levels should result in a corresponding change in the bulk absorption properties of the system. In this way, absorption characteristics in these organic materials could be significantly altered for specific applications, through the simple mixing of relatively straightforward components.

The potential of this design strategy for tuning absorption in DACLC materials is immediately recognized upon observing the bulk characteristics of the mixed-core donor series above.[43] As discussed in the sections above, all of the NDI–DAN mixtures previously studied exhibited the same deep-red color in the mesophase. This is not surprising, as changing the alkyl

Figure 4.15 Representation of the range in bulk colors achieved from simple mixing of different naphthalene donor components with N3. Samples were melted and slowly cooled on TLC slides to enhance the color contrast between materials.
Reprinted with permission from K. R. Leight *et al.*, *Chem. Mater.*, 2012, **24**, 3318. Copyright 2012 American Chemical Society.

derivatives on the 1,5-DAN should have little effect on the energy of the HOMO orbital. However, this is not true upon changing the core-substitution of the electron-rich naphthalene. The ability to significantly affect the CT absorption of DACLC materials, simply by mixing NDI with new aromatic donors, is immediately apparent in the vibrant array of colors produced by the series above, spanning the color spectrum (Figure 4.15).

Of significant interest, particularly with respect to the future design of DACLC materials for tailored outcomes and applications, is the ability to predict the onset of absorption in the studied DACLC materials, using relatively straightforward calculations on only the individual components.[43] In semiconducting materials, the HOMO/LUMO energy gap generally corresponds directly to the red-onset (inflection point) of the observed UV/Vis absorbance. Perhaps surprisingly, there is typically minimal perturbation of orbital energy levels upon donor–acceptor π-stacking (in contrast to the significant perturbation in covalently linked donor–acceptor polymer systems, so called 'push–pull' polymers). Therefore, in solvent-less systems, the observed red onset of the CT absorption should correlate with the energy of the relatively unperturbed $LUMO_{acceptor}$–$HOMO_{donor}$.

DFT-level calculations on the individual components above were used to find their corresponding HOMO and LUMO energy levels, which were simply subtracted to find a theoretical CT E_g for a given donor with **N3**. Remarkably, in *all* mixtures, this calculated CT E_g correlates extraordinarily well with the measured red-onset of CT absorption obtained from the asymptote method on UV/Vis spectra of thin films (Figure 4.16).[43] Regardless of substituent location, identity, or relative molecular shape, the energy of the calculated CT E_g is within 3% of that measured from the empirical CT onset (Table 4.4).

It is also important to note that the extinction coefficients (ε) of the measured CT absorbance in these materials are significant, ranging from 3000–6800 $M^{-1}\,cm^{-1}$. This is at least an order of magnitude more intense than that normally seen for similar types of aromatic donor–acceptor assemblies in solution, and begins to enter a practically useful range for applications of photo-conductive materials, particularly given the broad absorption observed in these materials. Overall, the precision and consistency of correlation of the calculated E_g to measured UV/Vis onset in these

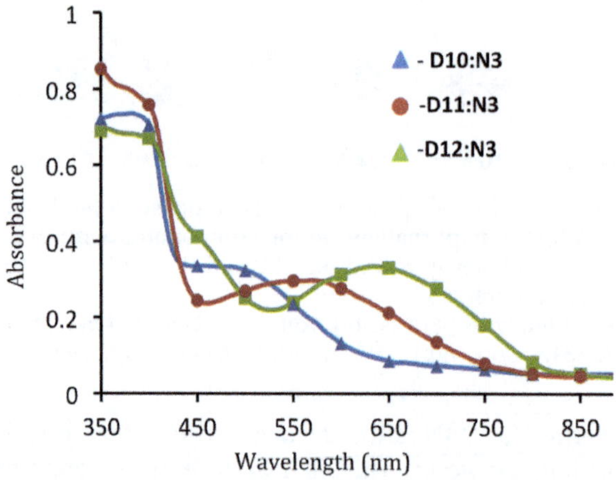

Figure 4.16 Representative UV/Vis absorption spectra of DACLC thin films. Reprinted with permission from K. R. Leight *et al.*, *Chem. Mater.*, 2012, **24**, 3318. Copyright 2012 American Chemical Society.

Table 4.4 Calculated and measured CT E_g, observed CT λ_{max}, and extinction coefficient. Reprinted with permission from K. R. Leight *et al.*, *Chem. Mater.*, 2012, **24**, 3318. Copyright 2012 American Chemical Society.

Mixture	$\lambda_{onset\ CT}$ (nm)	Obs E_g (eV)	Calc E_g (eV)	$\lambda_{max\ CT}$ (nm)	ε_{CT} ($M^{-1}\ cm^{-1}$)
D10–N3	656	1.89	1.86	486	6200
D11–N3	759	1.64	1.67	568	6000
D12–N3	812	1.53	1.52	636	6800
D13–N3	594	2.09	2.07	466	4200
D14–N3	705	1.76	1.74	491	3900
D15–N3	550	2.26	2.21	414	5100
D16–N3	613	2.04	2.01	450	6500
D17–N3	631	1.97	1.94	465	5500
D18–N3	826	1.50	1.53	606	3200
D19–N3	730	1.70	1.69	551	5200
D20–N3	728	1.70	1.70	572	6800
D21–N3	725	1.71	1.74	562	6400

NDI-based DACLCs are extraordinary, and suggest the possibility for targeted and efficient design of optical absorption characteristics in future materials.

4.3.6 Strongly Dichroic Thin-Films in NDI–Anthracene DACLC Materials

The exploration of new NDI-based DACLCs continues to provide a better understanding of structure property relationships in modular mixed-stack CLC materials, and results in the emergence of new functional properties for these systems. NDI compound **N3** was again used as the electron-poor component in a study exploring a series of DACLC materials using alkoxy-anthracene donor components (Figure 4.17).[44] Consistent with the work discussed above, all of the NDI–anthracene mixtures studied exhibited thermotropic mesophase behavior and all were deeply colored upon melting, showing a significant CT band in the UV/Vis spectra with an onset that matched with orbital calculations. Interestingly, the intensity of the CT band in mixtures of **N3** with the 1,3,5,7-substitued anthracene core (**D23a–d**) was significantly higher (~8000 M^{-1}) than those of the other anthracene derivatives. Modeling of the relative stacking geometries in this NDI–anthracene series shows that the **N3–D23a** complex is restricted with respect to rotational orientation, and the minimum orientation yields excellent orbital overlap between the two components. The restricted molecular orientation at the near optimal orbital overlap is likely the reason for the significant enhancement of CT band intensity in these materials.

The optical texture of films of the **N3–D23a–d** series is also distinct from those of the other materials in this study, showing large domains of narrow linear structures. Viewing these structures with linear polarized light (LPL) reveals a stark dichroism. Light oriented parallel to the long-axis of the film structures (I_0) is strongly absorbed in the CT band, while light oriented perpendicular (I_{90}) is almost completely transmitted in this region (Figure 4.18). This remarkable dichroic behavior results from the stacking orientation of the DACLC columns aligning perfectly with the propagation of the film structures. As with the relative intensity of the CT band noted above, this feature likely results from the relative rotational restrictions of the mesophase, which also result in these structures adopting a rare columnar lamellar phase (Col_L).[44] Importantly, two of these materials maintain this Col_L phase and dichroic properties at room temperature, making these NDI based DACLCs of potential interest as organic thin-films for display technologies.

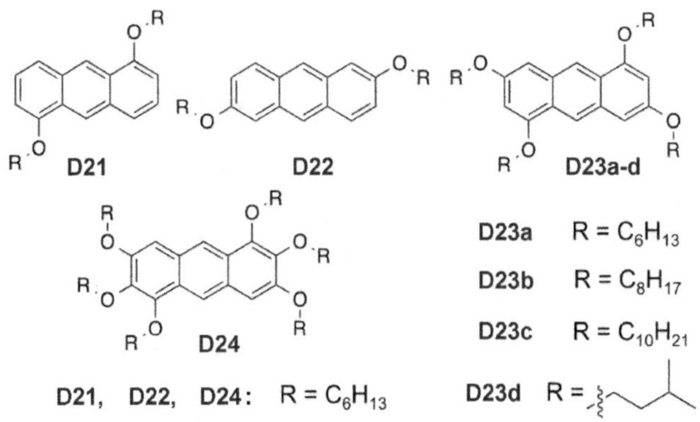

Figure 4.17 Series of anthracene donors combined with **N3**.

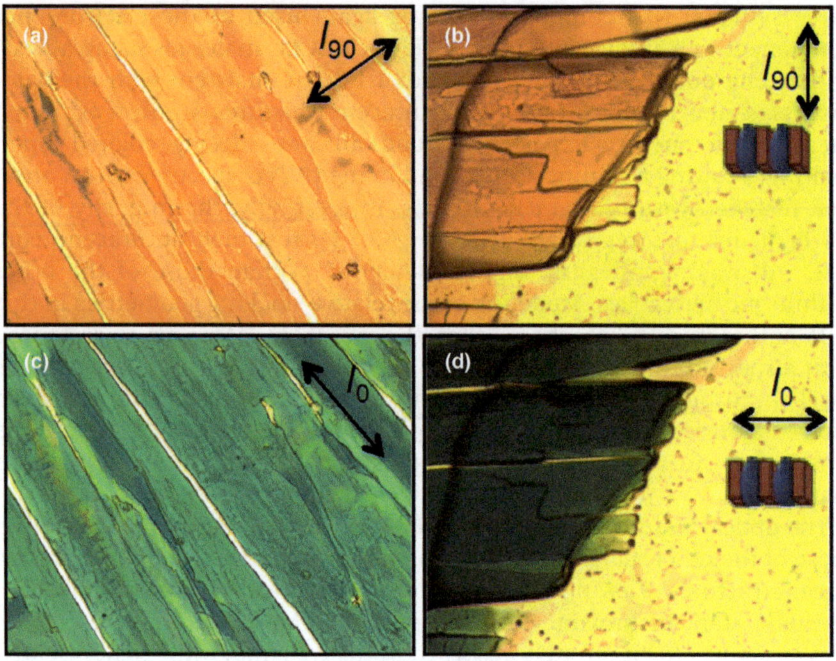

Figure 4.18 Optical microscopy images with LPL oriented perpendicular [(a), (b)] and parallel [(c), (d)] to the propagation direction of the linear structures. (a), (c) **D23b–N3**, 70 °C; (b), (d), **D23c–N3**, 120 °C. Reprinted with permission from A. Gray Bé *et al.*, *Org. Lett.*, 2015, **17**, 4834. Copyright 2015 American Chemical Society.

4.4 Conclusion

Columnar liquid crystals are a fascinating class of materials with the potential to significantly impact future technologies as organic electronic materials. Among the current types of CLCs, mixed-stack DACLCs based on designed donor–acceptor aromatic interactions between complementary components are an emerging class of modular materials with fascinating functional properties. Though not often liquid crystalline by itself, as a stable and moderately sized electron-poor aromatic, NDI is ideally suited for ADAIs and has played a key role in the development and study of DACLC materials. With proper design, these materials offer unprecedented tunability of physical and optical properties, often with regards to the structure of the donor *or* acceptor components independently. As such, they represent a highly modular class of materials from which new functional properties will continue to emerge. In particular, related materials based on CT co-crystals have recently enjoyed a resurgence of interest for their ambipolar charge-transport, bulk ferroelectric properties, and photoconductivity.[45–49] As with many liquid crystalline systems, well-designed DACLCs have the potential to mimic and outperform these crystalline counterparts for many applications. Important next steps in the study of the NDI-based CLCs include exploration of factors which affect their charge mobilities by the methods described above, development of methods to control columnar orientation in films, and continued investigation of other creative avenues for the application of these unique modular materials.

References

1. M. R. Fisch, *Liquid Crystals, Laptops and Life*, World Scientific, River Edge, NJ, 2004.
2. *Liquid Crystals Beyond Displays: Chemistry, Physics, and Applications*, ed. Q. Li, Copyright ©2012 John Wiley & Sons, Inc., Hoboken, N.J.
3. P. J. Collings, *Liquid Crystals: Nature's Delicate Phase of Matter*, Princeton University Press, Princeton, NJ, 2nd edn, 2002.
4. S. Chandrasekhar, B. K. Sadashiva and K. A. Suresh, *Pramana*, 1977, **9**, 471.
5. *Liquid Crystalline Semiconductors Materials, Properties, and Applications*, ed. R. J. Bushby, S. M. Kelly, and M. O'Neill, Springer, Dordrecht, The Netherlands, Springer Series in Materials Science 169.
6. B. R. Kaafarani, *Chem. Mater.*, 2011, **21**, 378–396.
7. S. Sergeyev, W. Pisula and Y. H. Geerts, *Chem. Soc. Rev.*, 2007, **36**, 1902–1929.
8. D. Adam, P. Schuhmacher, J. Simmerer, L. Haeussling, K. Siemensmeyer, K. H. Etzbach, H. Ringsdorf and D. Haarer, *Nature*, 1994, **371**, 141.
9. J. M. Warman and A. M. Van De Craats, *Mol. Cryst. Liq. Cryst.*, 2003, **396**, 41.
10. M. Garcia-Frutos, U. K. Pandey, R. Termine, A. Omenat, J. Barbera, J. L. Serrano, A. Golemme and B. Gomez-Lor, *Angew. Chem., Int. Ed.*, 2011, **50**, 7399.
11. A. M. Van de Craats and J. M. Warman, *Adv. Mater.*, 2001, **13**, 130.
12. A. M. Van de Craats, J. M. Warman, M. P. De Haas, D. Adam, J. Simmerer, D. Haarer and P. Schuhmacher, *Adv. Mater.*, 1996, **8**, 823.
13. N. Boden and R. J. Bushby, *J. Chem. Phys.*, 1993, **98**, 5920.
14. N. Boden, R. J. Bushby, J. Clements and R. J. Luo, *J. Mater. Chem.*, 1995, **5**, 1741.
15. P. G. Schouten, J. M. Warman, M. P. De Haas, J. F. Van der Pol and J. W. Zwikker, *J. Am. Chem. Soc.*, 1992, **114**, 9028.
16. C. R. Martinez and B. L. Iverson, *Chem. Sci.*, 2012, **3**, 2191.
17. C. A. Hunter and J. K. M. Sanders, *J. Am. Chem. Soc.*, 1990, **112**, 5525.
18. S. E. Wheeler, *Acc. Chem. Res.*, 2013, **46**, 1029.
19. R. M. Parrish and C. D. J. Sherrill, *J. Am. Chem. Soc.*, 2014, **136**, 17387.
20. M. Cubberley and B. L. Iverson, *J. Am. Chem. Soc.*, 2001, **123**, 7560.
21. E. A. Meyer, R. K. Castellano and R. Diederich, *Angew. Chem., Int. Ed.*, 2003, **42**, 1210.

22. M. J. Rashkin and M. L. Waters, *J. Am. Chem. Soc.*, 2002, **124**, 1860.
23. C. R. Patrick and G. S. Prosser, *Nature*, 1960, **188**, 1021.
24. H. Ringsdorf, R. Wustefeld, E. Zerta, M. Ebert and J. H. Wendorff, *Angew. Chem., Int. Ed. Engl.*, 1989, **28**, 914.
25. H. Bengs, M. Ebert, O. Karthaus, B. Kohne, K. Praefcke, H. Ringsdorf, J. Wendorff and R. Wüsterfeld, *Adv. Mater.*, 1990, **2**, 141.
26. K. Praefcke and D. Singer, *Handbook of Liquid Crystals*, ed. D. Demus, J. Goodby, G. W. Gray, H. W. Spiess and V. Vill, Wiley-VCH, Weinheim, 1998, vol. 2B, pp. 945–967.
27. L. A. Haverkate, M. Zbrir, M. R. Johnson, B. Deme, H. J. M. de Groot, F. Lefeber, A. Kotlewski, S. J. Picken, F. M. Mulder and G. J. Kearley, *J. Phys. Chem. B*, 2012, **116**, 13098.
28. M. Weck, A. R. Dunn, K. Matsumoto, G. W. Coates, E. B. Lobkovsky and R. H. Grubbs, *Angew. Chem., Int. Ed.*, 1999, **38**, 2741.
29. L. Y. Park, D. G. Hamilton, E. A. McGehee and K. A. McMenimen, *J. Am. Chem. Soc.*, 2003, **125**, 10586–10590.
30. W. Pisula, M. Kastler, D. Wasserfallen, J. W. F. Robertson, F. Nolde, C. Kohl and K. Müllen, *Angew. Chem., Int. Ed.*, 2006, **45**, 819.
31. J.-Y Wang, J. Yan, L. Ding, Y. Ma and J. Pei, *Adv. Funct. Mater.*, 2009, 1746.
32. F. Wurthner and M. Stolte, *Chem. Commun.*, 2011, **47**, 5109.
33. Y. Xu, S. Leng, C. Xue, R. Sun, J. Pan, J. Ford and S. Jin, *Angew. Chem., Int. Ed.*, 2007, **46**, 3896.
34. M.-A. Muth, G. Gupta, A. Wicklein, M. Carrasco-Orozco, T. Thurn-Albrecht and M. Thelakkat, *J. Phys. Chem. C*, 2014, **118**, 92.
35. Y. Zhang, H. Wang, Y. Xiao, L. Wang, D. Shi and C. Cheng, *ACS Appl. Mater. Interfaces*, 2013, **5**, 11093.
36. F. Würthner, C. R. Saha-Möller, B. Fimmel, S. Ogi, P. Leowanawat and D. Schmidt, *Chem. Rev.*, 2016, **116**, 962–1052.
37. M. Ichikawa, Y. Yokota, H.-G. Jeon, G. R. Banoukepa, N. Hirata and N. Oguma, *Org. Elect.*, 2013, **14**, 516.
38. T. Kakinuma, H. Kojima, M. Ashizawa, H. Matsumoto and T. Mori, *J. Mater. Chem. C*, 2013, **1**, 5395.
39. P. M. Alvey, J. J Reczek, V. Lynch and B. L. Iverson, *J. Org. Chem.*, 2010, **75**, 7682.
40. J. J. Reczek, K. R. Villazor, V. Lynch, T. M. Swager and B. L. Iverson, *J. Am. Chem. Soc.*, 2006, **128**, 7995.
41. E. Schab-Balcerzak, A. Iwan, M. Krompiec, M. Siwy, D. Tapa, A. Sikora and M. Palewicz, *Synth. Met.*, 2010, **160**, 2208.
42. R. S. Lokey and B. L. Iverson, *Nature*, 1995, **375**, 303.
43. K. R. Leight, B. E. Esarey, A. E. Murray and J. Reczek, *J.* Chem. Mater., 2012, **24**, 3318.
44. A. Gray Bé, C. Tran, R. Sechrist and J. J. Reczek, *Org. Lett.*, 2015, **17**, 4834.
45. K. P. Goetz, D. Vermeulen, E. Payne, C. Kloc, L. E. McNeil and O. D. Jurchescu, *J. Mater. Chem. C*, 2014, **2**, 3065.

46. Y. Su, Y. Li, J. Liu, R. Xing and Y. Han, *Nanoscale*, 2015, **7**, 1944.
47. A. S. Tayi, A. K. Shveyd, A. C.-H. Sue, J. M. Szarko, B. S. Rolczynski, D. Cao, T. J. Kennedy, A. A. Sarjeant, C. L. Stern, W. F. Paxton, W. Wu, S. K. Dey, S. C. Fahrenbach, J. R. Guest, H. Mohseni, L. X. Chen, K. L. Wang, J. F. Stoddart and S. I. Stupp, *Nature*, 2012, **488**, 485.
48. A. S. Tayi, A. Kaeser, M. Matsumoto, T. Aida and S. I. Stupp, *Nature Chem.*, 2015, **7**, 281.
49. W. Yu, X.-Y. Wang, J. Li, Z.-T. Li, Y.-K. Yan, W. Wang and J. Pei, *Chem. Commun.*, 2013, **49**, 54.

CHAPTER 5

Naphthalene-diimide (NDI) Nanofibre, Gel and Mesoscopic Material

ANINDITA DAS,[†] PRIYA RAJDEV[†] AND SUHRIT GHOSH*

Polymer Science Unit, Indian Association for the Cultivation of Science, 2A & 2B Raja S. C. Mullick Road, Kolkata 700032, India
*Email: psusg2@iacs.res.in

5.1 Introduction

Supramolecular assembly of extended aromatic compounds[1–3] leads to large numbers of mesoscopic materials with elegant structures and tunable photophysical properties.[4] For efficient self-organization, often a chromophoric building block is derivatized with different functional groups so that by virtue of multiple non-covalent forces, a desired structure can be formed. As many of these organic chromophores are semiconducting in nature and most of them exhibit excellent light absorption and emission properties, their organized assemblies over an extended length scale is intimately related to the production of flexible optoelectronic devices including field-effect transistors (FET),[5] light-emitting diodes (LED),[6] photovoltaic cells[7,8] and others. Bridging the molecules with materials is a highly challenging area of scientific research, which requires overcoming the difficulties associated with in-depth understanding of the inter-chromophoric interaction

[†]Contributed equally.

Monographs in Supramolecular Chemistry No. 21
Naphthalenediimide and its Congeners: From Molecules to Materials
Edited by G. Dan Pantoş

Published by the Royal Society of Chemistry, www.rsc.org

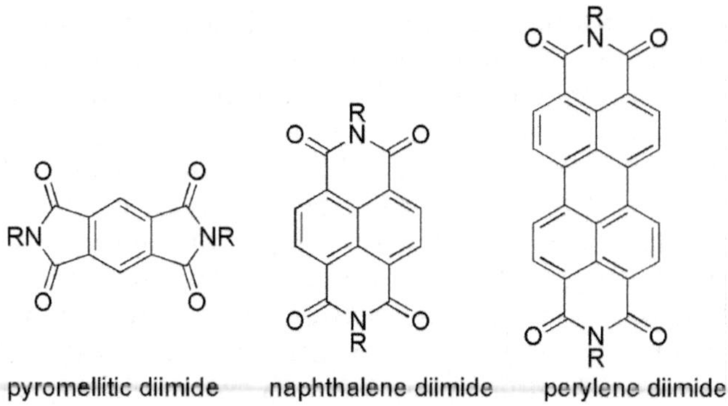

Figure 5.1 Structure of pyromellitic diimide, NDI and PBI.

in the self-organized state, implications in mobility and photophysical properties, inter-stack interactions, morphology, stability and related issues, which is only possible by a coherent effort from synthetic chemists, physicists and material scientists. In a sense, self-assembled π-systems can be compared to the conducting polymers[9] that generated enormous enthusiasm in the late 90s due to their potential in making plastic wires. However, these two are different areas and they have their own pros and cons in terms of applicability in material research. While the structure of a building block is precisely defined for small molecules and possibly structural fine-tuning is relatively easy, polymeric conjugated materials are often difficult to modify structurally. Macroscopic behavior, crystallinity, and inter-chain interactions are also difficult to predict for covalently linked conjugated polymers compared to the supramolecular polymers made out of the conjugated units. On the other hand, for small molecule materials, the control over solid state organization and processing remains a hard task, which has a paramount effect on device applications. While several p-type conjugated polymers have been studied over the last 30 years or so, relatively fewer examples of their n-type counterparts are known.[9] This could be a reason for the rapid progress in the research related to supramolecular assembly of n-type π-systems including pyromellitic diimides, 1,4,5,8-naphthalenediimide (NDI)[10] and perylene diimides (PBI) (Figure 5.1).[11]

Various NDI derivatives have been extensively studied in the recent past due to their excellent n-type semiconductivity.[12] Synthesis of symmetric derivatives of NDI can be achieved in a single step by condensing commercially available 1,4,5,8-naphthalenetetracarboxylic dianhydride with a primary amine at relatively high temperature (~90–120 °C). Synthetic ease allows access to a large number of NDI derivatives with different imide substitutions and examination of structural effects on supramolecular assembly. Unsymmetrical imide substitution is also possible by a two-step synthetic route,[13] which involves selective hydrolysis of one of the imides of a symmetric diimide followed by condensation in the presence of a different amine. On the other hand, treatment of 1,4,5,8-naphthalenetetracarboxylic dianhydride with urea produces the parent diimide,[14] which has also been studied in the context of supramolecular assembly. NDI is a planar, robust and redox-active compound. It exhibits excellent π-stacking ability in the solid state and also in non-polar solvents. It shows a strong absorption with fine structured bands below 400 nm and a very weak emission with relatively small Stoke's shift. However, in a non-polar solvent, it often exhibits an excimer-type emission at a longer wavelength. NDI shows a one-electron reduction with a potential of -1.10 V with respect to Fc/Fc^+ in a solvent such as dichloromethane, indicating its electron-deficient nature. It also forms very efficient donor–acceptor charge-transfer (CT)-complexes[15] in the ground state with electron-rich chromophores such as pyrene, dialkoxy-naphthalene and others. Although research on this particular chromophore was initiated way back in the 1970s,[16] it was only in the 90s that people identified NDI as an attractive building block for exploring supramolecular chemistry[10] and in the last 20 years, this has been one of the most extensively studied π-systems for non-covalent assembly, which include foldamers and organogels as well as hydrogels, vesicles, nanotubes, nanorods, catenanes, rotaxanes, synthetic ion channels, various supramolecular photosystems and so forth. Among this wide range of materials, in this chapter we focus on NDI-derivatized gelators, amphiphiles, other H-bonded and related mesoscopic materials and supramolecular systems based on core substituted NDIs.

5.2 Various Mesoscopic Materials from Self-assembled NDI

NDI-based building blocks show structure-dependent self-assembly and morphology by weak interactions such as H-bonding, hydrophobic effects and others. Sanders and coworkers demonstrated[17] that amino-acid-functionalized symmetrical NDIs can produce elegantly organized supramolecular chiral nanotubes in non-polar solvents and in the solid state by H-bonding among the carboxylic acid groups (Figure 5.2). The handedness of the nanotubes can be predicted by the stereochemistry of the amino acids. A structurally similar carboxylic-acid-appended NDI derivative without any chiral bias contrastingly exhibits rod-like assembly[18] by π-stacking and H-bonding reflecting the remarkable influence of structural variation on self-assembly.

Recent reports[19,20] by Ghosh and coworkers demonstrate very different modes of self-organization among mono-carboxylic-acid-functionalized unsymmetrical NDI derivatives (Figure 5.3). This series of chromophores shows facile aggregation of NDI chromophores by an unusual *syn–syn* catemer-type H-bonding through the acid groups.

Depending on the length of the methylene spacer segment, the morphology of self-assembly could vary from a 2D sheet to a 1D fibre. They also exhibit tunable emission color owing to *in situ* generation of two types of aggregates (J-aggregate and excimer) in a non-polar solvent as well as in the solid state. In some cases, even pure white light emission was observed (Figure 5.3), which is remarkable considering the emission originates from a

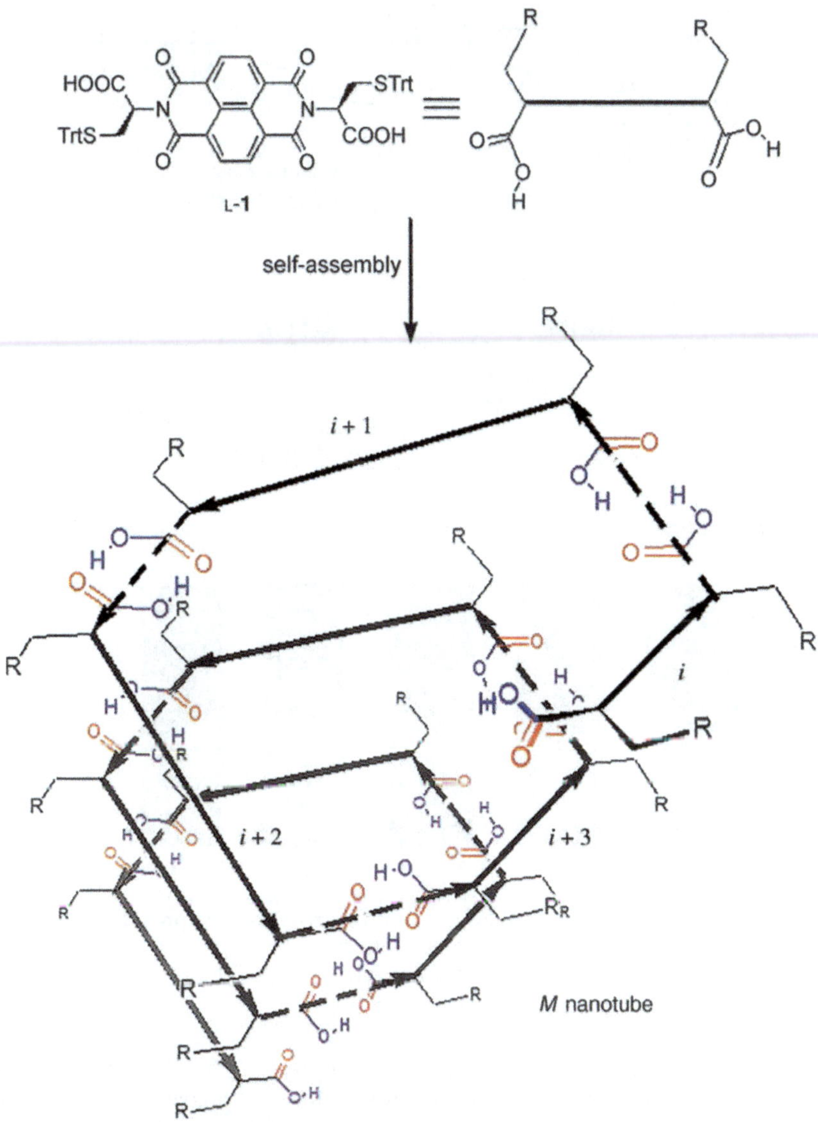

Figure 5.2 Nanotubular assembly of a NDI derivative by H-bonding. Reproduced by permission from ref. 17. Copyright © 2007 WILEY-VCH Verlag GmbH & Co. KGaA, Weinheim.

single chromophore. It is noteworthy that in the recent past, very similar systems have been shown to be effective candidates for organo-catalysis.[21]

In a more recent report,[22] the same group has shown a unique assembly behavior of a NDI building block (Figure 5.4) containing hydrazide (H1) and hydroxy (H2) groups by orthogonal H-bonding. It is organized to a reverse-vesicular structure by orthogonal H-bonding, which at 43 °C (lower critical solution temperature (LCST)) denatures due to the

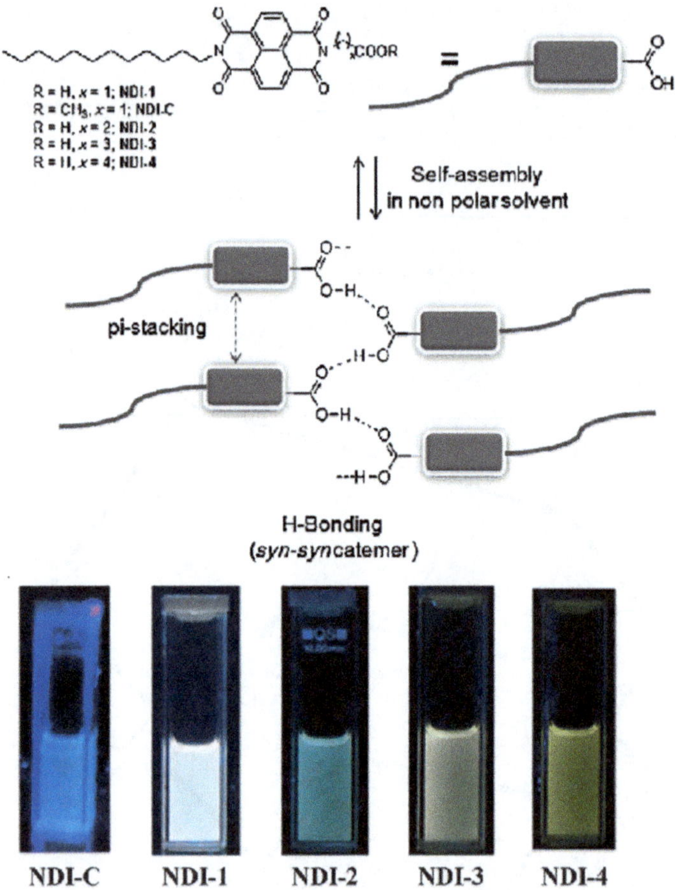

Figure 5.3 Mode of self-assembly of mono-carboxylic-acid-appended NDI and resulting emission properties (bottom).
Reproduced by permission from ref. 20. Copyright © 2014 WILEY-VCH Verlag GmbH & Co. KGaA, Weinheim.

dissociation of H2–H2 H-bonding. Intriguingly, at 70 °C (upper critical solution temperature (UCST)), it exhibits a second phase transition and adopts a reverse-micellar structure. A pyridine (H3)-appended pyrene (PY) forms a supramolecular dyad (NDI–PY) by H2–H3 H-bonding that gradually produces an alternate NDI–PY stack due to the charge-transfer interaction between NDI and PY. While the majority of examples of thermoresponsive phase transition are noticed in aqueous polymer solutions due to disruption of polymer–solvent interactions, the present examples set a new working principle for producing such thermoresponsive materials based on solute–solute interactions and thus offer a wider opportunity for property manipulation by structural fine tuning.

While these examples show self-assembly by H-bonding among abiotic units connected to NDI, there exist many examples of amino acid, short peptides or nucleic acid functionalized NDI building blocks (Figure 5.5)[23–28]

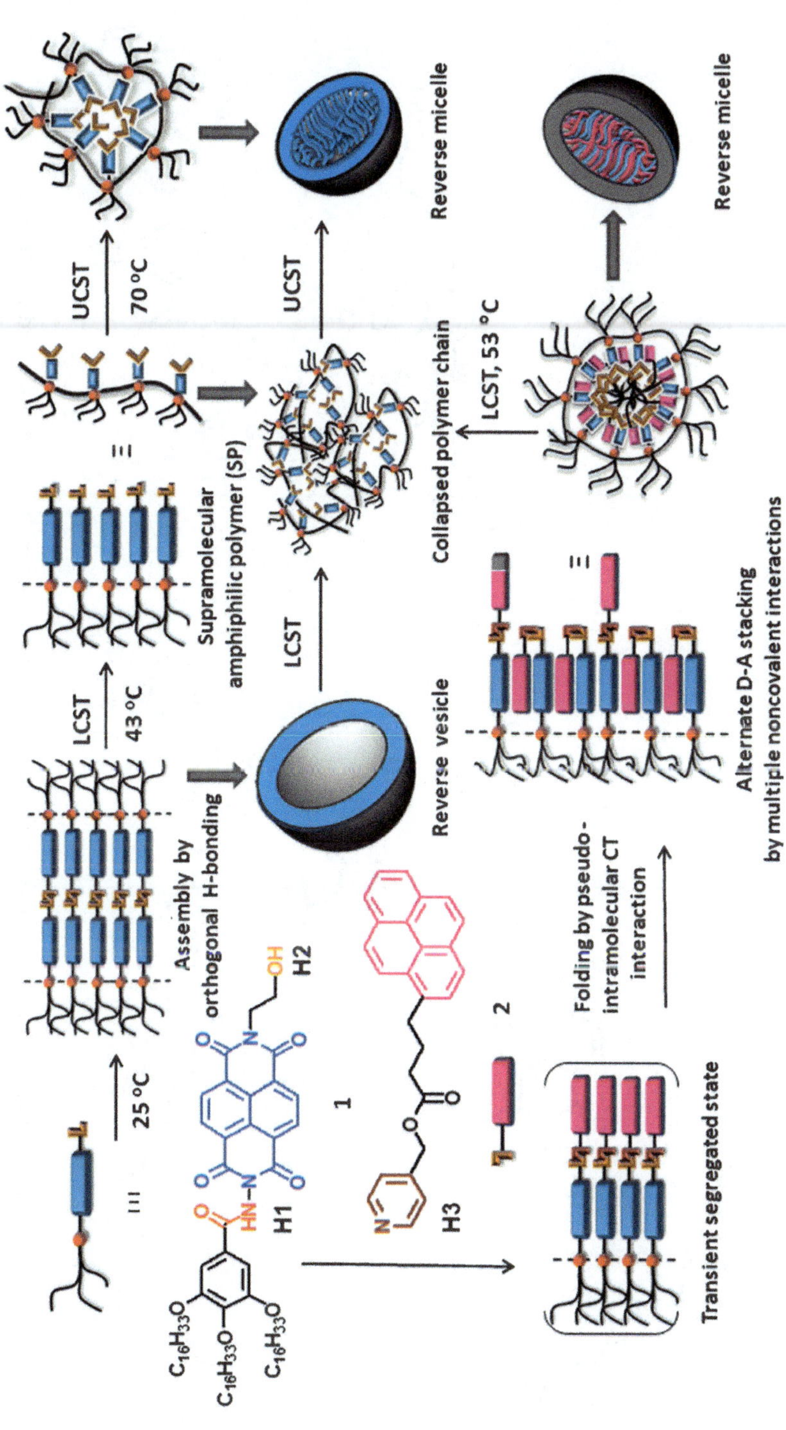

Figure 5.4 Thermoresponsive self-assembly of a NDI building block by orthogonal H-bonding and its co-assembly with a pyrene donor. Reproduced by permission from ref. 22. Copyright © 2014 WILEY-VCH Verlag GmbH & Co. KGaA, Weinheim.

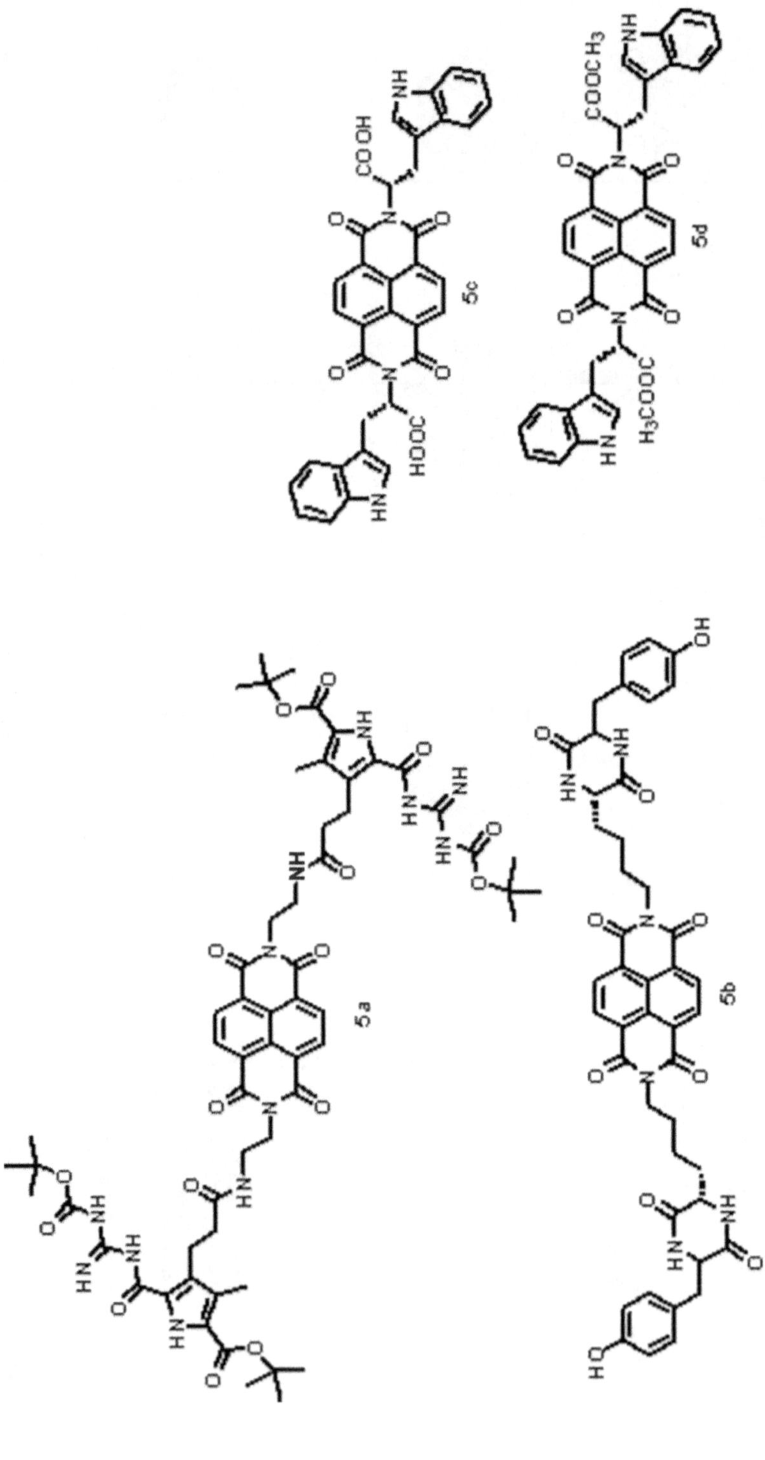

Figure 5.5 Representative examples of amino-acid-functionalized NDI units reported by Govindaraju and coworkers.

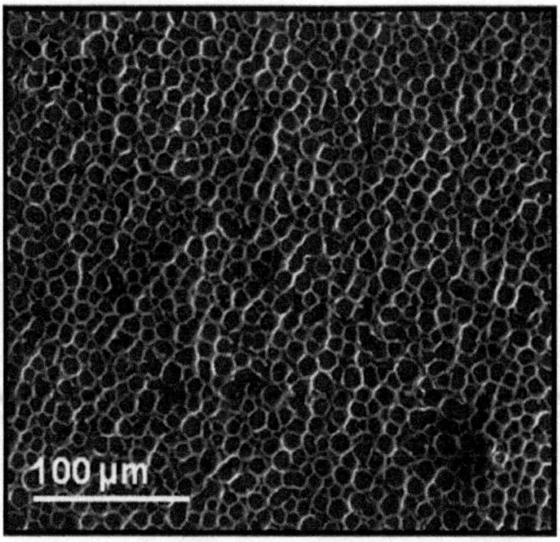

Figure 5.6 TEM images of compound **5a** showing a honeycomb-like structure formation.
Reproduced by permission from ref. 23. Copyright © 2012 WILEY-VCH Verlag GmbH & Co. KGaA, Weinheim.

for supramolecular assembly. Govindaraju, Schmuck and coworkers have shown[23] solvent-dependent morphology and self-assembly of **5a**. It produces self-assembled structures with film-like morphologies in acetonitrile, nanometer-to-micrometer length particulate-like aggregates in THF and DMSO, and discrete micro-wells in $CHCl_3$ and CCl_4, indicating the key role played by the solvent structure and polarity on self-assembly. Further in dichloromethane, it forms honeycomb-like microarrays (Figure 5.6). Interaction of the solvent molecules with the NDI carbonyl is believed to be responsible for this structure formation. Such supramolecular structures have helped to fabricate unprecedented self-cleaning surfaces with a very low slide angle (3°) and contact angle hysteresis (1°).

Compound **5b** (Figure 5.5) exhibits self-assembly with preferential supramolecular helical bias in organic solvents by π-stacking and H-bonding.[25] Intriguingly the helical assembly greatly depends on the solvent polarity and even chiroptical switching (M- to P-type helix) of these assemblies was established by solvent-dependent CD studies. Tryptophan-appended NDI building block **5c** (Figure 5.5) shows[26] solvent-dependent contrasting morphologies. It forms nanospheres (Figure 5.7(a),(b)), particles (Figure 5.7(c),(d)) and fractals (Figure 5.7(e),(f)) from acetonitrile, 90 : 10 acetonitrile–water and 10 : 90 acetonitrile–water (2 equiv. NaOH), respectively. In sharp contrast, the carboxylic-acid-protected derivative of the same building block (**5d** in Figure 5.5) forms nanospheres and bundles of nanobelts from acetonitrile and 40 : 60 acetonitrile–water, respectively.

This example illustrates the remarkable influence of solvent polarity and structural variation in morphology of supramolecular assemblies

Figure 5.7 SEM and TEM (b) images of compound **5c** (Figure 5.5) showing solvent-dependent morphology.
Reproduced from ref. 26 with permission from the Royal Society of Chemistry.

generated from the same or similar building units. Similar morphological variation was also observed with the self-assembly of adenine- and thymine-appended NDIs.[27] Amino-acid-functionalized NDI, even when devoid of any H-bonding functional groups, also appears to self-assemble in polar solvents[28] and aromatic solvents and exhibits solvent-dependent emission as depicted in Figure 5.8. Solvent-dependent excimer or exciplex formation could be a possible reason for such a drastic variation in photoluminescence in the aggregated state.

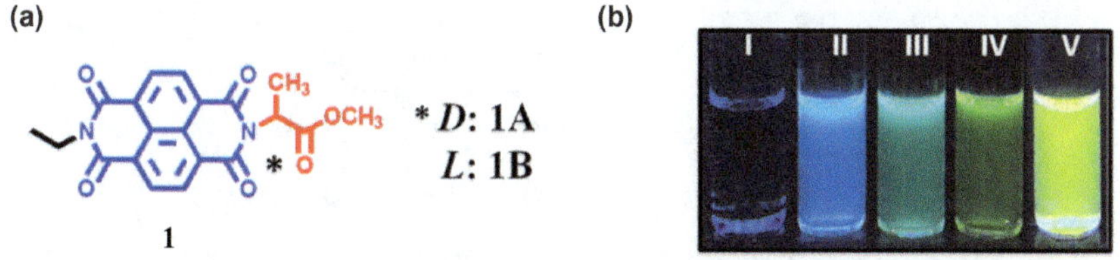

Figure 5.8 Structure (a) and emission color (b) of a simple NDI derivative in I: DMSO, II: benzene, III: toluene, IV: xylene and V: aqueous solution (H_2O–DMSO, 80 : 20) under UV light (365 nm).
Reproduced from ref. 28 with permission from the Royal Society of Chemistry.

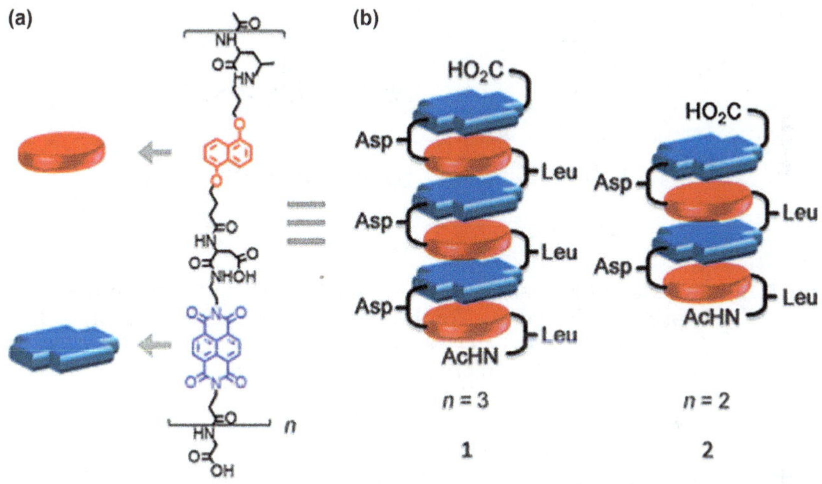

Figure 5.9 A representative example of D–A foldamers reported by Iverson and coworkers.
Reproduced by permission from ref. 31. Copyright © 2013 WILEY-VCH Verlag GmbH & Co. KGaA, Weinheim.

Iverson and coworkers and others have extensively used an NDI chromophore as an acceptor unit for donor–acceptor charge-transfer (CT)-interaction-promoted foldamer design[29–35] where the dialkoxy-naphthalene (DAN) and NDI units are alternately connected by a carboxylic acid attached to amino acid linkers.[29] The presence of a new CT band in the UV/vis absorption spectra indicates folded structure formation (Figure 5.9) for higher oligomers, which could also be confirmed by various NMR studies.

These foldamers exhibit irreversible structural denaturation[31] at an elevated temperature, a property similar to folded proteins. The folding pattern is also very sensitive to the sequence of connectivity of the DAN and NDI units in a given oligomer. For example, oligomers with DAD or DDA sequences exhibit pleated or intercalative folding, respectively.[34] As well as

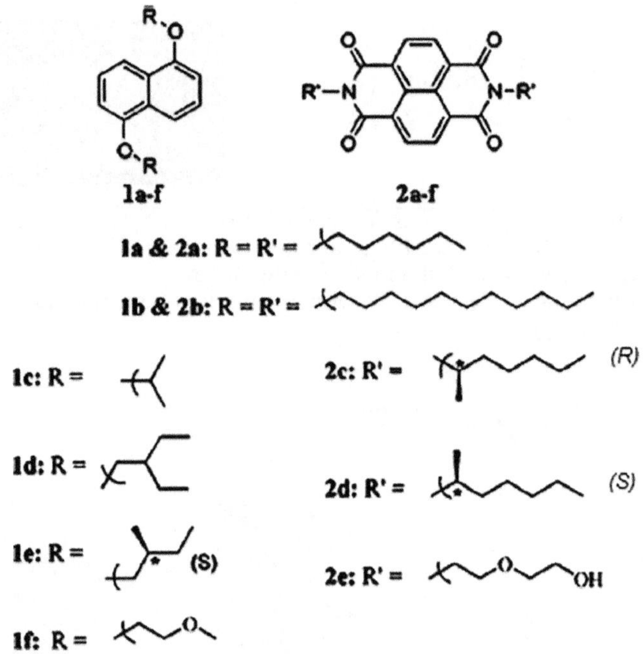

Figure 5.10 Structure of NDI and DAN compounds showing LC properties. Reprinted with permission from J. J. Reczek, K. R. Villazor, V. Lynch, T. M. Swager and B. L. Iverson, *J. Am. Chem. Soc.*, 2006, **128**, 7995. Copyright © 2006, American Chemical Society.

intra-chain CT interactions, inter-chain CT interactions among oligomers containing only NDI or DAN units help to create folded hetero-duplexes with extremely high stability constants by virtue of multi-valent CT interactions among the NDI and DAN units.[35] Inter-chain CT-interaction-promoted assembly has also been realized in polymeric systems. Iverson's group demonstrated[36] inter-chain CT interactions among two water soluble polymers with pendant NDI and DAN units, resulting in a purple viscous liquid owing to formation of a network.

The same group examined the liquid crystalline (LC) properties of a few DAN–NDI pairs (Figure 5.10) with different peripheral alkyl chains.[37,38] In most of these examples, columnar mesophases were obtained over a large temperature window owing to alternate NDI–DAN co-stacking and the phase transition temperatures could be correlated to the structure and thermal behavior of the particular NDI and DAN units. Depending on the nature of the peripheral long chain, few of the DAN–NDI pairs intriguingly showed a temperature-induced transformation from mesophases to a crystalline phase due to a reorganization of the alternate D–A stack to the segregated assembly of individual components. Similar studies on the LC properties of NDI- and DAN-based building blocks further highlight the scope of achieving tunable mesomorphic properties by structural variation.[39]

5.3 Assembly of Amphiphilic NDI

Supramolecular amphiphiles (SA) or superamphiphiles are those generated by dynamic non-covalent interactions, such as H-bonding, electrostatic interactions, charge-transfer (CT) interactions, metal–ligand co-ordination, π–π stacking, and host–guest recognition between the amphiphilic molecules, each containing a complementary functionality to the other.[40–42] NDI-containing amphiphiles self-assemble into well-defined structures in aqueous media with distinctly different photo-physical properties in the aggregated state. Substitutions at the imide position lead to the generation of various NDI amphiphiles with little effect on the inherent electronic and optical properties of the chromophore but significant effects on the solubility and molecular organization. The solubility and hence, the self-assembly of NDI amphiphiles, depend on the nature of the imide substitution, which has been exploited for structural manipulation and consequently the aggregation.

George and coworkers[43,44] recently examined the self-assembly of amphiphilic and bola-amphiphilic NDI (NDI-amphi and NDI-bola, respectively, in Figure 5.11). NDI-amphi forms a nano-tape-like structure in water. More interestingly, it exhibits reversible vapo- and thermo-chromism, which can be attributed to reorganization or molecular stacking (H- and J-type aggregates) (Figure 5.11) and consequently, different photo-physical properties in a given solvent vapor atmosphere. NDI-bola (Figure 5.11), on the other hand, self-assembles to produce nanoparticles or a micelle-like morphology by π–π stacking between the NDI chromophores, which exhibit enhanced green fluorescence owing to the emission from the pre-associated excimer. In another related report,[45] which is the first of its kind, they studied a pair of similar NDI-containing amphiphilic molecules functionalized with the dipicolylethylenediamine group (Figure 5.12), which self-assembles into a 1D nanofibre with reversible tunable chirality that can be induced by the specific molecular recognition of chiral guests.

Ghosh and coworkers have recently reported[46,47] the self-assembly of symmetric NDI bola-amphiphiles (NDI-2, Figure 5.13) with hydrogen-bonding hydrazide functional groups flanking both sides of the NDI-chromophore, which could further enhance the possibility of tuning the aggregation phenomenon in such π-conjugated amphiphiles. The hydrazide groups were strategically placed between the hydrophobic NDI core and benzyl ring to avoid competition from the bulk solvent molecules for H-bonding. NDI-2 shows J-type aggregation in water and spontaneously self-assembles into a vesicle-like morphology through the synergistic effect of π-stacking and hydrogen bonding (Figure 5.13). Furthermore, NDI-2 shows very effective intercalation ability for electron-rich pyrene donors and derivatives owing to CT interactions and the D–A mixture at higher concentration leads to formation of a hydrogel (*vide infra*).

Comparative self-assembly studies with NDI-1 (Figure 5.13) prove the necessity of the rigid placement of the hydrogen-bonding groups in the hydrophobic pocket of NDI-2. NDI-1, which lacks any flanking hydrazide

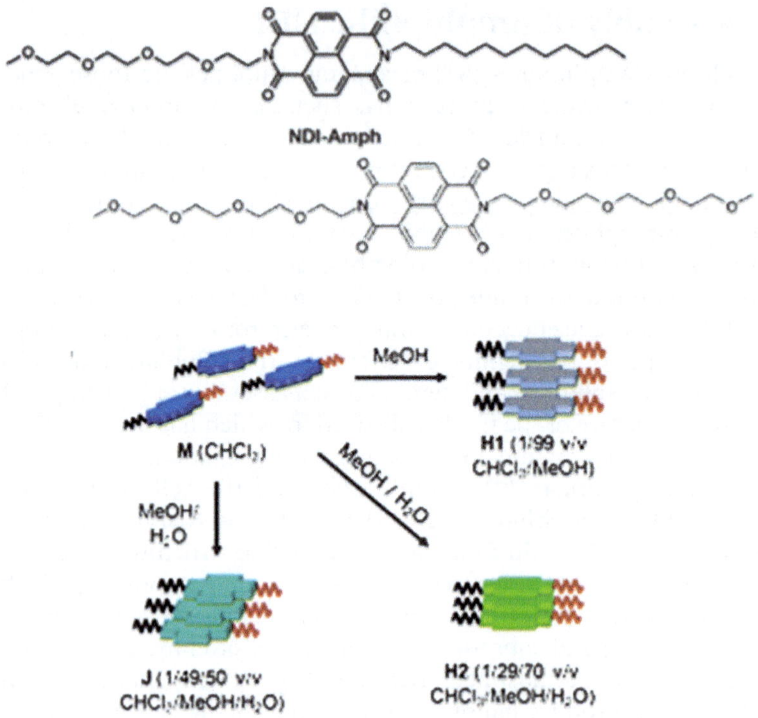

Figure 5.11 Structure of NDI amphiphilic units reported by George and coworkers and solvent-vapor-induced aggregation tuning of NDI-Amp (bottom). Reproduced by permission from ref. 43. Copyright © 2011 WILEY-VCH Verlag GmbH & Co. KGaA, Weinheim.

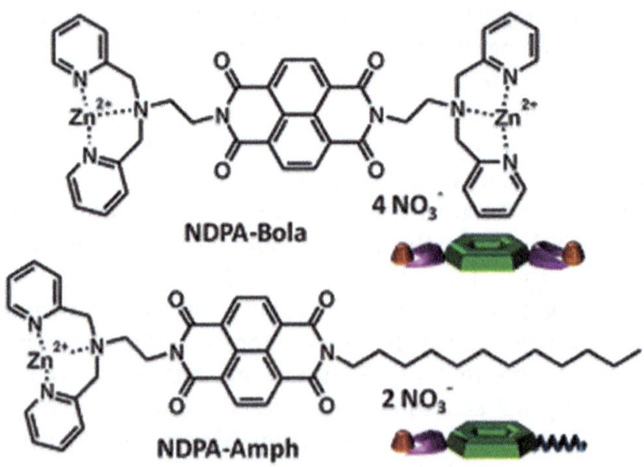

Figure 5.12 NDI amphiphile and bola-amphiphile functionalized with dipicolylethylenediamine.
Reproduced from ref. 45 with permission from the Royal Society of Chemistry.

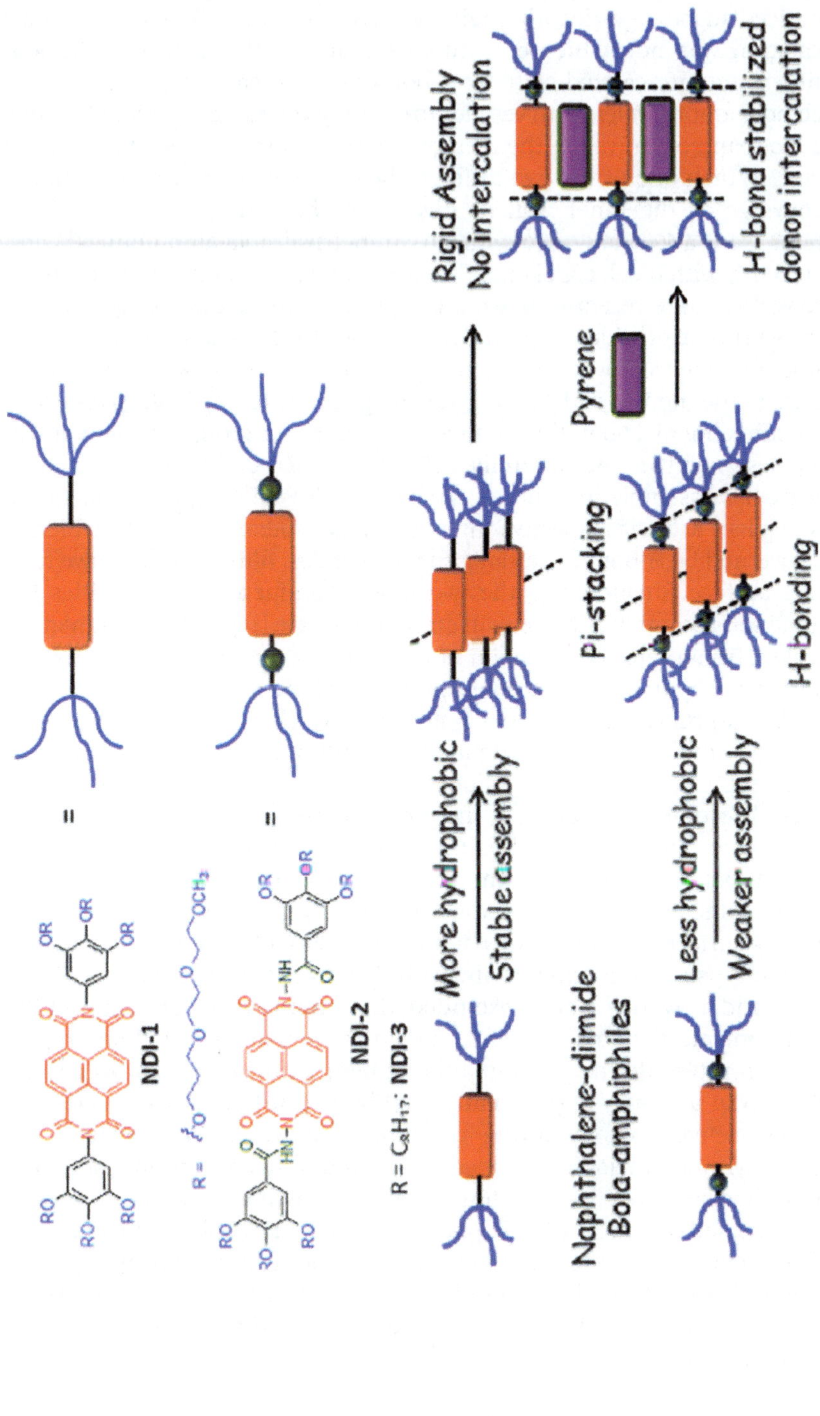

Figure 5.13 Schematic presentation showing contrasting self-assembly and pyrene co-assembly behavior of NDI-1 and NDI-2. Reprinted with permission from P. Rajdev, M. R. Molla and S. Ghosh, *Langmuir*, 2014, **30**, 1969. Copyright © 2014, American Chemical Society.

groups, also self-assembles to generate vesicular aggregates by face-to-face π-stacking between the NDI groups. But due to its more hydrophobic nature in the absence of polar hydrazide groups, the propensity to self-assemble was noticed to be significantly high for NDI-1. On the other hand, donor intercalation was not achieved in this case due to the lack of additional stability of the intercalated state by H-bonding as in case of NDI-2.

Such bola-amphiphiles form vesicles presenting the same functional groups at the inner and outer wall of the membrane. What if the two head-groups are different? Which one of the two different head groups or hydrophilic blocks will converge at the inner wall or diverge at the external surface? These questions were asked long ago with unsymmetrical bola-amphiphiles[48] and more recently with block copolymers[49] but it remains a non-trivial task. Ghosh and coworkers have recently shown a unique supramolecular design strategy (Figure 5.14) that provides a generalized answer to this classical problem by enabling full control over directionality in relative orientation as well as curvature of the monolayer by taking advantage of the strong hydrophobically-assisted directional H-bonding among the hydrazide groups present in rationally designed twin π-amphiphiles NDI-A and NDI-B.[50] Both exhibit identical vesicular assembly in aqueous media but with fully opposite functional group displays at their respective inner and outer membrane walls, which is solely governed by H-bonding among the hydrazides. Although they have fully opposite surface functionality, the two most prominent commonalities between the self-assembly of these twin amphiphiles are (i) parallel alignment in both cases, (ii) a preferred direction of curvature of the monolayer so that in either case, the hydrazides (and consequently the functional group attached to the same arm) remain at the inner wall of the vesicle. The motivation stems from the possibility of having extended H-bonding even at the cost of electrostatic repulsion among the anions in parallel orientation. Further, the preferred direction of curvature of the monolayer ensures that in both cases, the hydrazides remaining at the inner wall can benefit from a stronger interaction due to the relatively shorter inter-molecular distance than at the outer wall, which has a longer radius of curvature. As a direct outcome of contrasting surface functionality, NDI-A inhibits 80% of the enzymatic activity of α-chymotrypsin, but it remains almost fully active in the presence of NDI-B.

Ghosh and coworkers even extended this supramolecular amphiphilic design to polymeric domains[51] and studied the self-assembly of an amphiphilic macromolecule that contains the structure-directing hydrazide-functionalized NDI unit only at the chain end (Figure 5.15). It shows vesicle- and reverse-vesicle-type blue emitting assemblies in water and benzene, respectively, with appreciable dye encapsulation ability and thermal stability. Hydrophobically-assisted H-bonding among the hydrazide units is responsible for the self-assembly of this polymer. It is remarkable that the hydrophobic segment contributes only 13% by weight of the entire polymer, which contains only one hydrazide group in the entire polymer chain and still it shows such facile assembly indicating the power of directional supramolecular interactions.

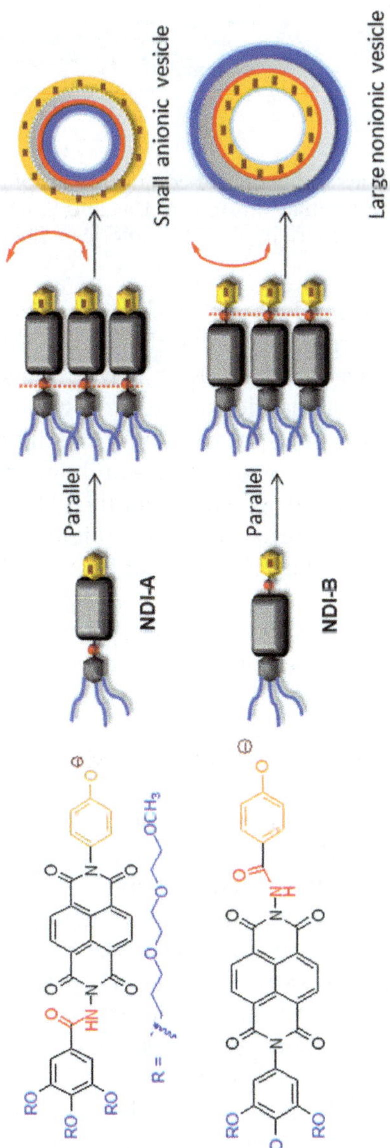

Figure 5.14 Vesicular assembly and distinct functional group display by unsymmetrical bola-amphiphiles. Reproduced by permission from ref. 50. Copyright © 2011 WILEY-VCH Verlag GmbH & Co. KGaA, Weinheim.

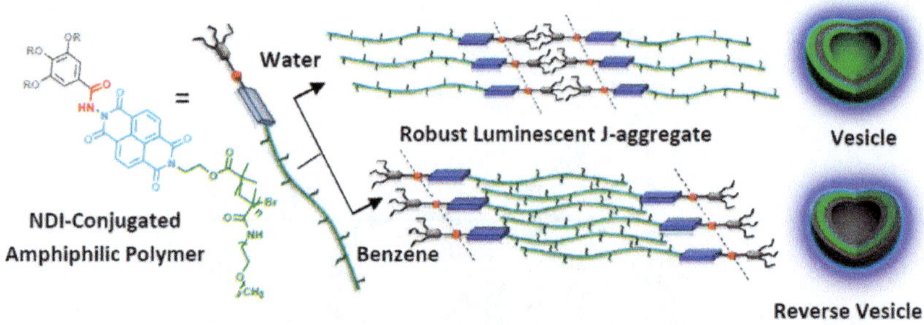

Figure 5.15 Adaptable self-assembly of NDI-conjugated amphiphilic polymer. Reprinted with permission from A. Das and S. Ghosh, *Macromolecules*, 2013, **46**, 3939. Copyright © 2013, American Chemical Society.

Parquette and coworkers studied peptide-functionalized NDI amphiphiles with either di-peptide-bearing NDI side chains or NDI bola-amphiphile appended to short peptides on both sides[52–54] with the aim of utilizing the β-sheet self-assembly of short peptides to achieve well-defined π-electronic nanostructures. Figure 5.16 shows the design for the di-peptide-bearing NDI and the model for the formation of nanofibres and nanoribbons,[52] driven by hydrophobic π–π association in water. In contrast, the peptide-appended bola-amphiphile organizes to form nano-rings (Figure 5.16), which eventually form a nanotube-like structure.[53] A similar bola-amphiphile comprising a central porphyrin chromophore flanked on both sides by NDI-lysine moieties (Figure 5.16) was constructed to explore the versatility of this strategy for the self-assembly of a donor (D)–acceptor (A) dyad into nanotube arrays with a D–A heterojunction.[54] The symmetrical design was intentional to exclude anti-parallel packing that may give rise to alternate D–A CT complexation and achieving charge separation within the same nanostructure *via* separate π-conjugated pathways.

Zhang and coworkers have reported many systems of the self-assembly of amphiphilic NDI derivatives by CT interactions with various donor-containing amphiphiles.[55–57] In a recent example, they showed J-aggregation of a NDI bola-amphiphile (Figure 5.17)[55] in an aqueous medium and formation of a nano-sheet-like structure by NDI–NDI stacking. In the presence of an electron-rich pyrene derivative (Figure 5.17), which interacts with the viologen head groups of the NDI bola-amphiphile, the sheet-like morphology transforms to ultra-long nanofibres (Figure 5.17).

The same NDI compound in the presence of an electron-rich DAN-containing bola-amphiphile (Figure 5.18) produces a H-shaped SA stabilized by CT interactions and the resulting DA complex results in a 2D sheet-like structure.[56] In the presence of another donor pyrene, a programmable transformation of the 2D nano-sheet to a 1D nano-fibre-like structure takes place because of the highly directional CT interactions between pyrene and the terminal viologen units of the SA. In another interesting study, they showed that a NDI bola-amphiphile forms different types of SA with DAN

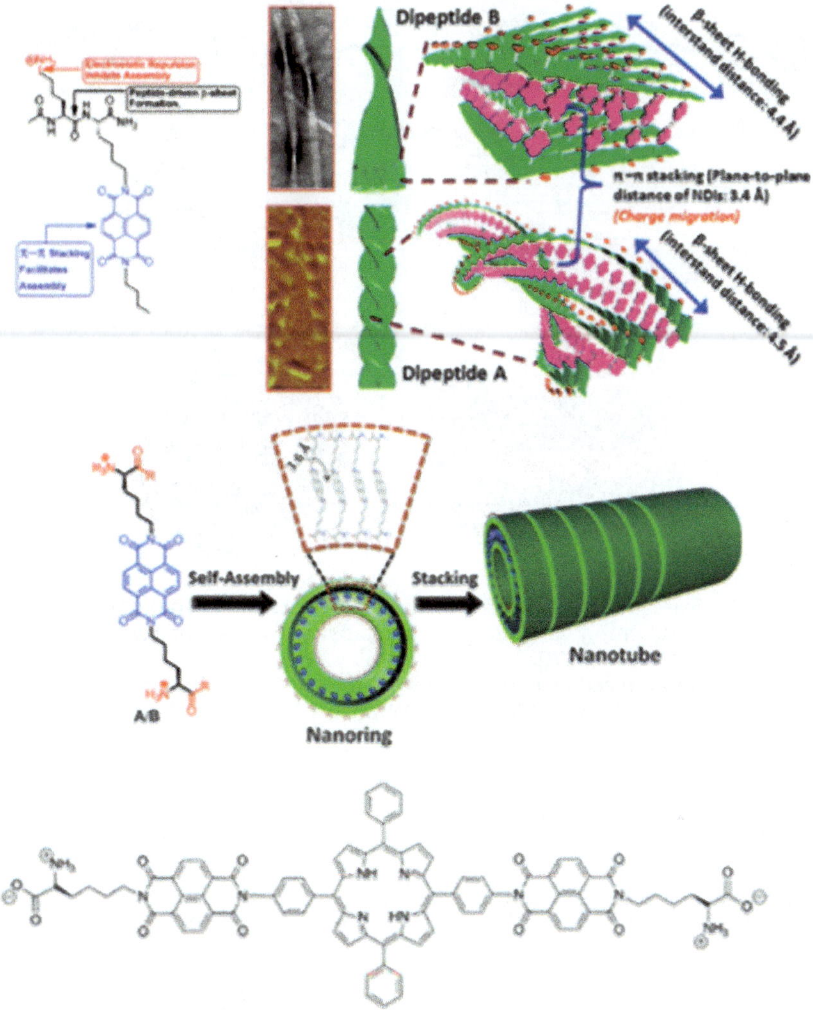

Figure 5.16 Self-assembly of a peptide-functionalized NDI-amphiphile (top), bola-amphiphile (middle) and ADA-type amphiphile reported by Parquette and coworkers.
Reprinted with permission from H. Shao, T. Nguyen, N. C. Romano, D. A. Modarelli and J. R. Parquette, *J. Am. Chem. Soc.*, 2009, **131**, 16374. Copyright © 2009, American Chemical Society. Reproduced by permission from ref. 53. Copyright © 2010 WILEY-VCH Verlag GmbH & Co. KGaA, Weinheim. Reprinted with permission from S. Tu, Se H. Kim, J. Joseph, D. A. Modarelli and J. R. Parquette, *J. Am. Chem. Soc.*, 2011, **133**, 19125. Copyright © 2011, American Chemical Society.

bola-amphiphiles depending upon the position of the two alkoxy substituents of the DAN chromophore and can produce different morphologies such as 2D nano-sheets and 1D nano-rods.[57] This was attributed to the different

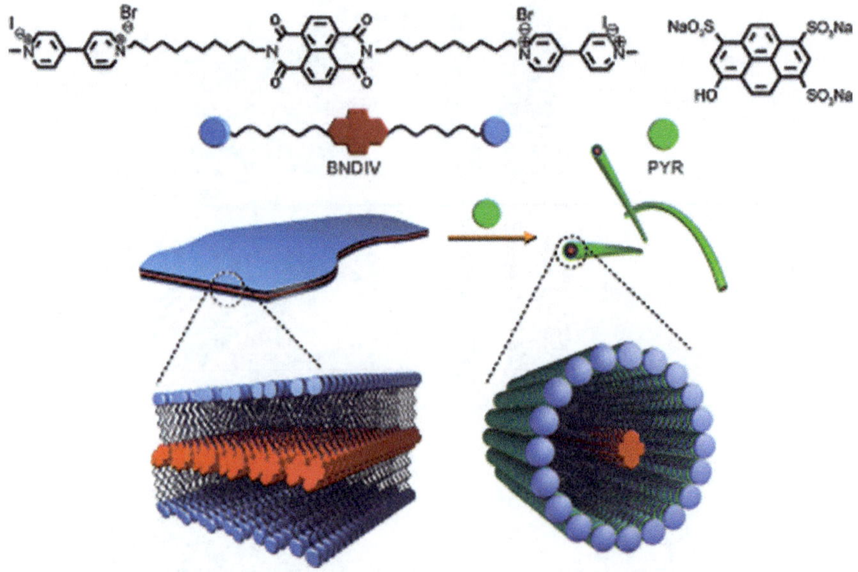

Figure 5.17 Self-assembly of a NDI-functionalized bola-amphiphile and its co-assembly and morphology transition in the presence of a pyrene derivative (PYR).
Reproduced by permission from ref. 55. Copyright © 2012 WILEY-VCH Verlag GmbH & Co. KGaA, Weinheim.

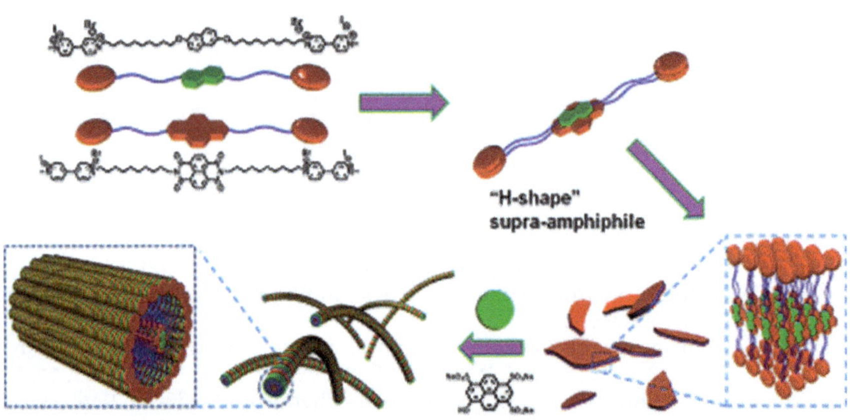

Figure 5.18 Co-assembly of NDI and DAN bola-amphiphiles.
Reprinted with permission from K. Liu, Y. Yao, Y. Liu, C. Wang, Z. Li and X. Zhang, *Langmuir*, 2012, **28**, 10697. Copyright © 2012, American Chemical Society.

geometry of the NDI–DAN SAs depending on the isomer of the DAN because the CT interaction that is responsible for formation of the SA is a highly directional one.

5.4 Self-assembled Core-substituted NDI

The previous discussion has illustrated a range of materials that can be generated by the self-assembly of NDI chromophores by incorporating different functional groups in the two-imide position.

In the recent past, excitement in this research area has also shifted to NDI derivatives that are substituted at the naphthalene ring, commonly known as core-substituted NDIs (cNDIs).[58,59] Much of the excitement in studying cNDIs stems from their tunable absorption and photoluminescence properties originating from a decreasing HOMO–LUMO gap with a concomitant increase in HOMO and LUMO energies as a function of the donating ability of the ring substituent. Figure 5.19 shows typical absorption and emission spectra of representative cNDIs, which cover almost the entire UV-Vis range. Further, their electrochemical properties also can be very different depending on the core substitution and, in selective examples, a cNDI may even work as a p-type semiconductor.

Research activity in this area has had a tremendous boost in the recent past due to the development of amicable synthetic routes,[60–63] which enable versatile ring substitution of the NDI core. A major focus in this area lies in examining the utilities of various push–pull type chromophores and main chain conjugated polymers derived by C–C linkages on the cNDI. Core substitution of NDI with an electron donating or withdrawing group such as aryl, thiophene, cyano *etc.* by Suzuki or Stille coupling leads to various oligomers and polymers[64–74]

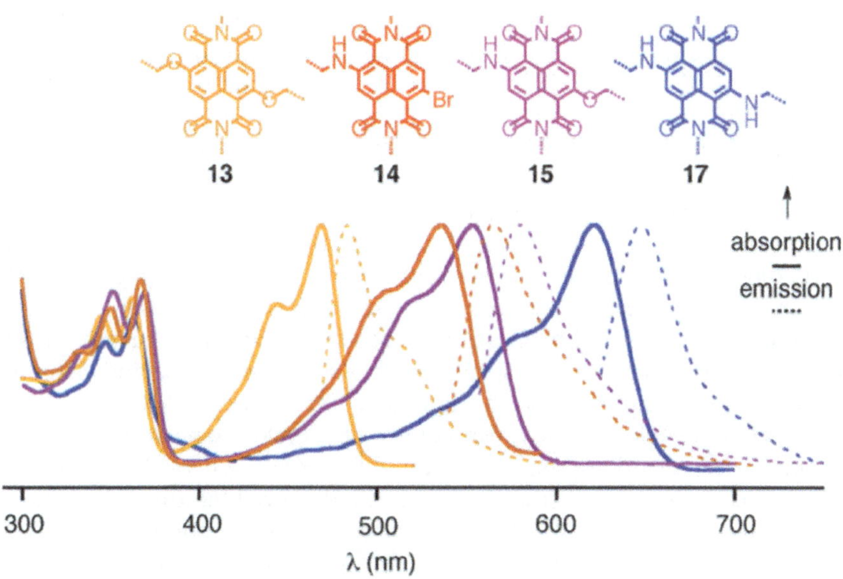

Figure 5.19 Typical absorption and emission spectra of a few major classes of cNDIs.
Reproduced from ref. 58 with permission from the Royal Society of Chemistry.

with potential applications in organic electronics and tunable photophysical properties. The utility of cNDI derivatives has been examined in the context of a diverse range of applications[58,59] including energy transfer, photo-induced electron transfer, sensing, ion-transport and biomedical applications. Nevertheless, examples of their supramolecular assemblies are still limited, presumably due to the lack of suitable synthetic methods, even until the recent past, for versatile ring substitution, which is essential for the preparation of structurally diverse supramolecular building blocks.

Würthner and coworkers showed[75] the supramolecular co-assembly of cNDI and OPV by H-bonding interactions leading to a rosette-like structure probed by STM (Figure 5.20). A melamine unit was covalently attached with two oligo(*p*-phenylenevinylene) chromophores, which showed complexation with a tetra-substitutes NDI by complementary H-bonding between the melamine and imide groups.

Bhosale and coworkers have examined[76] the self-assembly of a cNDI, which was derivatized by attaching long alkyl chains from the core (Figure 5.21). It exhibits well-defined worm-like aggregates in a MeOH–$CHCl_3$ solvent mixture owing to face-to-face π-stacking within the cNDI and a hydrophobic effect imparted by the alkyl chains. The worm-like morphology of the aggregates could be established unambiguously by atomic force microscopic images. Subsequently, they examined[77] the self-assembly of a different structurally related amphiphilic analogue of **1** (Figure 5.21) in an aqueous medium, which showed vesicular assembly.

Ghosh and coworkers have recently demonstrated[78] a straightforward design for the H-bonding driven self-assembly of an amine-substituted cNDI (cNDI-1, Figure 5.22) and shown J-aggregation in aliphatic hydrocarbons. They further revealed the remarkable impact of J-aggregation on excited state dynamics. Ultrafast electron transfer was noticed on a sub-ps timescale and, more importantly, the lifetime of the charge-separated state could be prolonged by an order of magnitude in J-aggregates compared to monomers owing to very efficient delocalization of the excited state charge-separated species. Unique solvent effects were realized on the macroscopic structure and morphology of the present J-aggregated cNDI. While J-aggregation with similar photophysical characteristics was noticed in all tested aliphatic hydrocarbons, the morphology strongly depended on the "structure" (NOT polarity) of the solvent. In linear hydrocarbons (*n*-hexane, *n*-octane, *n*-decane or *n*-dodecane), formation of an entangled fibrillar network led to macroscopic gelation while cyclic hydrocarbons (methylcyclohexane or cyclohexane), despite having a similar polarity, exhibited nanoscale spherical particles.

In a subsequent study,[79] they have identified a novel building block based on amine-substituted NDI (cNDI-2, Figure 5.23) for H-bonding-driven cooperative supramolecular assembly. High cooperativity originated from the synergistic operation of multiple weak interactions. NDI monomers initially undergo linear supramolecular polymerization (by H-bonding) as the nucleation step. Once the nucleus grows sufficiently long (degree of polymerization ~25), it propagates the self-assembly in an orthogonal direction

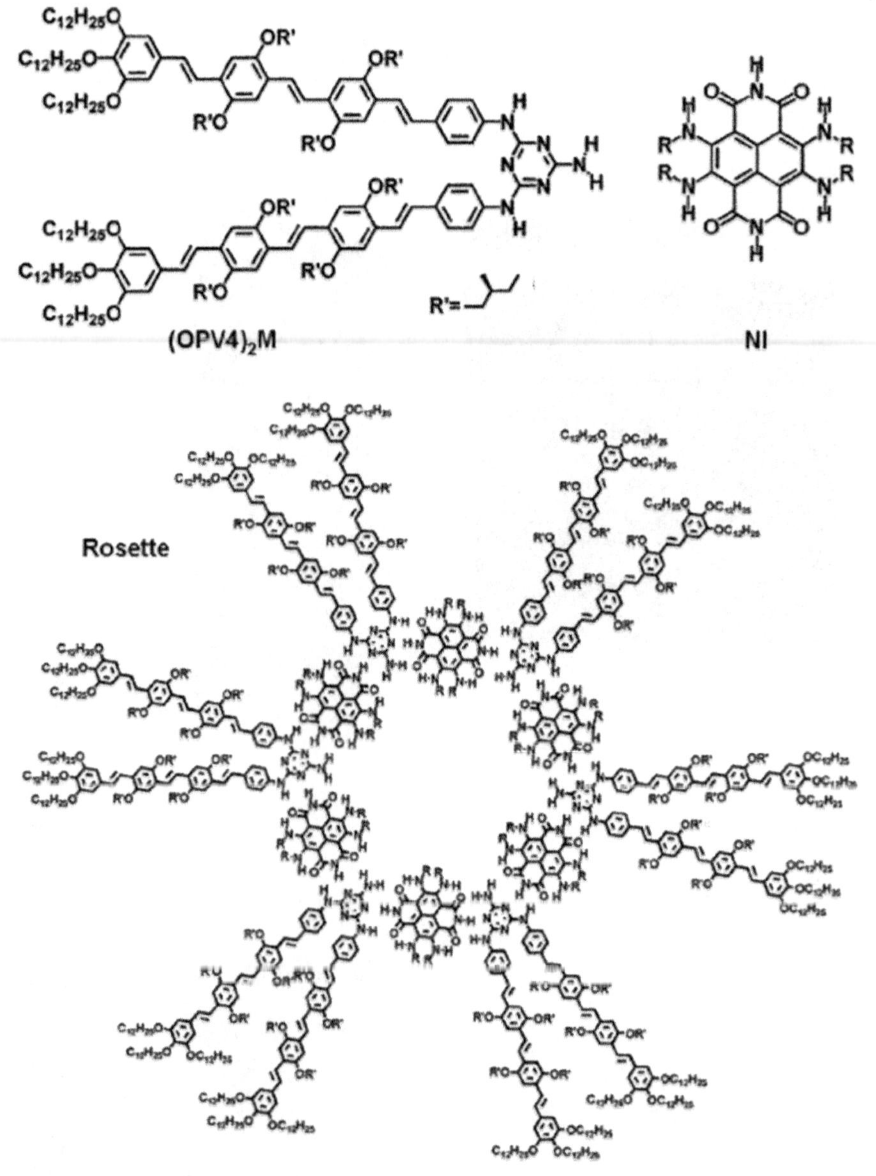

Figure 5.20 Co-assembly of OPV and cNDI by H-bonding.
Reproduced from ref. 75 with permission from the Royal Society of Chemistry.

by J-aggregation among the NDI chromophores leading to spontaneous formation of nanotubular structures. Further, the lifetime of the charge-separated state could be remarkably prolonged, extending beyond 100 μs in J-aggregates, owing to very efficient delocalization of the excited state over several chromophoric molecules assembled at the wall of the tube.

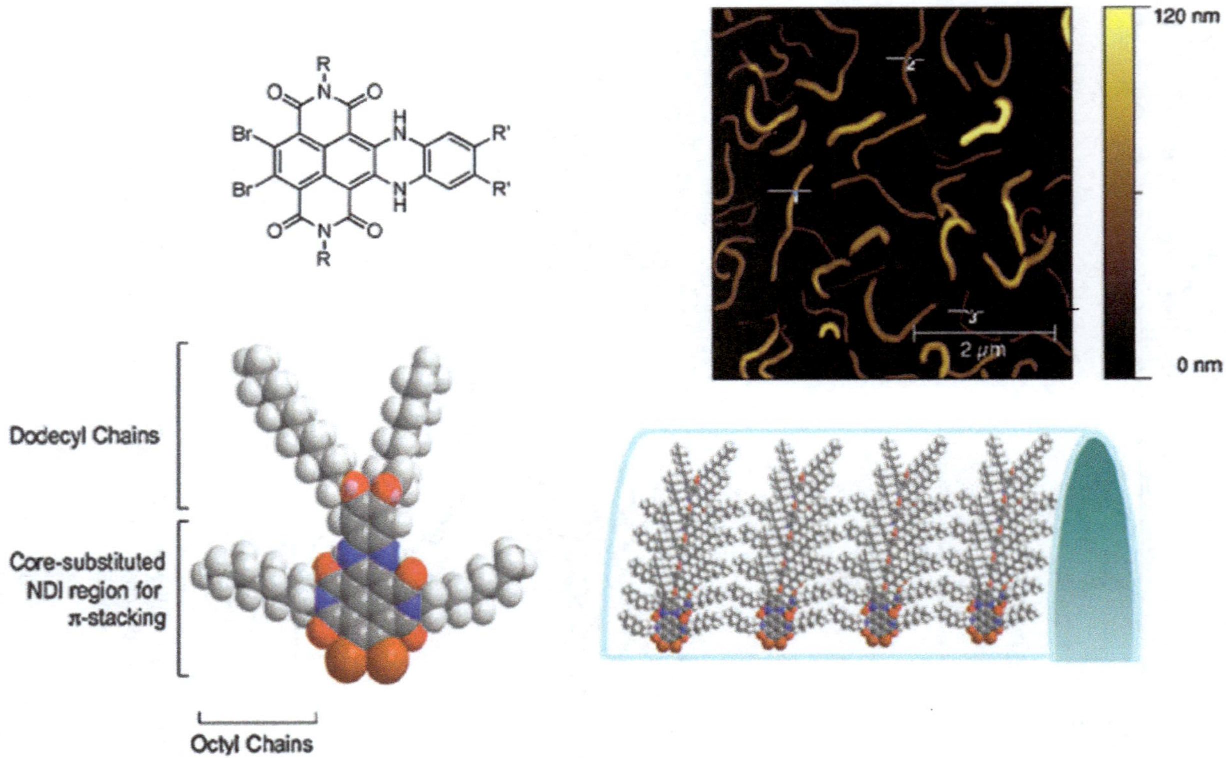

Figure 5.21 Solviphobically-driven supramolecular assembly of a cNDI (**1**). Reproduced from ref. 76 with permission from the Royal Society of Chemistry.

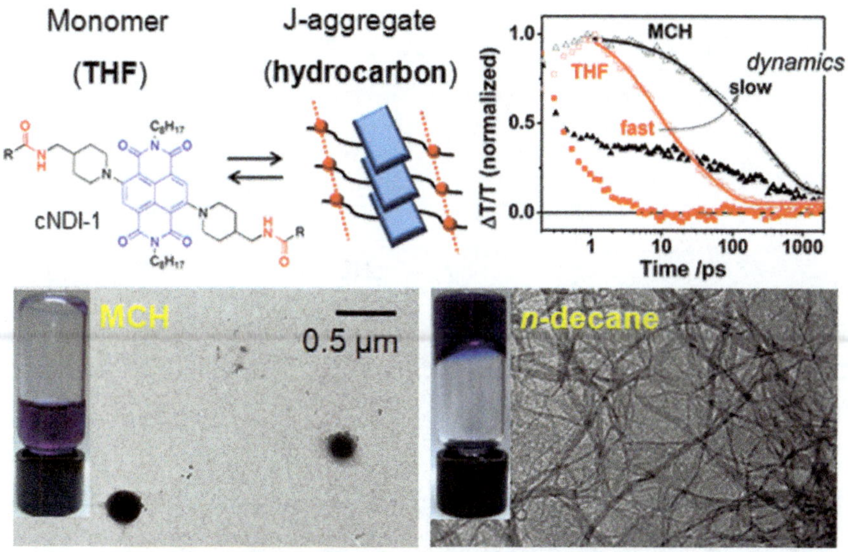

Figure 5.22 H-bonding-driven J-aggregation of cNDI-1 in hydrocarbons and impact on excited state dynamics. The bottom images show remarkable solvent geometry effects on gelation.
Reproduced from ref. 78 with permission from the Royal Society of Chemistry.

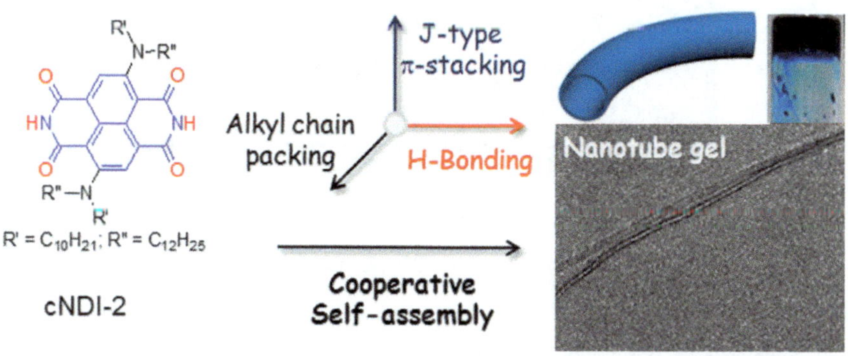

Figure 5.23 Cooperative supramolecular polymerization of cNDI-2 leading to tubular assembly and gelation.
Reproduced from ref. 79 with permission from the Royal Society of Chemistry.

More recently, supramolecular assembly of a sulfur-substituted NDI derivative (cNDI-3, Figure 5.24) has been reported,[80] leading to the formation of a very strong ($T_g > 90$ °C) organogel in aliphatic hydrocarbons due to the synergistic effect of H-bonding, aromatic and van der Waals interactions. Photophysical studies show explicit features of J-aggregation in aliphatic hydrocarbon solvents. J-aggregation restrains the non-radiative fluorescence

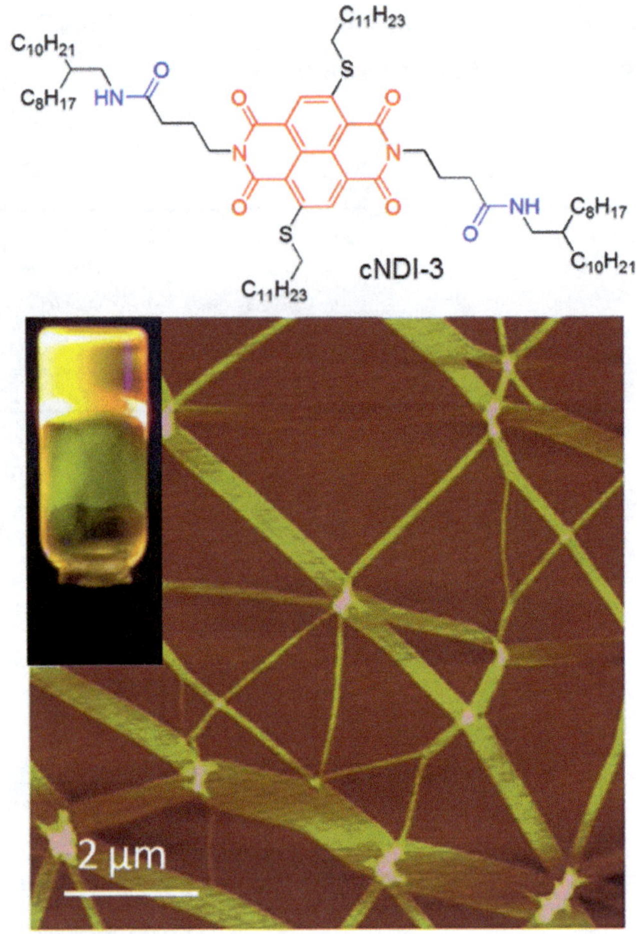

Figure 5.24 Emissive gel by cooperative J-aggregation of cNDI-3 and its tape-like morphology.
Reproduced from ref. 80 with permission from the Royal Society of Chemistry.

rate constant significantly and thus results in remarkable fluorescence enhancement (Φ_{PL} enhances from 1% to 30%), which is unprecedented in the entire NDI family, which rarely exhibits fluorescence due to fast inter-system crossing (ISC). The present dye is unique in a sense that its LUMO energy was already reduced due to sulfur substitution and further lowered by strong J-type excitonic coupling, which impacts on restraining the ISC.

5.4.1 Naphthalene-diimide-based Gelators

Over the years, chromophore-based gelators[1] have emerged as exciting soft materials owing to their potential for applications in various organic electronic devices[4,81–83] on account of their intrinsic electronic properties such as luminescence,[84] conductivity,[85] charge carrier mobility,[86,87] energy transfer *etc.*[88,89] The strategy for utilizing chromophore-based gels over their solution self-assembly stems from the fact that in the gel state, the chromophores are organized into highly ordered 1D nanostructures[90] that undergo 3D growth to form network structures efficient for solvent retention. Such discrete and ordered structure formation provides continuous pathways for enhanced charge transport and energy transfer essentially required for realization of these systems as efficient soft materials in photovoltaics. The electronic properties of π-gelators can be modulated by varying the extent of π stacking and the nature of inter chromophoric interactions.[91] Thus by employing various design strategies,[92,93] a library of tunable soft materials can be generated.

There are several reports of π-gelators based on p-type semiconducting chromophores such as porphyrins,[94,95] phthalocyanines,[96] oligophenylene-vinylenes,[89] oligothiophenes,[97,98] tetrathiofulvalene[99] and so forth. Reports on n-type semiconductor-based gelators are mostly limited to perylenebisimides (PBI),[3,92,100] naphthalene-diimide (NDI)[101–106] and pyromellitic diimide[117,118] derivatives. Among these, NDI gelators have been exploited more owing to their advanced optoelectronic properties[10] and better solubility, which enhances their ability to gelate a wide range of solvents depending upon the molecular design.

Organogel: The constant hunt for NDI-based gelators has prompted synthetic chemists to explore innumerable design strategies that utilize NDI stacking in conjunction with other non-covalent forces such H-bonding, hydrophobic interactions or electrostactic interactions to generate novel NDI-based gelators. The first report on NDI-based gel was documented by Shinkai and coworkers (Figure 5.25).[101]

The molecular design relied upon utilizing a combination of π-stacking between the NDI chromophores and H-bonding among the two amide groups to induce self-assembly. Long peripherial alkyl chains were introduced to solubilize the gelator in organic solvents by virtue of van der Waals interactions. Compound **1** was able to gelate a wide range of solvents of varying polarity with very low critical gelation concentrations (as low as 0.5 mg mL^{-1}). Involvement of hydrogen bonding in the gelation process was evident from attenuated total reflectance IR (ATR-IR) showing the characteristic peak for a H-bonded amide group (νN–H) at 3275 cm^{-1}. A morphological study of the xerogels from TEM showed bundles of elongated fibres forming a network structure essentially required to immobilise the solvent for gelation (Figure 5.25). The gel propensity was found to be dependent on the nature of the peripherial alkyl group. Compound **2** (Figure 5.25) with a 2-ethylhexyl unit at the periphery could not induce gelation in any tested solvent possibly due to the steric demand of the methyl groups from the adjacent alkyl chains thus illustrating the critical role played by the appendage in maintaining a subtle balance between solubility and gelation.

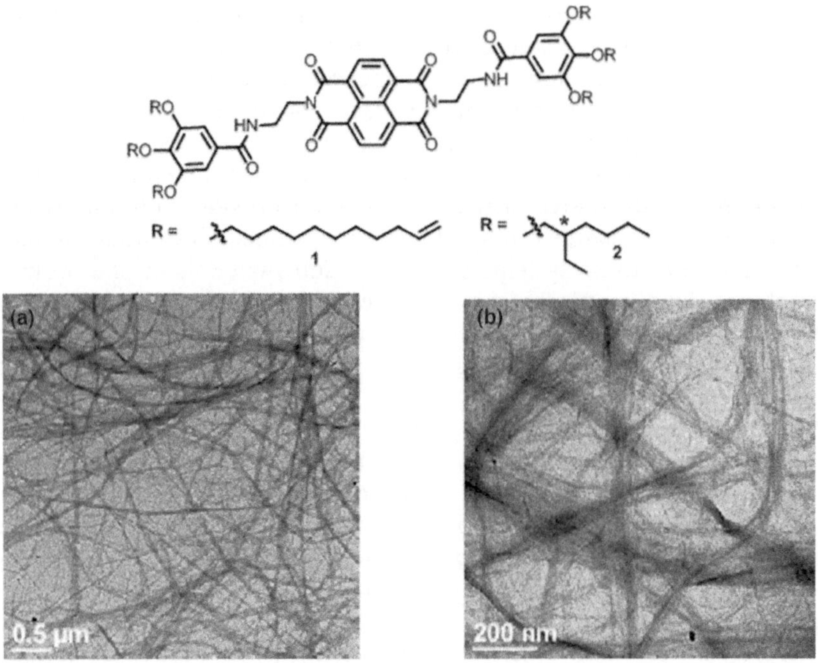

Figure 5.25 NDI-based gelators (top) and their TEM images (bottom) from cyclohexane at different resolutions reported by Shinkai and coworkes. Reprinted with permission from ref. 101. Copyright (2006) Wiley-VCH Verlag GmbH & Co. KGaA, Weinheim.

Ghosh and coworkers[46,102–105,112–115] have contributed significantly to this area by thoroughly investigating the effect of structural variations on gelation properties of diverse NDI-based gelators. In their first report on NDI gels,[102] they studied the gelation behaviour of a series of trialkoxybenzamide-appended NDI derivatives fabricated with different numbers of methylene units between the amide group and the π-chromophore (Figure 5.26). The propensity for solution self-assembly was found to follow the order **NDI-1** >> **NDI-2** > **NDI-3** > **NDI-4**. This was related to the poor H-bonding within the molecules due to enhanced flexibility with increasing numbers of methyl units (n). Surprisingly, the most strongly self-assembled **NDI-0** did not show gelation (Figure 5.26) in any tested solvent; rather, it formed lyotropic mesophases while the other three homologues maintained a similar trend, *i.e.* **NDI-2** >> **NDI-3** > **NDI-4** in the gel state as observed in the solution. The gelation propensity and gel stability were governed by the morphology of the self-assembled structures revealed in AFM-studies (Figure 5.26). **NDI-0** formed a short-range ordered assembly incapable of immobilising the solvent molecules for gelation whereas other NDI-derivatives formed long fibres that eventually produced gels in methylcyclohexane (Figure 5.26). The most robust gelation for **NDI-2** could be correlated to its morphology, which disclosed a network structure formed

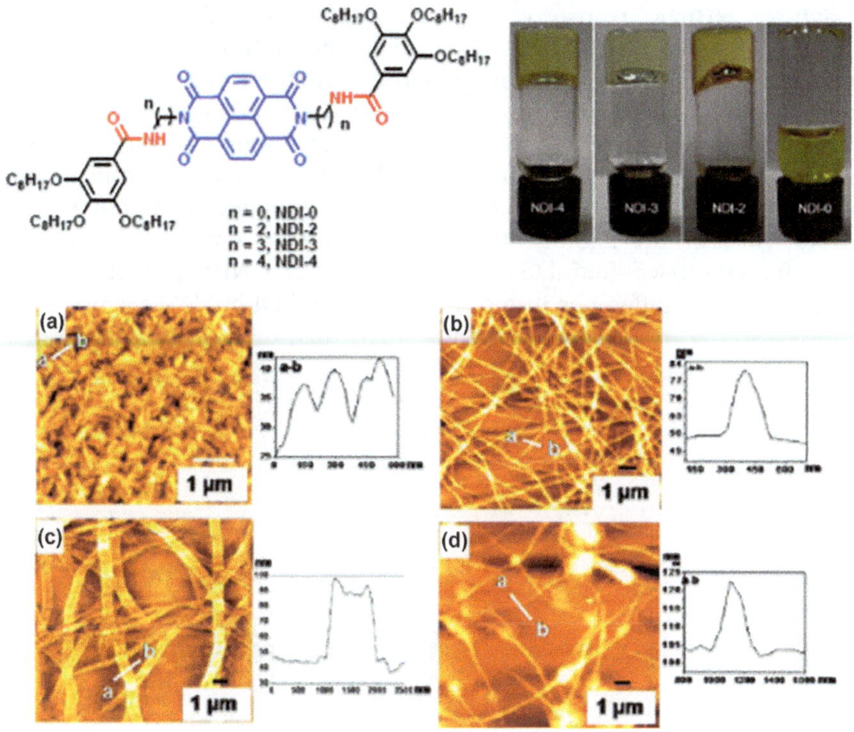

Figure 5.26 Top—structure of NDI-based gelators (left) and their gel–sol pictures in MCH; bottom—AFM cross-sectional analysis and height images of dilute solutions of (a) **NDI-0**, (b) **NDI-2**, (c) **NDI-3** and (d) **NDI-4** in MCH–$CHCl_3$ (95 : 5).
Reprinted with permission from M. R. Molla and S. Ghosh, *Chem. Mater.*, 2011, **23**, 95. Copyright (2011) American Chemical Society.

from continuous nanowires, which is the prerequisite for efficient gelation, unlike **NDI-3** and **NDI-4**, which formed flat nanoribbons and discontinuous nanofibres respectively.

In another study, they illustrated the role of H-bonding functionality in modulating the gelation property of two gelators, **NDI-2** (Figure 5.26) and **NDI-2a** (Figure 5.27) bearing different numbers of amide groups.[103] The higher thermal stability for **NDI-2** compared to **NDI-2a**, as determined from their T_{gel} (gel melting temperature) in two solvents, MCH and TCE respectively, was ascribed to the stronger H-bonding in the former case due to involvement of both the amide groups. The macroscopic properties in the gel state could be correlated with solution self-assembly. This was demonstrated by solvent variable UV-vis studies illustrating a greater propensity for self-assembly with **NDI-2** over **NDI-2a** as quantified from their critical solvent composition for aggregate formation in a binary solvent mixture (MCH–$CHCl_3$).

To validate that the difference in stability of the two gelators was indeed due to there being one less amide group and not due to the unsymmetrical design of **NDI-2a**, Ghosh *et al* compared[104] the gelation of **NDI-2b** (Figure 5.27) with **NDI-2** (Figure 5.26), which differ only in the position of the two-amide functionality. The unsymmetric **NDI-2b** formed thermally more stable self-assembled structures compared to the symmetric one as evident from their variable-temperature absorption spectroscopic studies; this was ascribed to the placement of both the amide groups in close proximity along one arm of **NDI-2b** thus exemplifying that the breaking of symmetry had no adverse effect on the stability of the nanostructures. Contrastingly, gelation ability for **NDI-2b** was found to be poorer compared to **NDI-2**, which could be explained by the difference in morphology as studied by TEM (Figure 5.28).

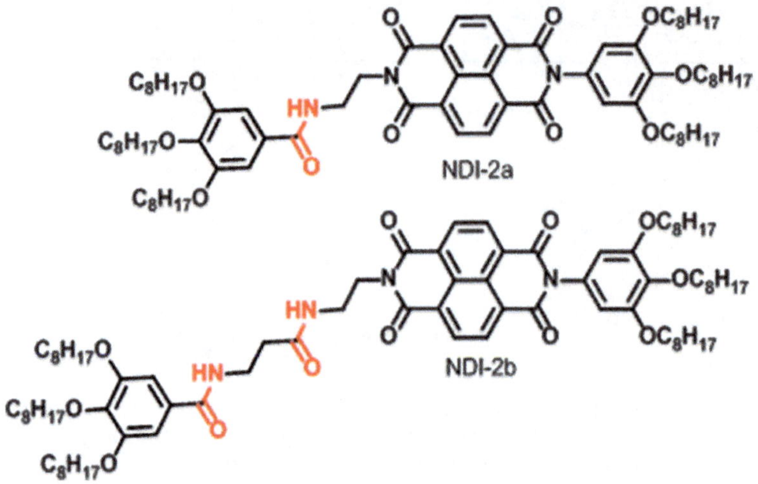

Figure 5.27 Structure of mono-amide- and bis-amide-functionalized NDI gelators reported by Ghosh and coworkers.[103,104]

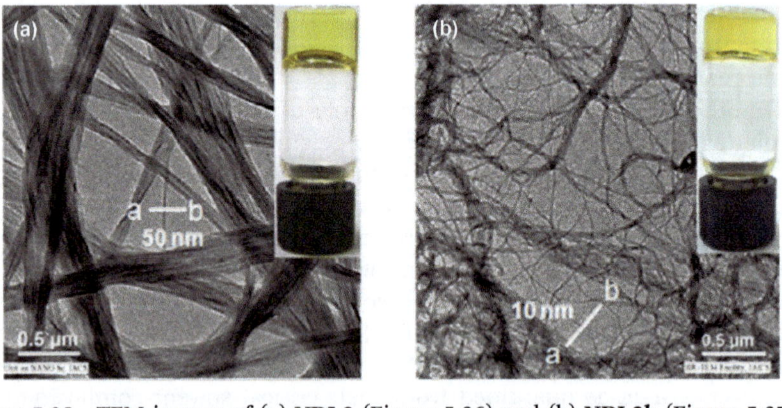

Figure 5.28 TEM images of (a) **NDI-2** (Figure 5.26) and (b) **NDI-2b** (Figure 5.27) in 95 : 5 MCH–$CHCl_3$. Inset shows gel pictures.

It revealed that a regular and continuous network formation was required for efficient gelation of **NDI-2**, while formation of undefined aggregates in **NDI-2b** led to irregular and discrete fibre formation that could not contribute to effective gelation. The superior viscoelastic property of **NDI-2** gel, as observed in rheological studies, further supported their hypothesis.

In another design, the same group showed gelation of a pyridine-functionalized NDI derivative (**NDI-Py**) using an external structural directing agent (**ESDA-1**) containing a carboxylic acid and urea functionality (Figure 5.29).[105] Utilizing orthogonal H-bonding involving complementary pyridine–carboxylic acid pairs and self-complementary urea groups, they could generate 1D assembly (Figure 5.29) for gelation. In the presence of a H-bond-breaking polar protic solvent such MeOH, the gel was transformed into sol (Figure 5.29) illustrating the crucial role of H-bonding in the gelation process. Control **ESDA-2**, lacking the urea functional group, could not induce gelation but formed spherical particles (Figure 5.29) confirming the necessity of H-bonding in orthogonal direction in concurrence with π-stacking for promoting 1D assembly for gelation. A superior conducting property for fibrillar gel compared to spherical assembly induced by the control ESDA ascertained the need for such soft materials in optoelectronic devices.

Wang *et al.* reported a NDI-based chiral gelator[106] made by appending a peripheral sugar moiety as the chirality induction group that could gelate more polar solvents such as butanol. In the aggregated state, the chirality was transferred from the sugar derivative to the chromophore as evident from the corresponding exciton-coupled CD spectrum in the NDI absorption region due to formation of more organised 1D helical columns in the gel state. NDI, being electrochromic in nature, can strongly absorb in the NRI region in its reduced electronic states (*i.e.* radical anion and dianion form).

Therefore, introducing a sugar moiety into the molecular design, a redox switchable chiroptical NDI gel was produced that retained its helical assembly over the entire redox cycle and could induce reversible CD signals between visible and NIR regions depending upon the electronic state. The birefringence character of the gel fibres, as evident from polarized optical microscopy (POM) (Figure 5.30), revealed hierarchical ordered structures essential for chirality transfer in the aggregated state.

In another report, George *et al.* designed another NDI-based chiral gelator (**C3-NDI**) utilizing a C3 symmetric disc-shaped H-bonding benzene-1,3,5-tricarboxamide (BTA) core that was covalently connected to three NDI chromophores with a benzene spacer (Figure 5.31).[107] The incorporation of (*S*)-(−)-3,7-dimethyloctyloxy side chains in the periphery induced chirotopic properties in the self-assembled structure. The chiral gel formed in MCH–toluene showed left-handed helical fibres as visualized from AFM studies (Figure 5.31). The appearance of bisignated CD signals for the NDI chromophore was in accordance with the chirality transfer from the alkyl wedges to the higher ordered assembly in the gel state. The unusual isodesmic mechanism for the self-assembly of such a BTA-functionalised NDI gelator

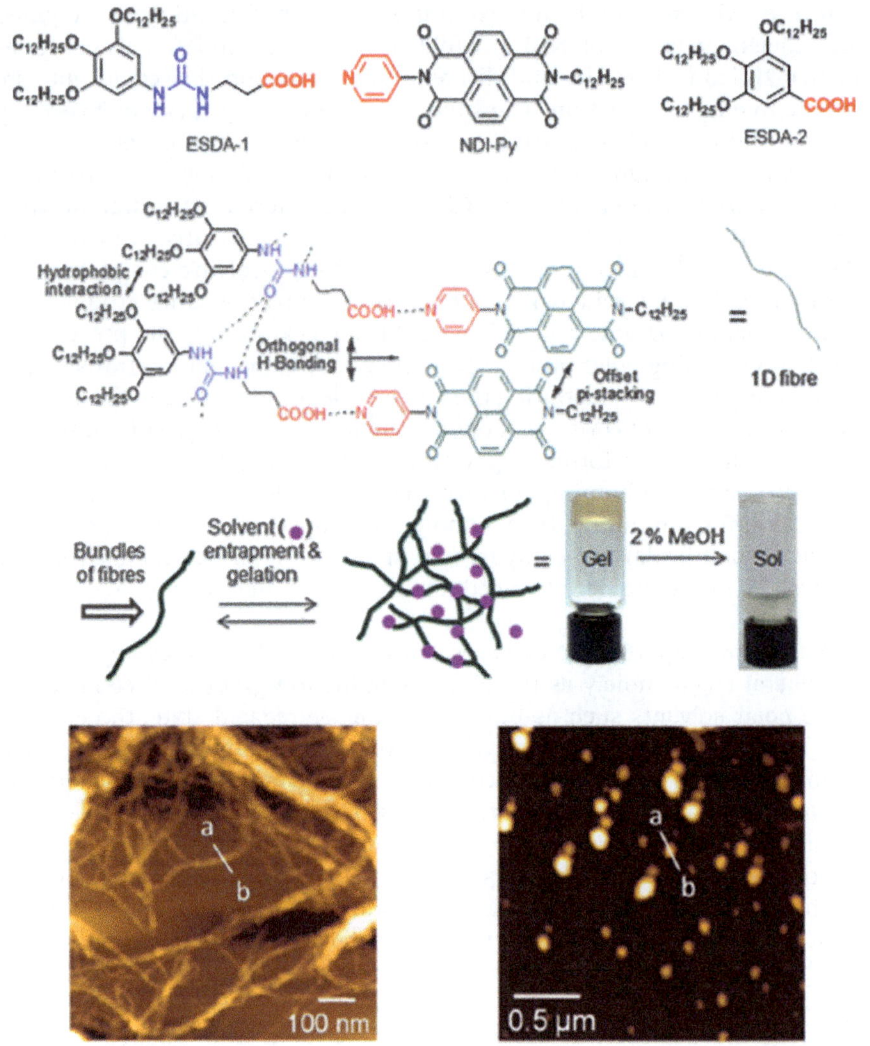

Figure 5.29 Top—structure of the ESDA and NDI building blocks; middle—proposed mode of assembly and MeOH-induced disassembly of gel; bottom—AFM images of equimolar **ESDA-1** + **NDI-Py** (left) and **ESDA-2** + **NDI-Py** (right) in MCH.
Reproduced from ref. 105 with permission of the Royal Society of Chemistry © 2013.

was probed by a solvent variable CD experiment, which was in line with the quantum chemical calculations. The involvement of all three NDI chromophores and the benzene spacer in the assembly process imparted different conformation to the three H-bonded amide bonds in the BTA spacer, which was thought to be the most plausible reason for the uptake of such an unusual isodesmic pathway.

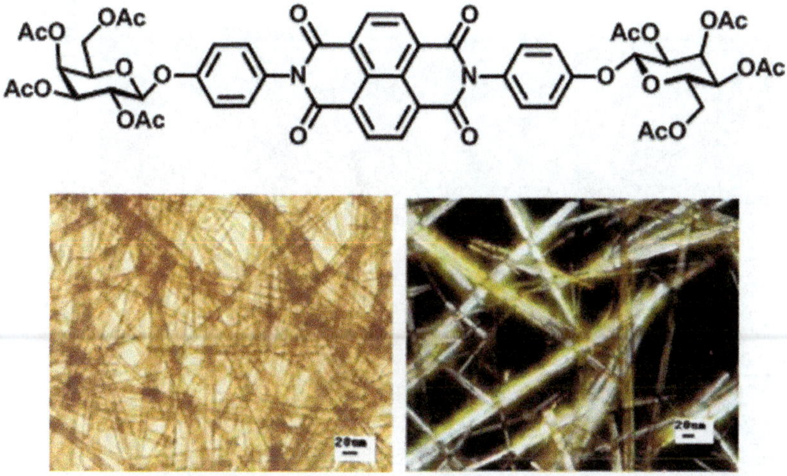

Figure 5.30 Structure of the sugar-functionalized NDI gelator (top) and images (bottom) of the xerogel in *n*-butanol under optical microscopy (left) and POM (right).
Reprinted with permission from J. Zheng, W. Qiao, X. Wan, J. P. Gao, and Z. Y. Wang, *Chem. Mater.*, 2008, **20**, 6163. Copyright (2008) American Chemical Society.

Banerjee and coworkers have reported luminescent organogels derived from terminal peptide grafted NDI-derivatives (Figure 5.32).[108] Turning fluorescence "on" and "off" could be demonstrated by a reversible sol gel phase transition, which showed yellow emission in the gel state and a feeble blue emission from the monomer in the sol state (Figure 5.32). The ß-sheet structure of the gel fibres induced by a cooperative effect of H-bonding within the peptide groups and π-stacking between the NDI chromophores was probed by both FTIR and XRD of the xero-gel. The semiconducting behaviours of the two gels formed from **NV** and **NF** (Figure 5.32) were compared by cyclic voltametry and I–V measurements. The electrical conductivity of **NV** was found to be better than **NF**, which was attributed to the difference in morphology of the two gelators.

Hydrogel: NDI-based hydrogels are also known in the literature.[50,109] Such soft materials are of biological importance for their impending applications in tissue engineering,[119] fluorescent reporters[120] and drug delivery.[121] Parquette and coworkers made NDI-based hydrogels by introducing polar dipeptide groups in the imide position (Figure 5.33).[109] Incorporation of a fluorenylmethoxycarbonyl (Fmoc) group at the N-terminal position of an NDI-dilysine peptide derivative induced ß-sheet formation by π–π interactions among both the NDI and Fmoc chromophores that promoted gelation in water with 1D nanobelt morphology (Figure 5.33).

Circular dichroism (CD) showed strongly negative excitonic Cotton effects at the NDI absorption region indicating M-type helicity of the nanobelts due to chirality transfer from the peptide to the chromophore in the gel state.

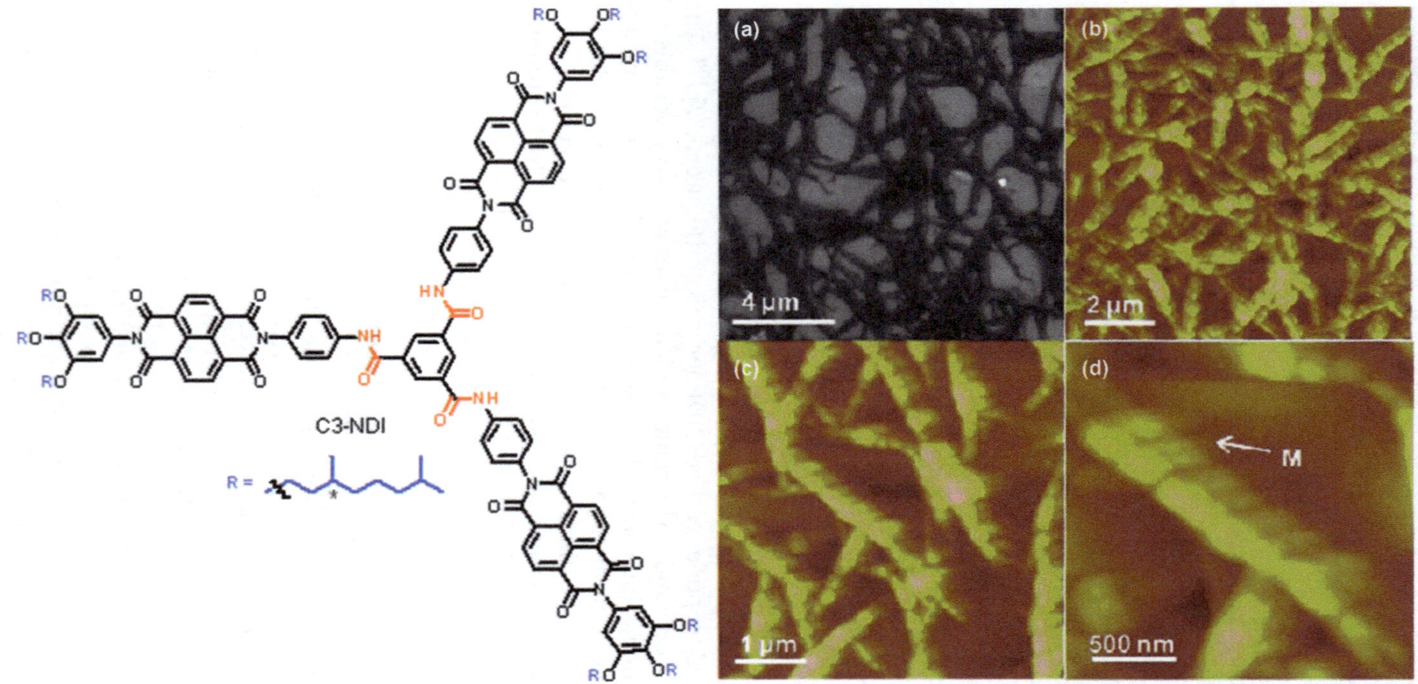

Figure 5.31 Images of the self-assembled structures of **C3-NDI** in toluene–TCE (90 : 10 v/v) solvent mixtures: (a) FE-SEM and (b)–(d) AFM images showing the presence of helically coiled bundles of fibres.
Reproduced from ref. 107 with permission of the Royal Society of Chemistry © 2013.

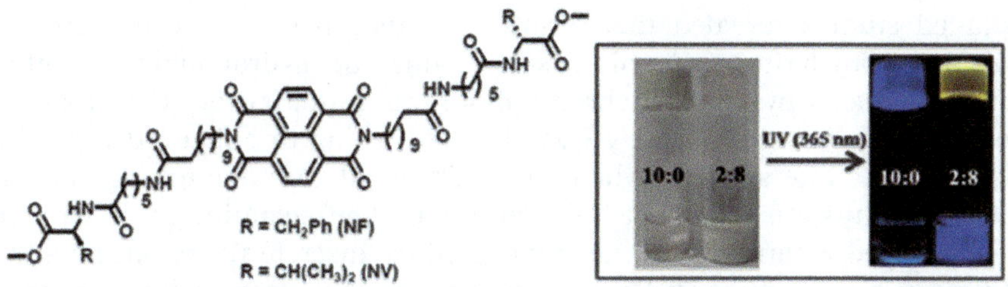

Figure 5.32 Gel and sol picture, in daylight and under UV light, of peptide-appended NDI gelator.
Reproduced from ref. 108 with permission of the Royal Society of Chemistry © 2013.

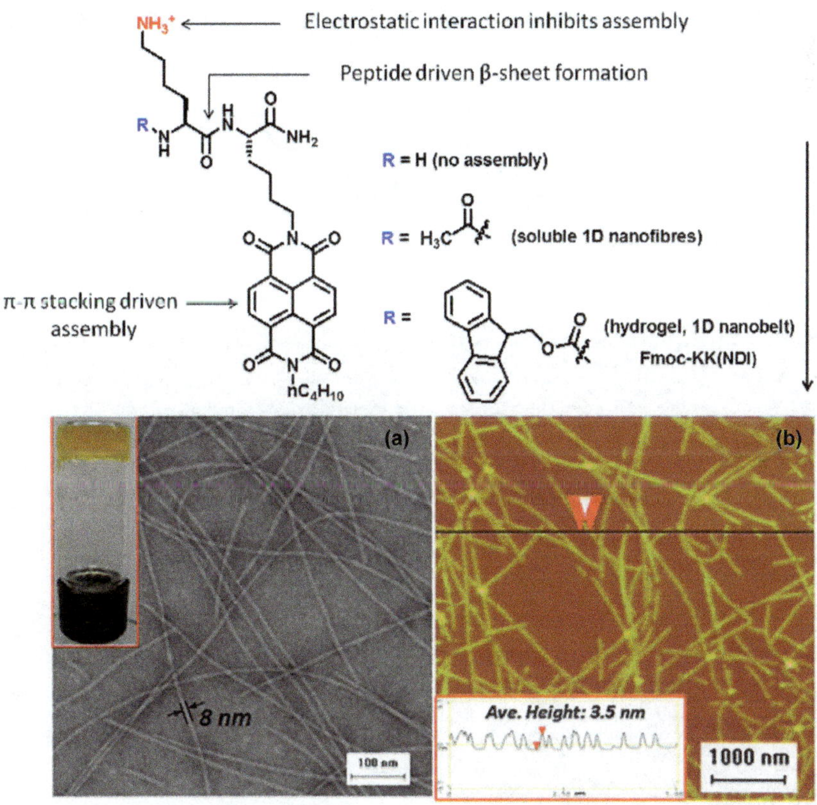

Figure 5.33 Top—structure of peptide-functionalized NDI building blocks; bottom—(a) TEM images of **Fmoc-KK (NDI)** in water; 2 wt% uranyl acetate as a negative stain. Inset: picture of the hydrogel. (b) AFM image of the gel fibres.
Reproduced from ref. 109 with permission of the Royal Society of Chemistry © 2010.

Detailed studies revealed that a subtle balance between electrostatic repulsions from lysine ammonium side chains and hydrophobic π–π interactions induced by the NDI chromophore was essential for gel formation. For example, no self-assembly was observed owing to enhanced repulsive electrostatic interactions when an additional ammonium group at the N-terminus was introduced. Replacement of a free amine group with an acetyl-capped N-terminal amine could produce discrete fibres but failed to initiate gelation, unlike an Fmoc-protected terminal amine, which provided better self-assembly for gelation due to increased hydrophobicity from the additional fluorenyl group.

In a slightly different design, the same group reported a hydrogel fabricated from an NDI-based bola-amphiphile symmetrically substituted with a polar L-lysine group at the imide position (Figure 5.34).[53] Both TEM and AFM images (Figure 5.34) revealed formation of nanotubes comprised of hydrophobic NDI stacks forming the wall of the tube shielded with polar lysine head groups both inside and outside the wall. XRD and magic-angle

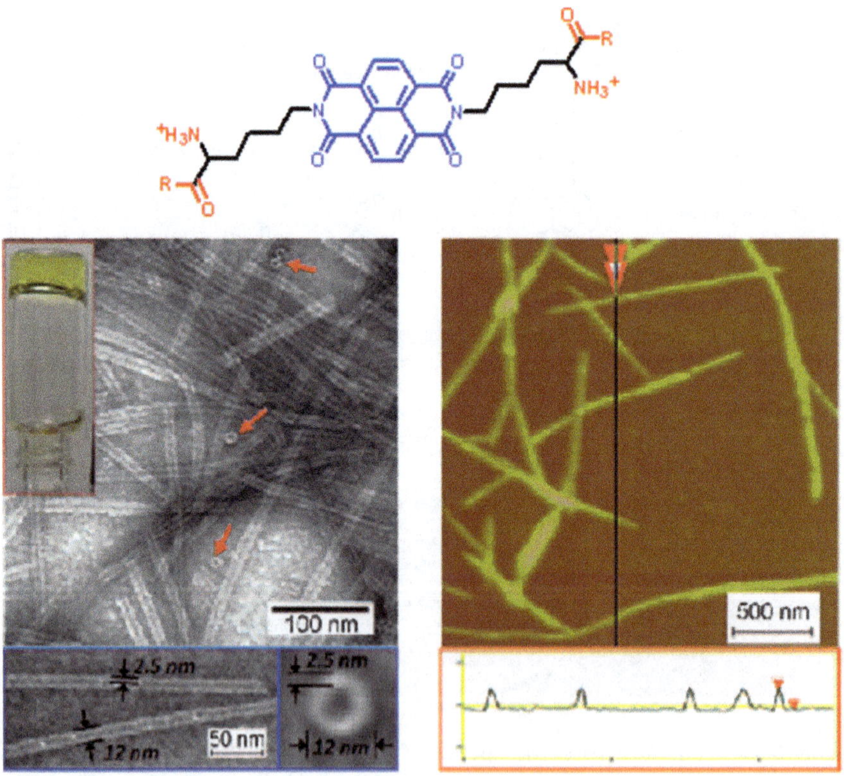

Figure 5.34 TEM (left) and AFM (right) images of NDI-bola-amphiphile (structure shown on the top) in water negatively stained with uranyl acetate. Reprinted with permission from ref. 53. Copyright (2006) Wiley-VCH Verlag GmbH & Co. KGaA, Weinheim.

spinning (MAS) solid-state NMR spectroscopy confirmed nanotube formation by interfusion of several of nano-rings together.

There are also reports of a polymeric gelator containing a NDI backbone.[110] Polymer **1** (Figure 5.35), synthesized by a Michael addition reaction, contained an alternate arrangement of a NDI chromophore and fluorinated alkyl chains and was found to efficiently gelate both aliphatic (THF and dioxane) as well as aromatic (toluene and chlorobenzene) solvents. However, a control polymer that contained alternate alkyl chains could not impart gelation, suggesting the need for fluorinated alkyl groups to induce gel formation. The gel could be irreversibly disassembled by addition of reducing agents such as hydrazide hydrate, F^- or CN^- ions and thus the gelator can be used as an efficient sensor for their detection (Figure 5.35). Also, the conductivity of the gel films increased many-folds on exposure to

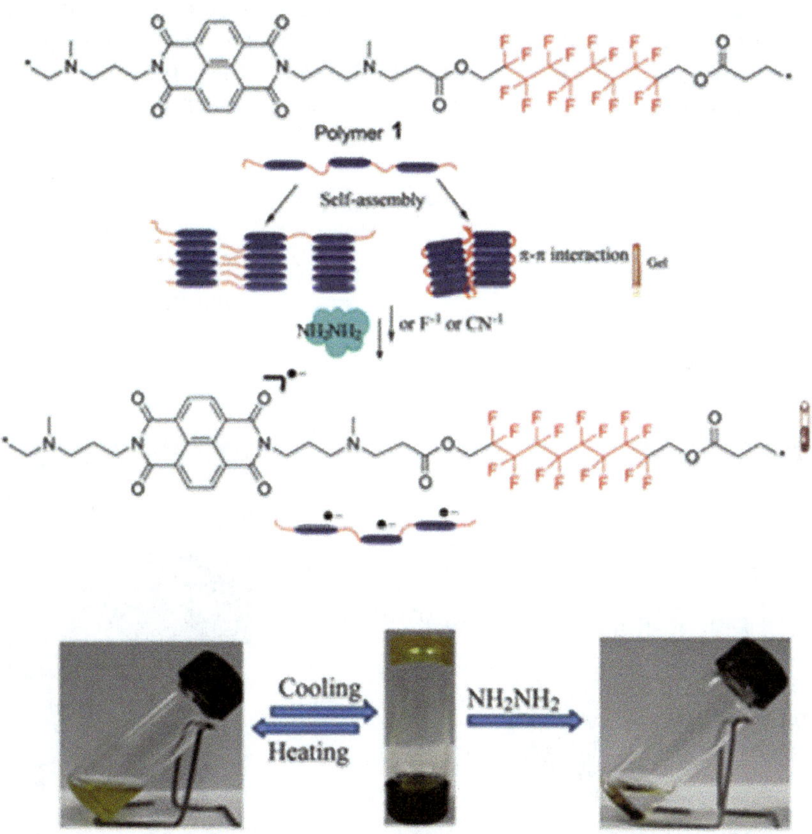

Figure 5.35 Top—structure of polymer **1** and a model showing its mode of assembly and disassembly in the presence of hydrazide hydrate; bottom—gel picture of polymer **1** in THF and its diassembly to sol on addition of hydrazine hydrate.
Reprinted with permission from ref. 110. Copyright (2012) Wiley-VCH Verlag GmbH & Co. KGaA, Weinheim.

hydrazide vapour due to the formation of a radical anion, which increased the carrier concentration in the presence of a reducing agent.

More recently, Ghosh and coworkers have reported[122] the self-assembly of NDI-appended 1,3-dihydroxy gelators, which show facile self-assembly in hydrocarbons by simultaneous intra- and inter-chain H-bonding among the 1,3-hydroxyl groups (Figure 5.36). The same building block was used for poly-condensation with hexa-methylene isocyanate, which produced novel polyurethane (NDI–PU, Figure 5.36) with pendant NDI units.[123] NDI–PU exhibits hierarchical assembly in hydrocarbons, which includes intra-chain folding by H-bonding followed by macroscopic assembly of folded structures to produce high aspect ratio nanotubes and green emitting gel. Placement of the n-type semiconducting NDI units in the wall of the nanotubes results in significant enhancement of charge carrier mobility.

Donor–Acceptor Gels: The high electron acceptability of NDI chromophores has tempted researchers to study their co-assembly with different electron-rich donor chromophores, which has created many organogels as well as hydrogels. Shinkai and coworkers showed[111] the thixotropic property of a NDI gel obtained from **1** (Figure 5.25), which disassembled to sol under

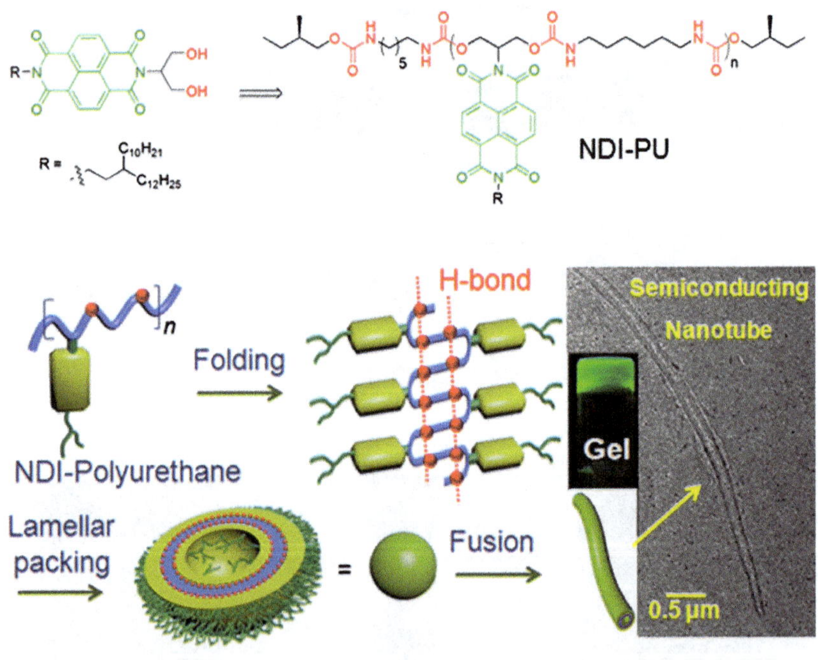

Figure 5.36 NDI-appended 1,3-dihydroxyl gelator and its utility in producing NDI–PU, which undergoes intra-chain folding followed by macroscopic assembly leading to semiconducting nanotubes producing a green emitting gel.
Reprinted with permission from T. Mondal, T. Sakurai, S. Yoneda, S. Seki and S. Ghosh, *Macromolecules*, 2015, **48**, 879. Copyright (2015) American Chemical Society.

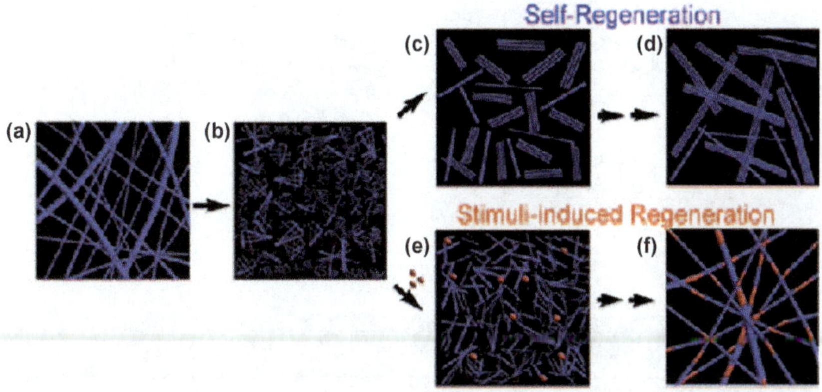

Figure 5.37 Schematic representation of the thixotropic process and its regulation with donor molecules: (a) original gel with long fibres; (b) short fibres after mixing by vortexing; (c),(d) self-regeneration of fibres and bundles; (e),(f) stimuli-induced regeneration of fibres in the presence of a molecular adhesive (red dots: 1,3-dihydroxynaphthalene).
Reprinted with permission from ref. 111. Copyright (2010) Wiley-VCH Verlag GmbH & Co. KGaA, Weinheim.

mechanical stress and reformed by a self-healing process (Figure 5.37). This was studied by monitoring the morphology of the solution by time-dependent TEM studies. The recovery time for reconstitution of the fibres was found to be dependent on the gel concentration and addition of a certain chemical stimulant, which acts as a "molecular adhesive". The disrupted gel solution undergoes a two-fold increase in the regeneration process in the presence of a donor chomophore, 1,3-dihydroxynaphthalene, that brings together the active end of the disintegrated gel fibres by D–A charge-transfer (CT) interactions producing a transparent violet gel.

The CT interaction was monitored by the appearance of a CT band at around 570 nm in the absorption spectrum. Morphological modulation of the gel fibre could be initiated by using another donor isomer, 1,6-dihydroxynaphthalene, which formed 2D sheet-like structures unlike the 1D fibres obtained in the former case.

They have also established a method for colorimetric detection of various positional isomers of a naphthalene donor (Figure 5.38) using the same NDI gelator (**1**) (Figure 5.25).[101] Interestingly, mixing of all positional isomers of dihydroxynaphthalene (**3a–g**) with **1** formed a CT gel with different characteristic colors that could be distinguished by the naked eye (Figure 5.38). However, the complex formed with **3f** was found to be the most intense with a red-shifted CT band. The energy minimized structure of the complex showed involvement of both the hydroxyl groups in H-bonding with the imide carbonyl of the NDI chromophore. Mixing of gelator **1** with various donor analogues such **4a, 4b, 4c, 4d** and pyrene showed no CT gelation as evident from yellow gel formation illustrating no D–A interaction in the gel matrix (Figure 5.38).

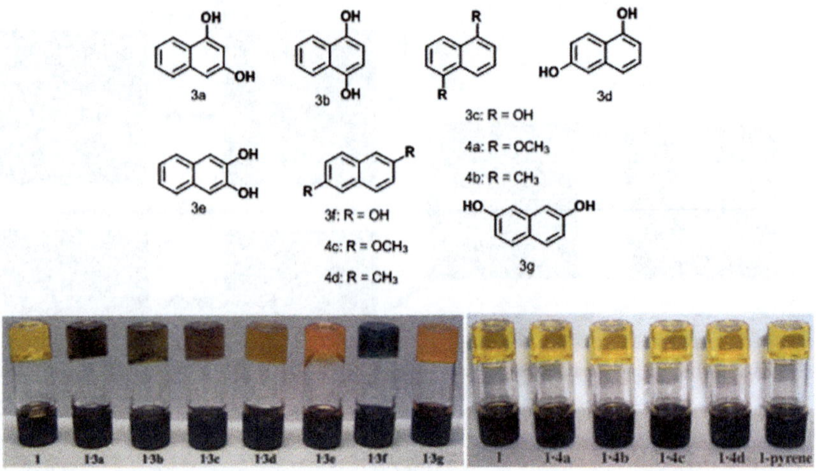

Figure 5.38 Gel pictures of NDI derivative **1** (Figure 5.25) with various naphthalene derivatives (top) showing structure-dependent CT complexation in the gel sate.
Reprinted with permission from ref. 101. Copyright (2006) Wiley-VCH Verlag GmbH & Co. KGaA, Weinheim.

Ghosh and coworkers have done in-depth studies to gain insight into the structure–property relationship in H-bonding-mediated supramolecular assembly and gelation of electron-rich dialkoxy-naphthalene (DAN) and naphthalene-diimide (NDI) building blocks. The primary motivation was to utilize various H-bonding synthons to gain control over the mode of co-assembly of NDI and DAN gelators in their mixtures. They demonstrated H-bonding-mediated self-sorted gelation of donor (**DAN-2**) and acceptor (**NDI-2**) chromophores with varying distances between the two amide groups in the donor and the acceptor chromophore (Figure 5.39).[112,124]

In a non-polar solvent (MCH), where H-bonding took a decisive role, the segregated assembly of two types of chromophores produced a yellow gel (Figure 5.40) in which all the amides were involved in H-bonding interactions, which could not be possible in an alternate D–A stack due to geometrical constraint. Further detailed circular dichroism (CD) studies with **NDI-1** (Figure 5.41) and **DAN-2** (Figure 5.39) demonstrated that molecularly-segregated NDI and DAN fibres did not undergo any macroscopic phase separation as expected due to the similar nature of the peripheral alkyl chains for both gelators.[113]

A control gelation experiment with a mixture of **NDI-1** and **DAN-2** (Figure 5.41) showed no chirality transfer to the donor due to immiscibility of the polar PEG chain of the donor with the hydrophobic alkyl chain of the acceptor. To demonstrate that self-sorting was indeed achieved due to proposed molecular design, the group made another control molecule, **DAN-4** (Figure 5.39),[114] in which the distance between the two amides was almost comparable to that in **NDI-2** so that H-bonding can be achieved between the

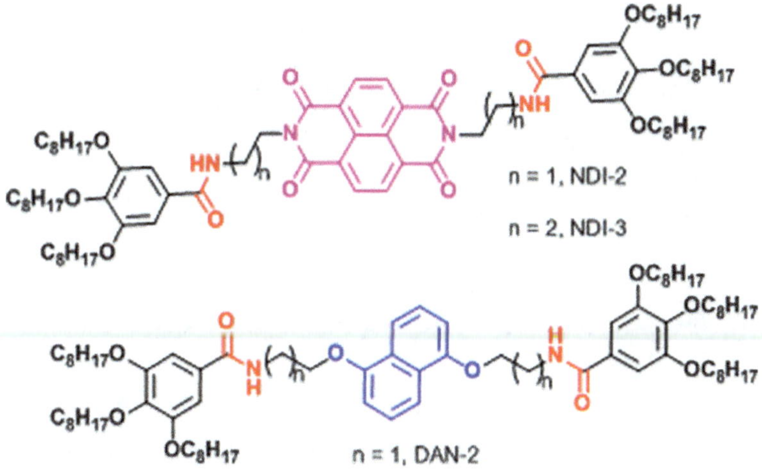

Figure 5.39 Strcutures of NDI- and DAN-based building blocks reported by Ghosh and coworkers for H-bonding controlled co-assembly and gelation.

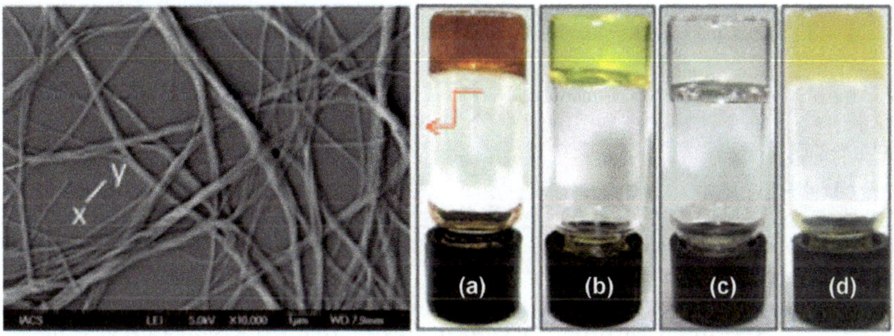

Figure 5.40 Left: SEM image of the CT gel (1 : 1 mixture of **NDI-2** + **DAN-4**) in TCE; right: gels in TCE from (a) **NDI-2** + **DAN-4** (1 : 1), (b) **NDI-2**, (c) **DAN-4**, (d) **NDI-2** + **DAN-2**.
Reprinted with permission from ref. 112. Copyright (2011) Wiley-VCH Verlag GmbH & Co. KGaA, Weinheim.

amide groups of **DAN-4** and **NDI-2** even in an alternate donor–acceptor stack. Indeed, as per proposition, a red gel was formed on mixing this D–A pair in TCE due to a H-bonding-induced CT interaction.[103] Further, they reported an unprecedented phenomenon of this CT gel. The red gel switched over to a yellow one (Figure 5.42) within a few hours due to the rearrangement of the donor and acceptor molecules from the CT state to the self-sorted state.[103]

Such an unusual switching phenomenon was attributed to the greater stability of the segregated gel owing to stronger H-bonding within the homoaggregates. As H-bonding is known to be stronger than π-stacking or

Figure 5.41 Structure of chiral NDI gelator and control DAN unit reported by Ghosh and coworkers.

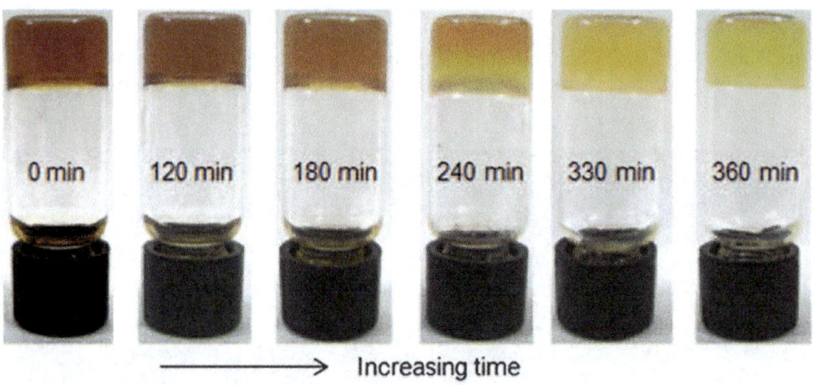

Figure 5.42 Gel picture of **NDI-2** + **DAN-4** (1 : 1) in MCH at various time intervals. Reprinted with permission from ref. 103. Copyright (2011) Wiley-VCH Verlag GmbH & Co. KGaA, Weinheim.

CT interactions in very non-polar solvents such as methylcyclohexane (MCH), the initially-formed kinetically-controlled CT gel switched to a thermodynamically more stable self-sorted gel by overcoming the weaker CT interaction and constrained H-bonding in the alternate D–A stacking by geometrically favored H-bonding in the homoaggregates. However, in a relatively more polarizable solvent such as tetrachloroethylene (TCE), switching did not occur and stable CT gel was observed. It was due to very poor self-assembly of the donor in TCE that the CT gel refrained from switching over to the self-sorted state. To gain further insight into this unique supramolecular switching phenomenon in D–A mixed systems, they studied[103] the co-assembly of various structurally related donor and acceptor chromophores by varying their spacer length (Figure 5.39). Gelation of **NDI-2**

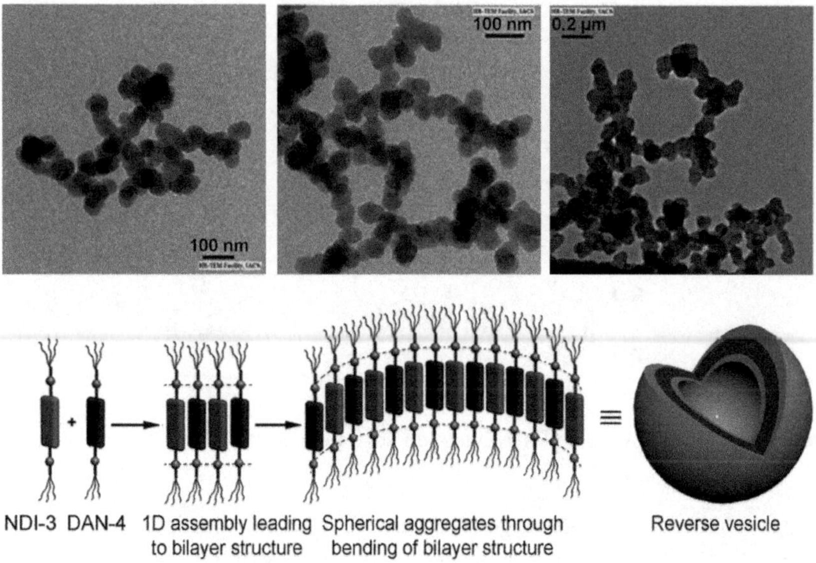

Figure 5.43 TEM images (taken from various sections of the grid) of the mixed assembly of **NDI-3** + **DAN-4** (1 : 1) in TCE; bottom: proposed model for reverse-vesicle-type assembly.
Reprinted with permission from ref. 103. Copyright (2012) Wiley-VCH Verlag GmbH & Co. KGaA, Weinheim.

with another structurally related donor molecule, **DAN-3**, in TCE revealed more stable alternate D–A type stacking than **NDI-2** + **DAN-4** due to the presence of fewer methylene units. Surprisingly, in this case, switching was also observed in TCE as the gelation ability of **DAN-3** was found to be stronger that **DAN-4.** Co-assembly was also tested with another D–A pair (**NDI-3** + **DAN-4**) in which a stable CT state was obtained without gelation even after a prolonged time. Microscopic studies revealed spherical reverse-vesicle-type aggregates (Figure 5.43) in this case, unlike the fibrillar network desired for gelation in the case of **NDI-2** + **DAN-4** (Figure 5.40).[103]

This was attributed to enhanced flexibility of the systems due to a maximum number of methylene units between the NDI and amide groups. Interestingly, they reported the drastic effect of stoichiometric imbalance on morphology transition for the **NDI-3** + **DAN-4** pair. With an excess of **NDI-3**, elongated fibres were noticed, resulting in gelation in contrast to red solution for a 1 : 1 ratio. In the presence of excess **DAN-4**, undefined aggregates were formed owing to poor donor–donor interactions that also failed to form a gel. Understanding the pitfall of their previous design for self-sorted gelation between **NDI-2** + **DAN-2**, which was too sensitive to the distance parameter,[103,112] they developed a more generalized strategy for fabricating segregated donor–acceptor gels. The strategy relied upon mixing an unsymmetrical bis-amide-functionalized NDI acceptor **NDI-1** with various symmetrical bis-amide substituted DAN gelators (Figure 5.44).[114] This

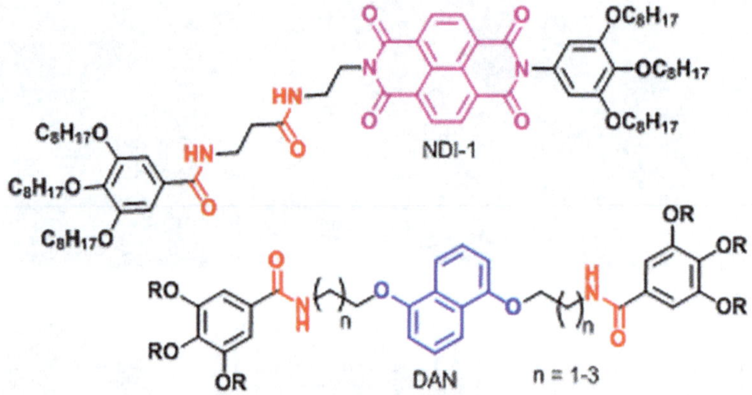

Figure 5.44 NDI–DAN pair for robust self-sorted gelation.

strategy ensured self-sorted gelation in all tested organic solvents, which was ascribed to the greater stability of the self-sorted yellow gel owing to participation of all the amide functionalities in intermolecular H-bonding with the neighbouring chromophores, which could not be achieved in alternate D–A stacking among symmetric DAN and unsymmetrical **NDI-1** (Figure 5.44).

They further explored a novel supramolecular design for CT gel formation between NDI and pyrene by utilizing an external structure directing agent (ESDA) (Figure 5.45).[115] **ESDA-A** self-assembled in a non-polar solvent by intermolecular H-bonding within the amide groups producing a supramolecular 1D polymer with pendant pyridine units, which acted as a proton receptor for alternate anchoring of carboxylic-acid-functionalized **NDI-1** and **Py-1** driven by CT interactions and pyridine–carboxylic acid H-bonding (Figure 5.45). The intense red colored CT gel formed in 10% $CHCl_3$ in MCH was characterized by the appearence of a new lower energy CT band in the UV-vis spectrum. The requirement of both CT interaction and orthogonal H-bonding for gel formation was verified by testing several other combinations possible from **NDI-1**, **Py-1** and **ESDA-A** where none of them produce gel except for **NDI-1** + **Py-1** + **ESDA-A** (1 : 1 : 2).

Ghosh and coworkers have further demonstrated that structural manipulation in the ESDA could modulate the gelation properties of the CT complex. For example, urea-functionalized **ESDA-U** could not induce gelation even at higher concentration even though it formed a stable CT solution. The morphology of the CT complex formed from **ESDA-U** showed ill-defined particles that failed to initiate gelation compared to fibrillar structures obtained from **ESDA-A**. Gel formation was found to be dependent on the stoichiometry of the D–A chromophores. When the composition of **NDI-1** and **Py-1** was varied from 1 : 1 to 3 : 1, **ESDA-U** could also induce CT gelation.

There are also reports where NDI–pyrene CT interactions have been used for hydrogel formation.[46] Ghosh and coworkers have documented a NDI-based non-ionic bola-amphiphile (**NDI-1**, Figure 5.46) that readily

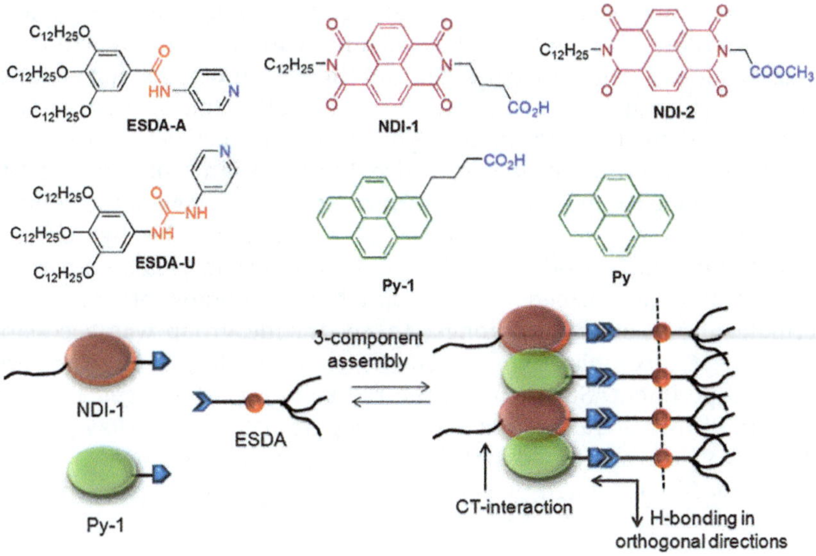

Figure 5.45 Top—structure of donor–acceptor building blocks and ESDA; bottom—schematic showing the three-component assembly.
Reproduced from ref. 115 with permission of the Royal Society of Chemistry © 2014.

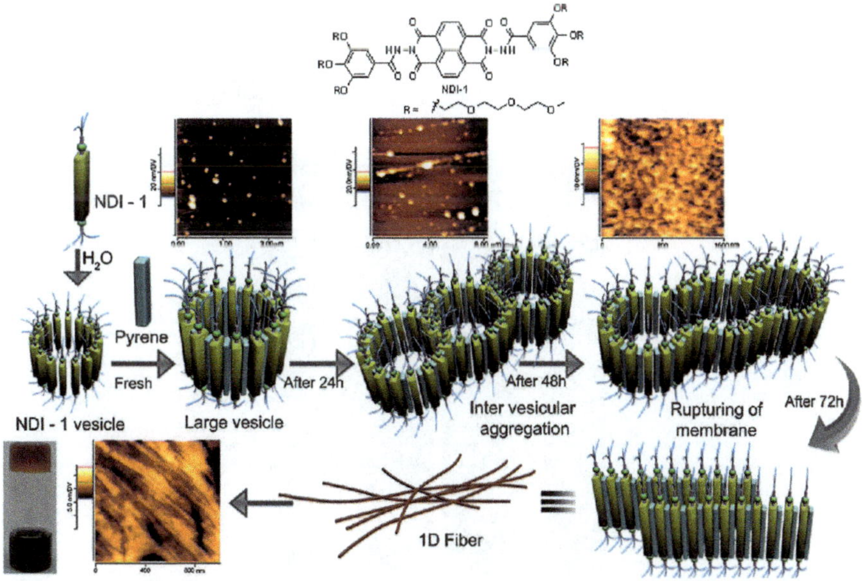

Figure 5.46 Proposed model for the guest-induced stepwise morphological transition from 2D vesicles to 1D fibres for **NDI-1**, thereby leading to gelation (gel image is shown in the bottom-left corner).
Reprinted with permission from ref. 46. Copyright (2012) Wiley-VCH Verlag GmbH & Co. KGaA, Weinheim.

formed a vesicular assembly in an aqueous medium by the synergistic effect of π-stacking between the NDI stack and hydrogen bonding within the hydrazide moiety. The molecular design relied upon using a site isolation strategy for H-bonding to be operative even in bulk water by incorporating the hydrazide functionality in the hydrophobic pocket between the NDI chromophore and the benzene derivative. In the presence of pyrene, **NDI-1** formed a D–A CT complex as evident from the color change of the solution from yellow to red in the presence of the donor. The hydrophobic pyrene donor, which is otherwise insoluble in water, intercalated between the acceptors that formed the hydrophobic membrane of the vesicle. Intermolecular H-bonding amongst the hydrazide group was accessible even in the CT complex state as elucidated by both FTIR and an urea addition experiment. Surprisingly, the red solution, when aged for three days, turned into a transparent red gel (Figure 5.46). Such slow gelation for **NDI-1** + **Py** was attributed to the morphology transition from 2D vesicles to 1D nanofibres, which was demonstrated by time-dependent AFM

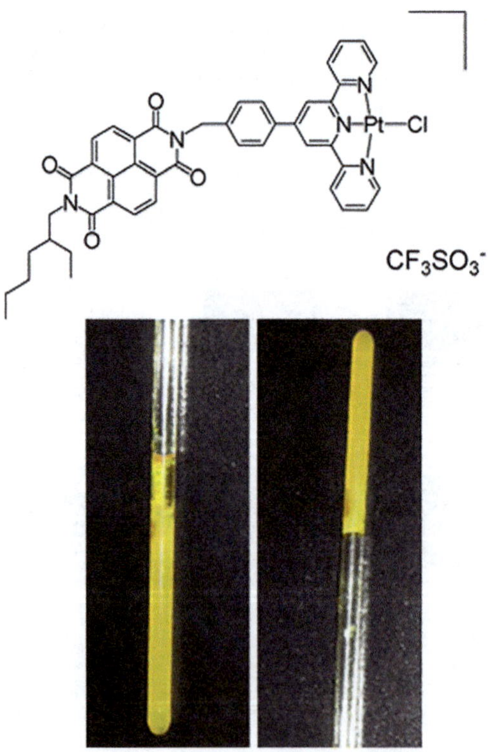

Figure 5.47 Top—structure of the D–A complex [(ND-tpy)PtCl](OTf); bottom—photograph of an NMR tube containing $[(\text{ND-tpy})\text{PtCl}]^+$ in DMSO following cooling after heating to 90 °C. The bottom image is the inverted tube showing the gel.
Reprinted with permission from ref. 116. Copyright © 2009, Elsevier.

studies (Figure 5.46). This showed a significant increase in the radius of curvature of the vesicle formed initially by an alternately stacked D–A assembly that led to intervesicular fusion and rupturing of the membrane to form 1D fibres for gel formation. However, this was not observed for **NDI-1** alone, which formed hollow spherical particles even at much higher concentrations. The gel formation was also found to be dependent on the nature of the intercalator. When the same experiment was done with the sodium salt of pyrene butyric acid, although there was CT complexation as evident from the formation of a green solution, gelation could not be induced even after keeping the solution for a prolonged time. This was acribed to the electrostatic repulsion between the carboxylate-functionalized vesicles that refrained from their interfusion to form a 1D assembly.

In another design, Schmehl and coworkers reported a covalently-linked terpyridyl Pt(II)–naphthalene-diimide (ND) donor–acceptor complex that formed a yellow gel in DMSO (Figure 5.47).[116] The gel was characterized by the formation of elongated tubular structures as evident from TEM studies. X-Ray powder diffraction studies revealed long-range order of the Pt centers. A gel-to-sol transition could be initiated by photolysis, which resulted in an electron transfer from the donor–Pt(II) complex to the NDI acceptor, leading to the formation of a charge-separated species that followed a slow decay process compared to that in the non-aggregated state in the solution phase.

5.5 Conclusion

This article highlights several examples of self-assembled nanofibres, gels and other mesoscopic materials from NDI building blocks by virtue of different weak interactions including π-stacking, H-bonding, CT interactions and solvophobic forces. Its ease of synthesis, the possibility of versatile imide substitutions, efficient π-stacking and the ability to form alternate stacking with electron-rich donor chromophores by CT interactions make NDI an extremely versatile building block for exploring supramolecular chemistry. Recent progress in the synthesis methods for making ring-substituted NDIs with very different photophysical and redox properties has added a new dimension to this research field. The motivation to use this building block for hierarchical structure generation originates from its well-known n-type semiconducting nature. However, most NDI derivatives that have been tested for making transistors are not supramolecularly assembled materials. Thus, it remains to be seen how self-assembly will actually affect device performance. The future prospects of this area will strongly depend on coherent efforts from chemists and device physicists to realize the impact of the knowledge that has already been developed in the area of self-assembly of NDIs on their application in supramolecular electronics.

Acknowledgements

AD thanks CSIR, New Delhi, India, for a research fellowship. PR thanks DST for a fellowship and funding (SR/WOS-A/CS-97/2011). SG thanks SERB for funding (SR/S1/OC-18/2012).

References

1. F. J. M. Hoeben, P. Jonkheijm, E. W. Meijer and A. P. H. J. Schenning, *Chem. Rev.*, 2005, **105**, 1491.
2. Z. Chen, A. Lohr, C. R. Saha-Möller and F. Würthner, *Chem. Soc. Rev.*, 2009, **38**, 564.
3. S. S. Babu, V. K. Praveen and A. Ajayaghosh, *Chem. Rev.*, 2014, **114**, 1973.
4. S. S. Babu, S. Prasanthkumar and A. Ajayaghosh, *Angew. Chem., Int. Ed.*, 2012, **51**, 1766.
5. C. D. Dimitrakopolous and P. R. L. Malenfant, *Adv. Mater.*, 2002, **14**, 99.
6. D. Braun, *Mater. Today*, 2002, **5**, 32.
7. C. J. Brabec, V. Dyakonov, J. Parisi and N. S. Sariciftci, *Organic Photovoltaics Concepts and Realization*, Springer-Verlag, London, 2003.
8. A. W. Hains, Z. Liang, M. A. Woodhouse and B. A. Gregg, *Chem. Rev.*, 2010, **110**, 6689.
9. A. O. Patil, A. J. Heeger and F. Wudl, *Chem. Rev.*, 1988, **88**, 183.
10. S. V. Bhosale, C. H. Jani and S. J. Langford, *Chem. Soc. Rev.*, 2008, **37**, 331.
11. F. Würthner, *Chem. Commun.*, 2004, 1564.
12. F. Würthner and M. Stolte, *Chem. Commun.*, 2011, 5109.
13. W. S. Horne, N. Ashkenasy and M. R. Ghadiri, *Chem. – Eur. J.*, 2005, **11**, 1137.
14. J. L. Sessler, C. T. Brown, D. O'Connor, S. L. Springs, R. Wang, M. Sathiosatham and T. Hirose, *J. Org. Chem.*, 1998, **63**, 7370.
15. A. Das and S. Ghosh, *Angew. Chem., Int. Ed.*, 2014, **53**, 2038.
16. S. Hunig, J. Grob, E. F. Lier and H. Quast, *Liebigs Ann. Chem.*, 1973, 339.
17. G. D. Pantoş, P. Pengo and J. K. M. Sanders, *Angew. Chem., Int. Ed.*, 2007, **46**, 194.
18. M. Tomasulo, D. M. Naistat, A. J. P. White, D. J. Williams and F. M. Raymo, *Tetrahedron Lett.*, 2005, **46**, 5695.
19. M. R. Molla, D. Gehrig, L. Roy, V. Kamm, A. Paul, F. Laquai and S. Ghosh, *Chem. – Eur. J.*, 2014, **20**, 760.
20. M. R. Molla and S. Ghosh, *Chem. – Eur. J.*, 2012, **18**, 1290.
21. Y. Zhao, Y. Domoto, E. Orentas, C. Beuchat, D. Emery, J. Mareda, N. Sakai and S. Matile, *Angew. Chem. Int. Ed.*, 2013, **52**, 9940.
22. A. Das and S. Ghosh, *Angew. Chem., Int. Ed.*, 2014, **53**, 1092.
23. M. B. Avinash, E. Verheggen, C. Schmuck and T. Govindaraju, *Angew. Chem., Int. Ed.*, 2012, **51**, 10324.

24. M. Pandeeswar, M. B. Avinash and T. Govindaraju, *Chem. – Eur. J.*, 2012, **18**, 4818.
25. S. Manchineella, V. Prathyusha, U. D. Priyakumar and T. Govindaraju, *Chem. – Eur. J.*, 2013, **19**, 16615.
26. M. B. Avinash and T. Govindaraju, *Nanoscale*, 2011, **3**, 2536.
27. N. Narayanaswamy, M. B. Avinash and T. Govindaraju, *New J. Chem.*, 2013, **37**, 1302.
28. M. Pandeeswar and T. Govindaraju, *RSC Adv.*, 2013, **3**, 11459.
29. R. S. Lokey and B. L. Iverson, *Nature*, 1995, **375**, 303.
30. J. Q. Nguyen and B. L. Iverson, *J. Am. Chem. Soc.*, 1999, **121**, 2639.
31. C. Peebles, R. Piland and B. L. Iverson, *Chem. – Eur. J.*, 2013, **19**, 11598.
32. A. J. Zych and B. L. Iverson, *J. Am. Chem. Soc.*, 2000, **122**, 8898.
33. V. J. Bradford and B. L. Iverson, *J. Am. Chem. Soc.*, 2008, **130**, 1517.
34. G. J. Gabriel, S. Orey and B. L. Iverson, *J. Am. Chem. Soc.*, 2005, **127**, 2637.
35. G. J. Gabriel and B. L. Iverson, *J. Am. Chem. Soc.*, 2002, **124**, 15174.
36. J. J. Reczek and B. L. Iverson, *Macromolecules*, 2006, **39**, 5601.
37. J. J. Reczek, K. R. Villazor, V. Lynch, T. M. Swager and B. L. Iverson, *J. Am. Chem. Soc.*, 2006, **128**, 7995.
38. P. M. Alvey, J. J. Reczek, V. Lynch and B. L. Iverson, *J. Org. Chem.*, 2010, **75**, 7682.
39. K. R. Leight, B. E. Esarey, A. E. Murray and J. J. Reczek, *Chem. Mater.*, 2012, **24**, 3318.
40. X. Zhang and C. Wang, *Chem. Soc. Rev.*, 2011, **40**, 94.
41. K. Liu, Y. Kang, Z. Wang and X. Zhang, *Adv. Mater.*, 2013, **25**, 5530.
42. J.-M. Lehn, *Science*, 2002, **295**, 2400.
43. M. Kumar and S. J. George, *Chem. – Eur. J.*, 2011, **17**, 11102.
44. M. Kumar and S. J. George, *Nanoscale*, 2011, **3**, 2130.
45. M. Kumar, N. Jonnalagadda and S. J. George, *Chem. Commun.*, 2012, **48**, 10948.
46. M. R. Molla and S. Ghosh, *Chem. – Eur. J.*, 2012, **18**, 9860.
47. P. Rajdev, M. R. Molla and S. Ghosh, *Langmuir*, 2014, **30**, 1969.
48. J.-H. Fuhrhop and D. Itsch, *Acc. Chem. Res.*, 1986, **19**, 130.
49. R. Vyhnalkova, A. H. E. Müller and A. Eisenberg, *Langmuir*, 2014, **30**, 5031.
50. A. Sikder, A. Das and S. Ghosh, *Angew. Chem., Int. Ed.*, 2015, **54**, 6755.
51. A. Das and S. Ghosh, *Macromolecules*, 2013, **46**, 3939.
52. H. Shao, T. Nguyen, N. C. Romano, D. A. Modarelli and J. R. Parquette, *J. Am. Chem. Soc.*, 2009, **131**, 16374.
53. H. Shao, J. Seifert, N. C. Romano, M. Gao, J. J. Helmus, C. P. Jaroniec, D. A. Modarelli and J. R. Parquette, *Angew. Chem., Int. Ed.*, 2010, **49**, 7688.
54. S. Tu, S. H. Kim, J. Joseph, D. A. Modarelli and J. R. Parquette, *J. Am. Chem. Soc.*, 2011, **133**, 19125.
55. K. Liu, Y. Yao, C. Wang, Y. Liu, Z. Li and X. Zhang, *Chem. – Eur. J.*, 2012, **18**, 8622.

56. K. Liu, Y. Yao, Y. Liu, C. Wang, Z. Li and X. Zhang, *Langmuir*, 2012, **28**, 10697.
57. K. Liu, C. Wang, Z. Li and X. Zhang, *Angew. Chem., Int. Ed.*, 2011, **50**, 4952.
58. N. Sakai, J. Mareda, E. Vauthey and S. Matile, *Chem. Commun.*, 2010, **46**, 4225.
59. S. V. Bhosale, S. V. Bhosale and S. K. Bhargava, *Org. Biomol. Chem.*, 2012, **10**, 6455.
60. F. Würthner, S. Ahmed, C. Thalacker and T. Debaerdemaeker, *Chem. – Eur. J.*, 2002, **8**, 4742.
61. C. Thalacker, C. Röger and F. Würthner, *J. Org. Chem.*, 2006, **71**, 8098.
62. R. S. K. Kishore, V. Ravikumar, G. Bernardinelli, N. Sakai and S. Matile, *J. Org. Chem.*, 2008, **73**, 738.
63. B. A. Jones, A. Facchetti, M. R. Wasielewski and T. J. Marks, *J. Am. Chem. Soc.*, 2007, **129**, 15259.
64. X. Guo and M. D. Watson, *Org. Lett.*, 2008, **10**, 5333.
65. P. M. Alvey and B. L. Iverson, *Org. Lett.*, 2012, **14**, 2706.
66. S.-L. Sararu and F. Würthner, *Synthesis*, 2009, **11**, 1841.
67. C. Gu, W. Hu, J. Yao and H. Fu, *Chem. Mater.*, 2013, **25**, 2178.
68. Y. Wei, Q. Zhang, Y. Jiang and J. Yu, *Macromol. Chem. Phys.*, 2009, **210**, 769.
69. Y. Kim, J. Hong, J. H. Oh and C. Yang, *Chem. Mater.*, 2013, **25**, 3251.
70. E. Ahmed, G. Ren, F. S. Kim, E. C. Hollenbeck and S. A. Jenekhe, *Chem. Mater.*, 2011, **23**, 4563.
71. X. Guo, F. S. Kim, M. J. Seger, S. A. Jenekhe and M. D. Watson, *Chem. Mater.*, 2012, **24**, 1434.
72. M. M. Durban, P. D. Lazarinoff and C. K. Luscombe, *Macromolecules*, 2010, **43**, 6348.
73. K. Naakabayashi and H. Mori, *Macromolecules*, 2012, **45**, 9618.
74. J. Wang, M. Ueda and T. Higashihara, *ACS Macro Lett.*, 2013, **2**, 506.
75. I. D. Cat, C. Röger, C. C. Lee, F. J. M. Hoeben, M. J. Pouderoijen, A. P. H. J. Schenning, F. Würthner and S. De Feyter, *Chem. Commun.*, 2008, 5496.
76. S. V. Bhosale, C. Jani, C. H. Lalander and S. J. Langford, *Chem. Commun.*, 2010, **46**, 973.
77. S. V. Bhosale, C. H. Jani, C. H. Lalander, S. J. Langford, I. Nerush, J. G. Shapter, D. Villamaina and E. Vauthey, *Chem. Commun.*, 2011, **47**, 8226.
78. H. Kar, D. Gehrig, F. Laquai and S. Ghosh, *Nanoscale*, 2015, **7**, 6729.
79. H. Kar, D. W. Gehrig, N. K. Allampally, G. Fernández, F. Laquai and S. Ghosh, *Chem. Sci.*, 2016, **7**, 1115.
80. H. Kar and S. Ghosh, *Chem. Commun.*, 2016, **52**, 8818.
81. R. G. Weiss and P. Terech, *Molecular Gels: Materials with Self-Assembled Fibrillar Networks*, Springer, 2006.
82. N. M. Sangeetha and U. Maitra, *Chem. Soc. Rev.*, 2005, **34**, 821.

83. J.-L. Bredas, J. E. Norton, J. Cornil and V. Coropceanu, *Acc. Chem. Res.*, 2009, **42**, 1691.
84. L. Maggini and D. Bonifazi, *Chem. Soc. Rev.*, 2012, **41**, 211.
85. *Handbook of Conducting Polymers*, ed. T. A. Skotheim, R. L. Elsenbaumer and J. R. Reynolds, Marcel Dekker Inc., New York, 1998.
86. J. Puigmart-Luis, V. Laukhin, A. P. Pino, J. Vidal-Gancedo, C. Rovira, E. Laukhina and D. B. Amabilino, *Angew. Chem., Int. Ed.*, 2007, **46**, 238.
87. D. A. Stone, A. S. Tayi, J. E. Goldberger, L. C. Palmer and S. I. Stupp, *Chem. Commun.*, 2011, **47**, 5702.
88. A. Ajayaghosh, V. K. Praveen and C. Vijayakumar, *Chem. Soc. Rev.*, 2008, **37**, 109.
89. A. Ajayaghosh and V. K. Praveen, *Acc. Chem. Res.*, 2007, **40**, 644.
90. R. E. Melendez, A. J. Carr. B. R. Linton, A. D. Hamilton, in *Controlling Hydrogen Bonding: From Molecular Recognition to Organogelation, Struct. Bonding*, 2000, vol. 96 pp. 31–61.
91. B. J. Schwartz, *Annu. Rev. Phys. Chem.*, 2003, **54**, 141.
92. S. Ghosh, X.-Q. Li, V. Stepanenko and F. Würthner, *Chem. – Eur. J.*, 2008, **14**, 11343.
93. S. Yagai, S. Kubota, T. Iwashima, K. Kishikawa, T. Nakanishi, T. Karatsu and A. Kitamura, *Chem. – Eur. J.*, 2008, **14**, 5246.
94. S. I. Tamaru, M. Nakamura, M. Takeuchi and S. Shinkai, *Org. Lett.*, 2001, **3**, 3631.
95. M. Shirakawa, S. I. Kawano, N. Fujita, K. Sada and S. Shinkai, *J. Org. Chem.*, 2003, **68**, 5037.
96. H. Engelkamp, S. Middelbeek and R. J. M. Nolte, *Science*, 1999, **284**, 785.
97. A. Mishra, C.-Q. Ma and P. Bauerle, *Chem. Rev.*, 2009, **109**, 1141.
98. F. S. Schoonbeek, J. H. Van Esch, B. Wegewijs, D. B. A. Rep, M. P. De Haas, T. M. Klapwijk, R. M. Kellogg and B. L. Feringa, *Angew. Chem., Int. Ed.*, 1999, **38**, 1393.
99. T. Akutagawa, K. Kakiuchi, T. Hasegawa, S.-I. Noro, T. Nakamura, H. Hasegawa, S. Mashiko and J. Becher, *Angew. Chem., Int. Ed.*, 2005, **44**, 7283.
100. K. Sugiyasu, N. Fujita and S. Shinkai, *Angew. Chem., Int. Ed.*, 2004, **43**, 1229.
101. P. Mukhopadhyay, Y. Iwashita, M. Shirakawa, S.-I. Kawano, N. Fujita and S. Shinkai, *Angew. Chem., Int. Ed.*, 2006, **45**, 1592.
102. M. R. Molla and S. Ghosh, *Chem. Mater.*, 2011, **23**, 95.
103. A. Das, M. R. Molla, B. Maity, D. Koley and S. Ghosh, *Chem. – Eur. J.*, 2012, **18**, 9860.
104. A. Das, M. R. Molla and S. Ghosh, *J. Chem. Sci.*, 2011, **123**, 963.
105. H. Kar, M. R. Molla and S. Ghosh, *Chem. Commun.*, 2013, **49**, 4220.
106. J. Zheng, W. Qiao, X. Wan, J. P. Gao and Z. Y. Wang, *Chem. Mater.*, 2008, **20**, 6163.
107. B. Narayan, C. Kulkarni and S. J. George, *J. Mater. Chem. C*, 2013, **1**, 626.
108. S. Basak, J. Nanda and A. Banerjee, *Chem. Commun.*, 2013, **49**, 6891.

109. H. Shao and J. R. Parquette, *Chem. Commun.*, 2010, **46**, 4285.
110. L. Hewei, L. Zitong, C. Zhengxu, W. Liusuo, Z. Guanxin, L. Chenyang and Z. Deqing, *Chin. J. Chem.*, 2012, **30**, 1453.
111. P. Mukhopadhyay, N. Fujita, A. Takada, T. Kishida, M. Shirakawa and S. Shinkai, *Angew. Chem., Int. Ed.*, 2010, **49**, 6338.
112. A. Das, M. R. Molla, A. Banerjee, A. Paul and S. Ghosh, *Chem. - Eur. J.*, 2011, **17**, 6061.
113. M. R. Molla, A. Das and S. Ghosh, *Chem. Commun.*, 2011, **47**, 8934.
114. A. Das and S. Ghosh, *Chem. Commun.*, 2011, **47**, 8922.
115. H. Kar and S. Ghosh, *Chem. Commun.*, 2014, **50**, 1064.
116. D. Kumaresan, K. Lebkowsky and R. H. Schmehl, *J. Photochem. Photobiol., A*, 2009, **207**, 86.
117. Z. Yu, H. Wang, B. Bai, S. Qu, F. Li, X. Ran, J. Sun, G. Jin and M. Li, *Mater. Sci. Eng., C*, 2010, **30**, 699.
118. P. Stefan, S. Anke, S. Dana, B. Erica, V. Michael and K. Edwin, *New J. Chem.*, 2013, **37**, 169.
119. A. W. Feinberg, A. Feigel, S. S. Shevkoplyas, S. Sheehy, G. M. Whitesides and K. K. Parker, *Science*, 2007, **317**, 1366.
120. M. He, J. Li, S. Tan, R. Wang and Y. Zhang, *J. Am. Chem. Soc.*, 2013, **135**, 18718.
121. M. D. Burke, J. O. Park, M. Srinivasarao and S. A. Khan, *J. Controlled Release*, 2005, **104**, 141.
122. T. Mondal, D. Basak, A. Al-Ouahabi, M. Schmutz, P. Mésini and S. Ghosh, *Chem. Commun.*, 2015, **51**, 5040.
123. T. Mondal, T. Sakurai, S. Yoneda, S. Seki and S. Ghosh, *Macromolecules*, 2015, **48**, 879.
124. M. R. Molla, A. Das and S. Ghosh, *Chem. - Eur. J.*, 2010, **16**, 10084.

CHAPTER 6

Polymeric Materials Based on NDI and its Congeners

LEWIS R. HART,[a] WAYNE HAYES[a] AND BARNABY W. GREENLAND*[b]

[a] Department of Chemistry, University of Reading, Whiteknights, Reading RG6 6AD, UK; [b] The Reading School of Pharmacy, University of Reading, Whiteknights, Reading RG6 6AP, UK
*Email: b.w.greenland@reading.ac.uk

6.1 Introduction

The structure of naphthalene diimide (NDI) has many useful chemical and electronic properties that have resulted in wide ranging research interest into small molecules containing either the NDI skeleton or congeners of NDI. In recent years, polymer chemistry featuring the use of NDI has also flourished, affording materials with high value properties including conducting polymers, sensors and healing materials.[1–3] This chapter uses selected examples from the literature to illustrate how NDI-based structures have been utilised within polymeric structures to generate materials with fascinating properties.

6.1.1 The Structures of NDI-based Polymers

The polymer chemistry surrounding NDI has focussed principally on polymers that have three distinct architectures (Figure 6.1). These macromolecules can be broadly categorised as those that have (i) NDI in the polymer main chain connected through the aromatic core (Figure 6.1(a)),

Monographs in Supramolecular Chemistry No. 21
Naphthalenediimide and its Congeners: From Molecules to Materials
Edited by G. Dan Pantoş

Published by the Royal Society of Chemistry, www.rsc.org

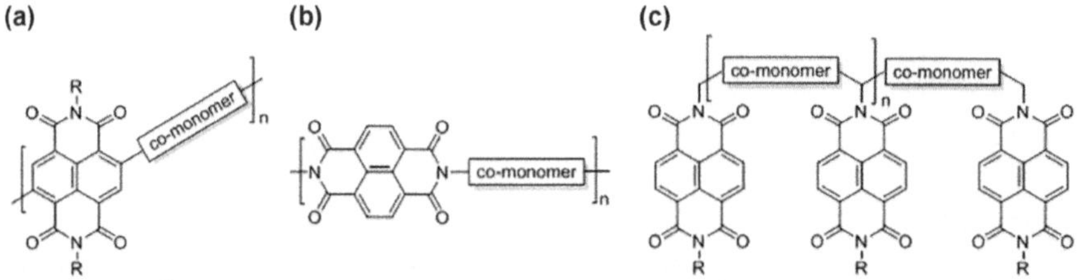

Figure 6.1 Generic structures of polymers containing: (a) NDI in the polymer main chain connected through the aromatic core. (b) NDI connected through the imide nitrogen into the polymer main chain. (c) NDI residues as pendent side groups.

(ii) NDI connected through the imide nitrogen into the polymer main chain (Figure 6.1(b)) or (iii) NDI as pendent groups on an extended polymer backbone (Figure 6.1(c)).

The addition of two imide units to a naphthalene ring renders NDI an extremely π-electron-deficient species. This results in HOMO/LUMO energy levels that are favourable for electronic applications, particularly as n-type semiconductors (electron accepting species), which have traditionally been difficult materials to access.[4–7] A large body of literature has detailed functionalisation of the naphthalene core of NDI in order to tune its electronic band gap[8,9] and has enabled rapid progress to be made in producing polymeric materials with naphthalene residues embedded within the main chain. Judicious choice of the co-monomer allows the production of materials with highly delocalised structures that frequently retain the high thermal and oxidative stability of the parent NDI monomer. The synthesis and evaluation of this class of polymer is discussed in the following Section 6.2.

The extremely π-electron-deficient nature of NDI, combined with its planar geometry allows it to form highly defined and predictable π–π stacked complexes with π-electron-rich species including 1,5-dimethoxy naphthalene and pyrene.[10–12] This phenomenon has resulted in a rich literature surrounding the use of NDI in small molecule (non-polymer) supramolecular chemistry.[13,14] In more recent years, inserting NDI into a polymer backbone or as pendent side groups (Figure 6.1(b) and (c) respectively), using simple condensation chemistries, has resulted in the realisation of sensors, self-assembling structures and healable polymers, which are covered in the latter sections of this chapter (6.3, 6.4 and 6.5).

6.2 Conjugated Polymers Containing NDI for Organic Electronic Devices

Research into plastic electronics has been one of the most rapidly advancing fields over the last 25 years. The goal of this body of research is to replace the fragile inorganic semiconducting components of electronic devices with polymeric materials that have lower raw material and manufacturing costs.[15,16] The ultimate aim of plastic electronics research is to be able to use the solution processing characteristics of polymers to 'print' large area semiconductor components including transistors, light emitting diodes and solar cells in a continuous process. Achieving these goals would enable these high value commodities to be both mechanically flexible and inexpensive enough to cover every roof with photovoltaic cells or produce display screens that can cover walls and then be rolled up and carried away.

6.2.1 NDI Polymers for Field-Effect Transistors

Field-effect transistors (FETs) are the basis of all modern electronics; they act as electronic amplifiers or on–off switches that form the heart of integrated circuits and computer processors. Replacing inorganic transistors with alternatives derived from organic polymeric materials would have a transformative effect on society.[7,17–20]

Two typical organic field-effect transistor (OFET) architectures are shown in Figure 6.2, consisting of a gate electrode, a dielectric insulating layer, and an organic semiconductor (OSC) sandwiched between a source and drain electrode. In many devices, heavily doped silicon is used as the gate electrode. In top-contact bottom-gate geometries (Figure 6.2(a)), the organic film is deposited on a uniform dielectric surface, before a shadow mask is used to apply the source and drain electrodes on top by metal evaporation. This process reduces contact resistance as a result of the intimate contact between the semiconductor and the electrodes.

In bottom-contact devices (Figure 6.2(b)), the source and drain electrodes are lithographically pre-patterned on the substrate before the organic layer is deposited. Bottom-contact devices typically exhibit less than half the

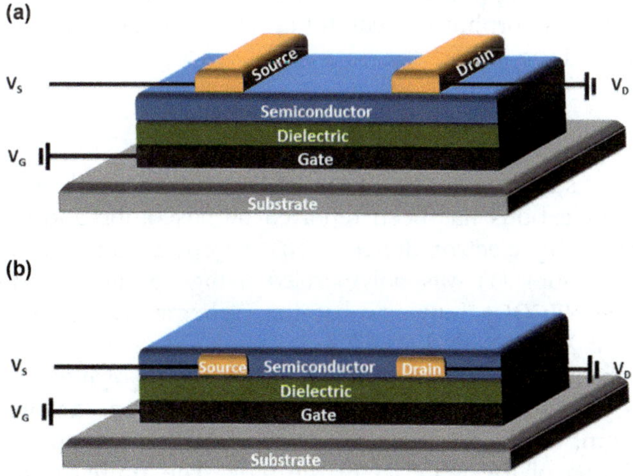

Figure 6.2 Schematic of top-contact bottom-gate (a) and bottom-contact bottom-gate (b) geometries in OFET devices.

effective drive current of top-contact devices as a result of contact resistance, although bottom-contact devices are integrated easily into low-cost manufacturing processes. In both geometries, the OSC is deposited to bridge the source and drain electrodes, and is itself separated from the gate electrode by an insulating dielectric layer.

OFETs operate *via* the reversible application of two electric fields, one laterally between the source and drain as well as a second vertically between gate and semiconductor. The gate field is able to induce a layer of charge at the interface between the OSC and dielectric, termed the 'conducting channel', the magnitude of which is dependent on the capacity of the dielectric and the gate field. The mobility within the OSC ideally reflects charge migration across a continuous material supporting the channel, induced by the drain field. In practice, the measured mobility incorporates various barriers and series resistances that are extrinsic to the OSC, some of which are gate-dependent. These include interface dipoles, morphological interruptions near contacts, charge traps on the dielectric surface, grain boundaries, chemical traps and dopants, and environmental influences. Thus, the measured efficiency of a device is highly dependent on not just the chemical structures of the materials but also on the fabrication of the device. Depending on the semiconductor used in the transistor, OFETs can function as a p-channel (hole-transporter) or n-channel (electron-transporter) switch.

The key device performance parameters include the field-effect mobility (μ_e) and the current on–off ratio (I_{on}/I_{off}). Holes and electrons should be able to be induced at applied electric fields, as dictated by the energy levels of the highest occupied/lowest unoccupied molecular orbital (HOMO/LUMO) energies of the individual molecules. The molecular scale packing within the material should allow sufficient overlap of frontier orbitals to provide efficient charge migration between neighbouring (macro)molecules.[6] To enable OFET-based applications to reach market, these parameters should surpass those of amorphous silicon for devices fabricated/functioning under ambient conditions.

6.2.1.1 NDI Homopolymers for OFETs

Perhaps the structurally simplest NDI-containing polymer developed for OFET applications has been reported by Luscombe and co-workers in 2010.[21] The highly electron-deficient NDI acceptor unit with octyldodecyl-solubilising groups (**1**) was polymerised using Yamamoto coupling conditions using $Ni(COD)_2$/bipy as a catalyst (Scheme 6.1) to yield polymer **2** with an M_n of 71.1 kDa and $Đ$ of 2.9.

Optical absorption spectra of films of the homopolymer **2**, spin-coated from a $CHCl_3$ solution onto glass, showed an absorption band $\lambda_{max} = 426$ nm, whilst the X-ray diffraction (XRD) spectrum did not reveal visible ordering of the material in the bulk. The similarity in the UV-vis spectra of the monomer **1** and resultant polymer **2** suggested that the conjugation length of the homopolymer was similar to that of the starting

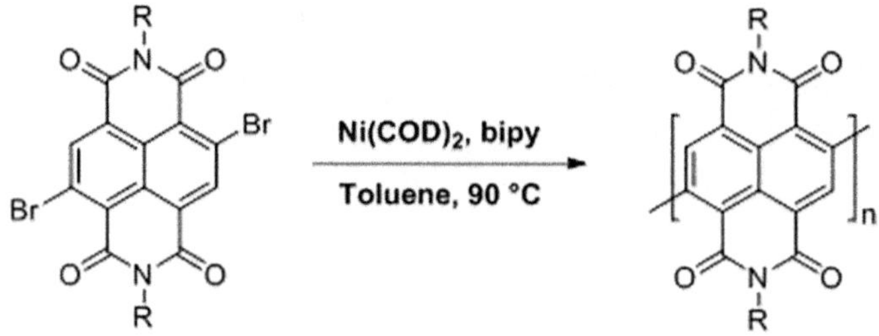

1 2

R = Branched alkyl chain

Scheme 6.1 Synthesis of NDI polymer **2** by Luscombe *et al.* using Yamamoto coupling conditions.[21]

material. It was postulated that the bulky branched alkyl groups used to aid solubility forced the polymer into a twisted conformation in the solid state, resulting in an extremely low persistence length. The LUMO energy levels were calculated from the onset of the first polymer reduction peak using cyclic voltammetry (CV) and demonstrated to be 3.76 eV as a result of the intrinsic properties of NDI.[22,23] The HOMO energy level (6.50 eV) was estimated using the optical band gap ($E_{g\,(opt)}$) from the calculated LUMO.

OFET devices were fabricated in the typical top-contact bottom-gate geometry. Typical n-type behaviour was observed at positive gate-source biases ($V_g = +100$ V), whilst ambipolar or p-type responses with low current flow and on–off ratios below 10 when a negative gate-source bias was applied ($V_g = -100$ V) was not evident.

The on–off ratios were also calculated for the homopolymer **2** ($I_{on/off} = 10^3$) as well as threshold voltage ($V_t = 40$ V). AFM, XRD and UV-vis spectroscopic data indicated that the polymers were not crystalline in nature, which accounted for the observed low electron mobility (6.0×10^{-4} cm^2 V^{-1} s^{-1}) as well as the relatively low $I_{on/off}$ and high V_t values.

6.2.1.2 Core-functionalised NDI–Thiophene Copolymers for OFETs

A series of NDI–thiophene copolymers have been studied by several groups, including Luscombe and co-workers,[21] Facchetti and co-workers,[24,25] as well as Jenekhe and Watson *et al.*[26] and Sommer and co-workers[27] (**3–5**, Figure 6.3).

Facchetti *et al.* first reported the NDI–dithiophene copolymer **4** in 2009 *via* a Pd-catalysed Stille-coupling polymerisation in high yields (>90%) with high molecular weight (M_n 11.0 kDa, $Đ = 2.90$) for applications in OFETs.[24] Thin film optical polymer absorption spectra for the NDI–dithiophene copolymer **4** revealed two significant absorptions at $\lambda_{max} = 697$ and 391 nm,

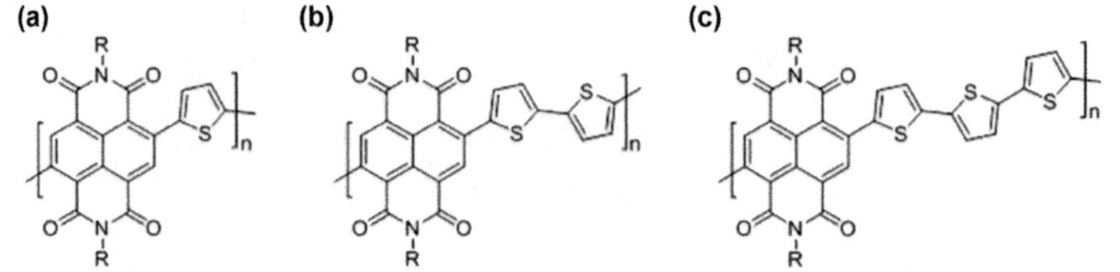

3 4 5

Figure 6.3 NDI–thiophene copolymers synthesised by Luscombe and co-workers[21] (a)–(c), Jenekhe and Watson *et al.*[26] (a),(b), Facchetti and co-workers[24,25] and Sommer and co-workers[27] (b) *via* Pd-catalysed Stille polymerisation.

respectively, with a corresponding $E_{g(opt)}$ of 1.45 eV. This data confirmed the extended π-conjugated electronic structure of this semiconductor as a consequence of the regioregular polymeric backbone. The corresponding LUMO energy of ~4.0 eV is among the lowest reported to date for a semiconducting polymer, approaching those of strongly electron-depleted core-cyanated rylenes.[28]

CV analysis revealed two reversible reductions with the first and second reduction potentials located at −0.49 and −0.96 V, respectively. The HOMO (−5.36 eV) and LUMO (−3.91 eV) energy levels were then calculated, which revealed the LUMO energy to be borderline for the ambient operation of OFETs containing this material. Bottom-gate top-contact OFETs were fabricated by spin coating polymer solutions ($<$10 mg mL^{-1} in $CHCl_3$–dichlorobenzene) onto n^{2+}-Si/SiO_2/OTS substrates to afford films of *ca.* 100 nm in thickness. The films were annealed at 110 °C for four hours before Au source/drain vapour deposition. Current–voltage plots were created to calculate charge mobility, under high vacuum and at ambient pressures. These experiments showed the material to be an n-type semiconductor with electron motilities of 6.0×10^{-2} cm^2 V^{-1} s^{-1} under vacuum, 100 times greater than that of the homopolymer **2**.[21] However, when the same OFET array was assessed under ambient conditions, the NDI–dithiophene-copolymer (**4**) device functioned effectively as an n-type semiconductor for a prolonged period of 14 weeks.

The device architectures containing the NDI–dithiophene copolymer **4** were optimised by Facchetti and co-workers.[25] A top-gate bottom-contact architecture OFET, with glass or PET substrate and Au source-drain contacts, was fabricated in contrast to the previous studies. A typical staggered (top-gate) architecture shows superior injection characteristics as well as facile channel miniaturisation for bottom-contact OFETs, which could lead to high-frequency circuits.[29] The OFET devices were created by depositing the NDI–dithiophene copolymer **4** either by spin-coating, gravure, flexographic or inkjet printing, whilst the dielectric layer was deposited by spin-coating before testing in ambient conditions. Comparable electron motilities of greater than 1.0×10^{-1} cm^2 V^{-1} s^{-1} were demonstrated by all of these devices, which was an increase of at least one order of magnitude when compared to the previous iteration of devices. Current on–off ratios were also measured to be above 10^6 in most instances and 10^8 in the case of the best performing device, demonstrating a substantial increase when compared to other thiophene co-polymers,[21] whilst V_t values were found in the 5–40 V range. The optimum OFET device, in which the NDI–dithiophene copolymer **4** was spin-coated on a glass substrate, demonstrated electron mobility of $\mu_e = 8.5 \times 10^{-1}$ cm^2 V^{-1} s^{-1}, an almost 20-fold increase in performance when compared to the previous iteration of the device, as well as $I_{on/off} = 10^7$ and V_t values of 5–10 V.

To explore the stability of the top-gate bottom-contact OFET on glass, three different devices, each containing different dielectrics (PS, D2200, and PMMA) were monitored for performance over several weeks after fabrication and revealed minimal statistical variation in key performance parameters such as μ, $I_{on/off}$, and V_{on}. Furthermore, the stability of these devices in air was monitored under atmospheres of increased relative humidity (0–98%), which revealed minimal erosion of the device characteristics below 70% relative humidity.

More recently, in 2015, Noh, Yang and co-workers produced[30] a hybrid material containing a NDI–dithiophene copolymer which had been modified with terminal hybrid siloxane hexyl chains (SiC_6), appended *via* the imide, analogous in structure to polymer **4** (Figure 6.4). The polymers were successfully synthesised by condensation of the NDI residue **1** with 6-amino-1-hexene, obtained *via* a Gabriel synthesis, followed by hydrosylation of the terminal alkene in the presence of Karstedt catalyst. This hybrid monomer could then undergo Stille-coupling polymerisation to afford the

(a)

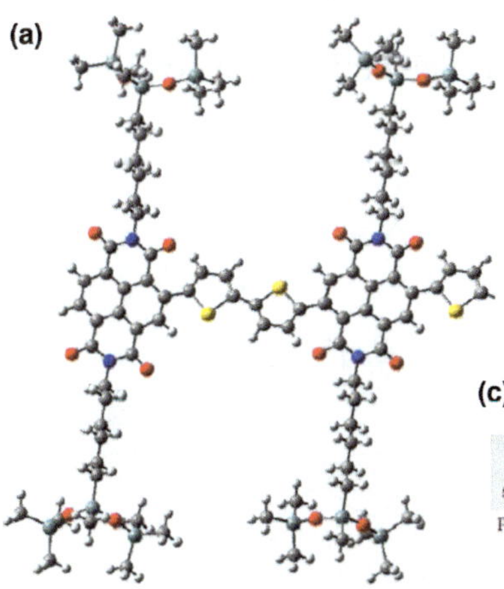

(b)

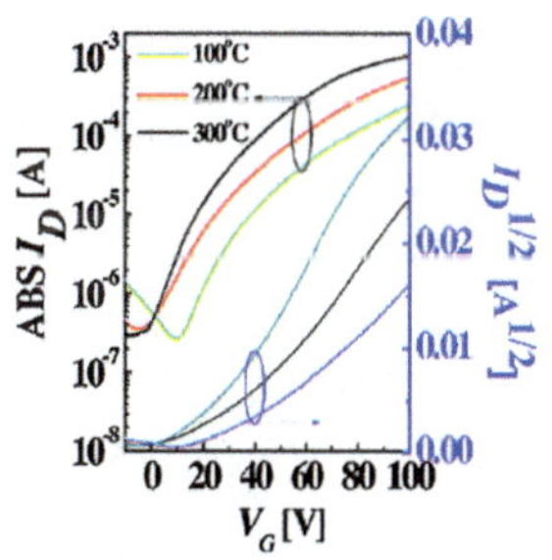

(c)

semiconductor[a]	annealing temp/time [°C/min]	$\mu_{e,avg}$ [cm^2/V·s][b]	V_{Th} [V]	$I_{on/off}$
P(NDI2SiC6-T2) (CF solvent)	100/30	0.43 (±0.14)	25.1 (±4.00)	10^3
	200/30	0.79 (±0.25)	22.4 (±3.50)	10^3
	300/30	1.04 (±0.21)	21.6 (±4.17)	10^3

Figure 6.4 (a) Molecular model of the imide-modified hybrid polymer **4**. (b) Current–voltage (*I–V*) characteristics of the OFETs fabricated and (c) table of the electronic parameters with different annealing temperatures. Adapted with permission from ref. 30. Copyright 2015 American Chemical Society.

desired, highly soluble, polymer ($M_n = 32$ kDa, $Đ = 2.02$) through facile purification steps.

UV-vis analysis of the polymer showed a broad absorption peak in the low-energy region (500–950 nm), corresponding to the strong charge transfer (CT) transition, and another in the high-energy region (330–470 nm) arising from π–π* transitions. Electronic energy levels were measured by cyclic voltammetry (CV), demonstrating two reversible reduction waves. The HOMO energy level (−5.27 eV) was estimated using the optical band gap ($E_{g(opt)} = 1.44$ eV) from the LUMO (−3.83 eV). It was noted that the dihedral angles between NDI and adjacent thiophene residues ($\theta = 36.6$–$37.4°$) were similar to NDIs functionalised with alkyl side-chains. Top-gate, bottom-contact OFETs fabricated *via* the polymer thin films (20–22 nm) were prepared by spin-coating of the NDI-polymer, yielding films of *ca.* 20 nm. The OFET of the hybrid NDI–dithiophene copolymer exhibited typical n-channel OFET characteristics and the electron mobility of the device was progressively improved by increased annealing temperature. After optimised annealing at 300 °C, the maximum mobility was determined to be as high as 1.04 $cm^2 V^{-1} s^{-1}$. The on–off ratios were also calculated for the hybrid polymer ($I_{on/off} = 10^3$) as well as threshold voltage ($V_{Th} = 22$ V). Despite the very similar LUMO energy levels between the hybrid NDI–dithiophene copolymer and reference polymer with simple branched alkane functionality, the improved charge transport properties suggest that other factors, such as molecular packing and crystallinity, can strongly influence the properties.

Luscombe *et al.* have introduced fused-thiophene moieties to the polymer backbone in increasing numbers **6–8** (Figure 6.5).[21] In doing so, it was demonstrated that the electronic structure could not only be tuned, but also the charge mobility within the active layer of the bottom-gate top-contact OFET could be altered with increasing fused-thiophene content from materials deposited by spin-coating from dichlorobenzene. Typical n-type

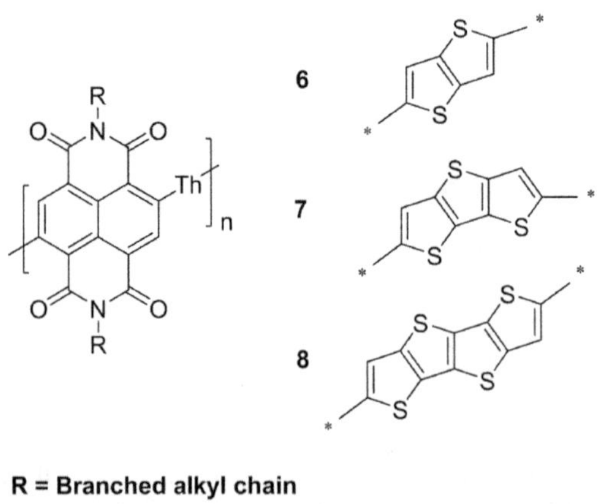

Figure 6.5 NDI-fused-thiophene polymers **6–8** as reported by Luscombe and co-workers.[21]

behaviour was exhibited by all copolymers within the series, as a result of low lying LUMO energy levels (~3.30 eV) with a positive gate-source bias ($V_g = +100$ V) and they did not display an ambipolar or p-type response with a low current flow and on–off ratios below 10 when a negative bias was applied ($V_g = -100$ V). Electron mobilities and on–off ratios as high as 1.2×10^{-2} $cm^2 V^{-1} s^{-1}$ and $I_{on/off}$ values greater than 10^5 were measured from n-channel thin-film transistors using the NDI-fused-tetrathiophene copolymer **8** with a V_t value as low as 1 V, further demonstrating that increasing the π-conjugated length of the fused-thiophene donor moiety yields a measurable improvement in the electron transport properties of the copolymer as larger π conjugated systems experience stronger intermolecular π–π stacking interactions.

A high performance NDI-based conjugated polymer for use as the active layer of n-channel organic field-effect transistors has also been reported by Noh, Kwon, Kim and co-workers.[31] An extended π-conjugated polymer backbone was developed by affixing a thiophene–vinylene–thiophene moiety to the NDI core to yield a NDI–thiophene–vinylene–thiophene copolymer **9** (Figure 6.6) of high molecular weight ($M_n = 70$ kDa, $Đ = 1.98$) which did not increase the LUMO energy level (−4.00 eV) but would allow for increased rigidity within the polymer backbone, preventing adjacent thiophenes from rotating which would reduce persistence length.[32] In addition, the strong donor–acceptor interaction could enhance intermolecular interactions as a consequence of the alternating NDI (electron acceptor) and thiophene–vinylene–thiophene (electron donor) residues.

Density functional theory (DFT) calculations revealed the coefficients of the HOMO orbital were located predominantly on the thiophene–vinylene–thiophene moiety, whilst the central NDI unit possessed mainly LUMO orbital characteristics (Figure 6.7). The calculated energy gap of the NDI–thiophene–vinylene–thiophene copolymer **9** was narrower than that of the NDI–dithiophene copolymer **4** (Figure 6.3(b)), which was consistent with the optical band gap recorded for the system ($E_{g\,(opt)} = 2.02$ eV and 2.30 eV, respectively). These results suggest that the donor unit dictates the electron

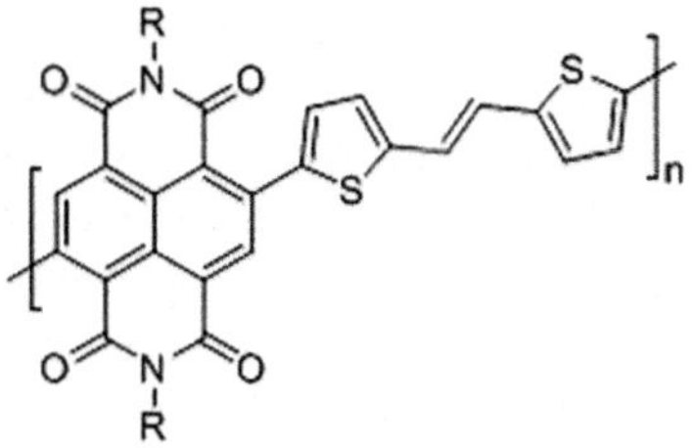

9

R = Branched alkyl chain

Figure 6.6 NDI–thiophene–vinylene–thiophene copolymer **9** reported by Noh, Kwon, Kim and co-workers.[31]

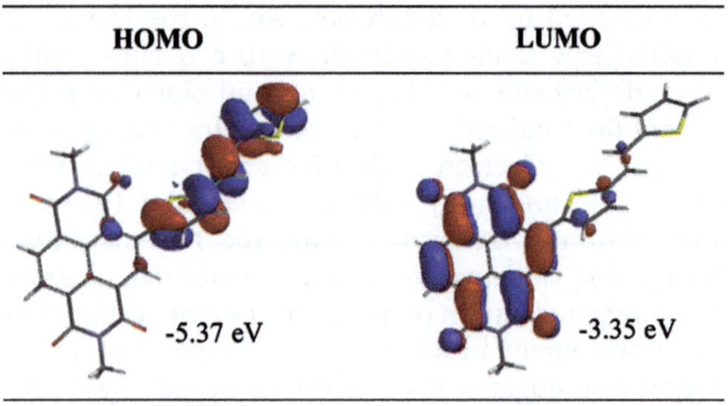

Figure 6.7 Calculated stereostructures and frontier orbitals of copolymer **9** and HOMO/LUMO orbital energies.
Reproduced with permission from ref. 31. Copyright © 2013 WILEY-VCH Verlag GmbH & Co. KGaA, Weinheim.

density of the HOMO, whilst the acceptor (NDI residue) controls the LUMO electron density. CV measurements confirmed both the LUMO orbital energy (−4.00 eV) and HOMO orbital energy (−5.41 eV), which enable the use of NDI–thiophene–vinylene–thiophene copolymer **9** OFETs under ambient conditions.

The NDI–thiophene–vinylene–thiophene copolymer **9** was used to produce OFET devices in the top-gate bottom-contact orientation, which demonstrated n-type semiconductor behaviour. The most effective device exhibited an electron mobility of $\mu_e = 1.8$ cm^2 V^{-1} s^{-1}, and an $I_{on/off}$ value of 10^6 with a Cs_2CO_3-treated Au electrode, after annealing at 150–210 °C, which demonstrated an order of magnitude increase when compared to the NDI–dithiophene copolymer **4** (Figure 6.3(b)). Good air and operational stabilities were demonstrated by minimal drop off in electrical characteristics over a period of five days, as well as after 100 cycling scans at ambient conditions.

6.2.1.3 Core-functionalised NDI–Selenophene Copolymers

n-Type conjugated copolymers containing NDI and various selenophene derivatives have been synthesised by Jenekhe and co-workers[33] using Stille coupling copolymerisation of 4,9-dibromo-naphthalene diimide **1** with 5,5′-bis(trimethylstannyl)-2,2′-biselenophene **10** (Scheme 6.2). The high molecular weight NDI–diselenophene copolymer **11** ($M_n = 40.1$ kDa, $Đ = 2.6$) was reported to have LUMO and HOMO energy levels of 3.9 eV and 5.90 eV, respectively. Although the LUMO energy level is comparable with the analogous dithiophene–copolymer **4**, HOMO energy levels were increased by *ca.* 0.55 eV.

The optical absorption spectra of these polymers revealed two distinct absorption bands, a higher energy band (388 nm) which can be assigned as a π–π* transition band and an intramolecular charge transfer (ICT) band in a lower energy range (675 nm). These values are red-shifted with respect to the

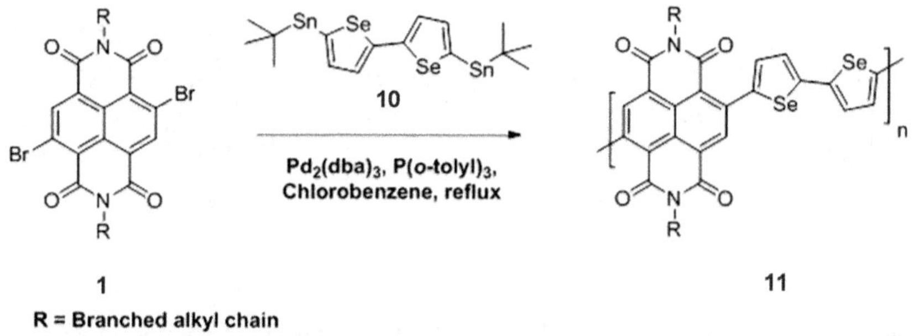

Scheme 6.2 Copolymer synthesised by Jenekhe and co-workers *via* Stille coupling copolymerisation, utilising a selenophene co-monomer.[33]

NDI homopolymer **2** (Scheme 6.1) as a result of the stronger electron-donating nature of the biselenophene moiety.

The field-effect charge transport properties of the NDI–diselenophene copolymer **11** were investigated by fabricating and testing OFETs with bottom gate/top contact architecture. A charge carrier mobility (μ_e) value of 2.4×10^{-1} cm^2 V^{-1} s^{-1} was reported, as well as a $I_{on/off}$ of 10^6, similar to those reported by Facchetti *et al.* for the analogous optimised OFETs of the NDI–bisthiophene copolymer **4.**[25]

6.2.1.4 Core-functionalised NDI–Aryl Copolymers

Oh, Yang and co-workers have produced a family of copolymers with various aryl moieties (Figure 6.8), including a NDI–benzene copolymer **12**, NDI–naphthalene copolymer **13**, and NDI–pyrene copolymer **14**, which can act as alternative electron donating residues to copolymers between NDI and thiophenes (**3–5**) or selenophenes (**11**).[34] All of the polymers feature a number average molecular weight below 31 kDa with *Đ* values less than 3.3. Photophysical measurements revealed characteristic absorption bands of high energy (330–410 nm) as well as a broad low energy ICT band (430–630 nm), which were significantly blue shifted when compared to the thiophene **3–5** and selenophene **11** analogues as a result of the weaker donating abilities of the aryl residues. CV analysis demonstrated that LUMO energy levels were *ca.* 3.8 eV for all of the polymers produced whilst HOMO energy levels showed a strong dependency on the nature of the co-monomer used. HOMO energies decreased with increasing number of fused aromatic systems (6.1 eV (**12**)–5.73 eV (**14**)), indicating that destabilisation was a consequence of the electron donating ability of the aryl residue. The relative high-lying HOMO levels (5.73 eV) of the NDI–pyrene copolymer **14** suggested that the material could hold favourable properties of charge transport for holes, possibly being useful as semiconductors for ambipolar OFETs.

The electrical properties of NDI–aryl copolymers **12–14** (Figure 6.8) were investigated by fabricating bottom-gate top-contact configuration

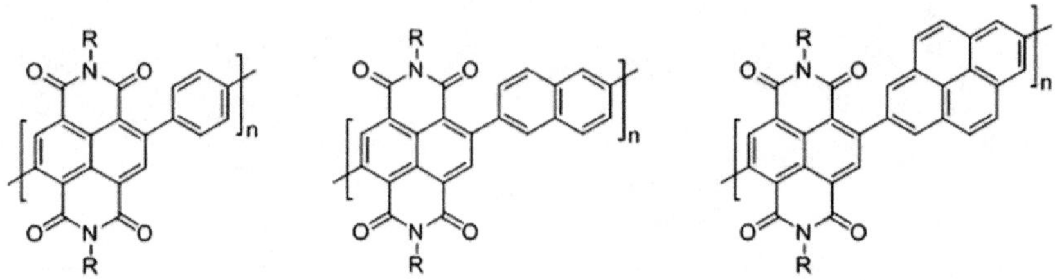

12 13 14

Figure 6.8 A series of NDI–copolymers **12–14** with aromatic co-monomers as synthesised by Oh, Yang and co-workers.[34]

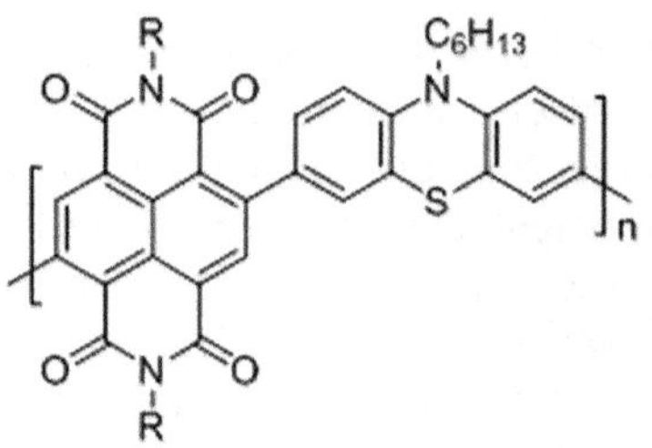

15

R = Branched alkyl chain

Figure 6.9 NDI and phenothiazine co-polymer synthesised by Liu, Zhan and co-workers.[35]

OFET devices, which all exhibited unipolar n-channel field-effect behaviours with electron mobilities and current on–off ratios ranging from $\mu_e = 8.0 \times 10^{-3}$ cm^2 V^{-1} s^{-1}, and $I_{on}/I_{off} = 10^5$ (**12**), $\mu_e = 5.6 \times 10^{-2}$ cm^2 V^{-1} s^{-1}, and $I_{on}/I_{off} = 10^5$ (**13**), and $\mu_e = 7.6 \times 10^{-3}$ cm^2 V^{-1} s^{-1}, and $I_{on}/I_{off} = 10^6$ (**14**), whilst all of the thiophene and selenophene copolymers analysed possessed μ_e values of magnitude $\sim$1.0–3.0 $\times 10^{-2}$ cm^2 V^{-1} s^{-1}. The NDI–pyrene copolymer **14** was also tested as a p-type field-effect transistor, and exhibited field-effect mobility of *holes* (μ_h) in excess of 5×10^{-5} cm^2 V^{-1} s^{-1} and $I_{on}/I_{off} = 10^1$, indicating that the materials had ambipolar transport characteristics.

Copolymers between NDI and phenothiazine **15** (Figure 6.9) have also been synthesised by Liu, Zhan and co-workers *via* a Pd-catalysed Suzuki coupling polymerisation approach,[35] yielding both relatively high ($M_n = 12$ kDa) and low ($M_n = 7$ kDa) molecular weight polymers. Both polymers exhibited characteristic absorptions centred at 328 and 628 nm, respectively, corresponding to high (π–π* transition) and low energy (ITC) absorptions. HOMO (−5.8 eV) and LUMO (−3.7 eV) levels of the polymers were measured by CV and bottom-gate top-contact geometry OFETs were fabricated. The NDI–phenothiazine copolymers **15** exhibit electron mobilities as high as 5×10^{-2} cm^2 V^{-1} s^{-1} and current on–off ratios as high as 10^5

in nitrogen atmospheres, which are comparable to other rylene diimide-based polymers (such as perylenediimides and pyromellitic diimide) under the same test conditions.

6.2.1.5 *Ladder-NDI–Acene Copolymers for Use in OFETs*

An interesting structural variation on the NDI core-functionalised copolymers thus far reported has featured a ladder-based NDI structure.[36–38] In 2003, Jenekhe *et al.* reported polymers where functionality was introduced though the imide moiety by means of an imine to produce novel polymer architectures, termed ladder polymer FETs (**16** and **17**, Figure 6.10).[36]

The double-stranded (ladder) polymer benzobisimidazobenzophenanthroline (BBL) **16** and its non-ladder derivative BBB **17** were synthesised by polymerisation in polyphosphoric acid at 185 °C.[39,40] The high nitrogen and oxygen heteroatom content in both polymers gave rise to excellent electron-accepting properties, which were demonstrated by OFETs in bottom-contact architectures. Both polymers exhibited an n-type channel characteristic with field-effect mobility of electrons as high as 1×10^{-1} $cm^2 V^{-1} s^{-1}$ in samples of BBL **16** which was vapour deposited on top of SiO_2. BBB **17** devices demonstrated much more modest electron mobilities in the order of 10^{-6} $cm^2 V^{-1} s^{-1}$, whilst blends of BBL/BBB **16** + **17** also demonstrated relatively poor mobilities. Current on–off ratios were also observed to be in excess of 10^5 (**16**) and 10^3 (**17**). The finding that the electronic properties of **16** with a ladder architecture were enhanced by many orders of magnitude when compared to its non-ladder analogue **17** was instructive in tailoring intermolecular π-interactions for maximising electron transport.

6.2.2 NDI Polymers for Photovoltaic Devices

Solution-processed organic photovoltaic cells (OPVC) offer a promising low cost solution for solar energy technologies.[41,42] Although the structures of OPVC have been investigated intensively to maximise solar conversion efficiency, the design rationale OPVCs can be understood by considering the structure of a simple bi-layer type solar cell (Figure 6.11).

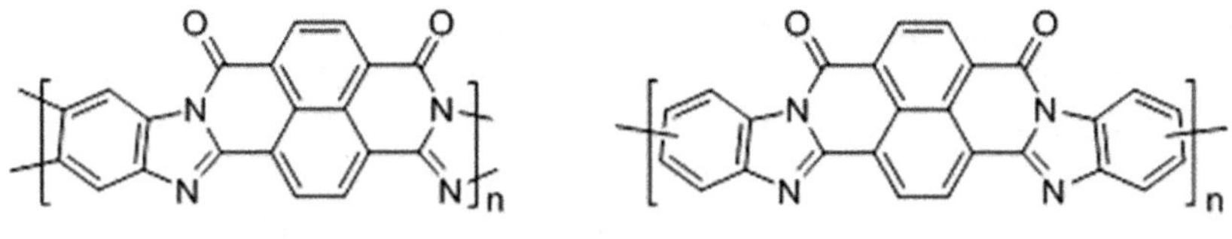

16 **17**

Figure 6.10 The molecular structures of ladder BBL **16** and non-ladder BBB **17** polymer semiconductors as reported by Jenekhe and co-workers.[36–38]

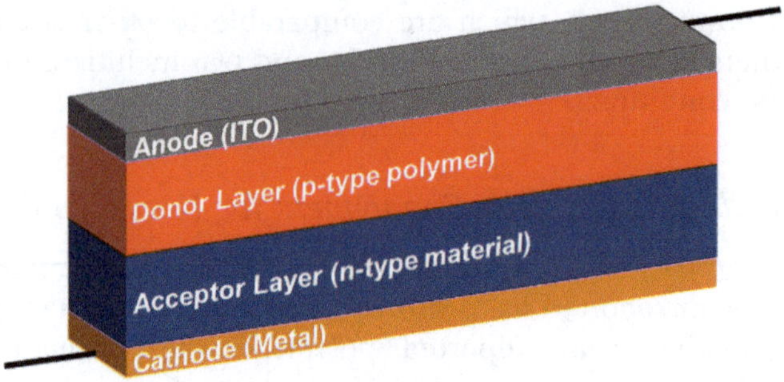

Figure 6.11 Schematic of a simple bi-layer OPVC.

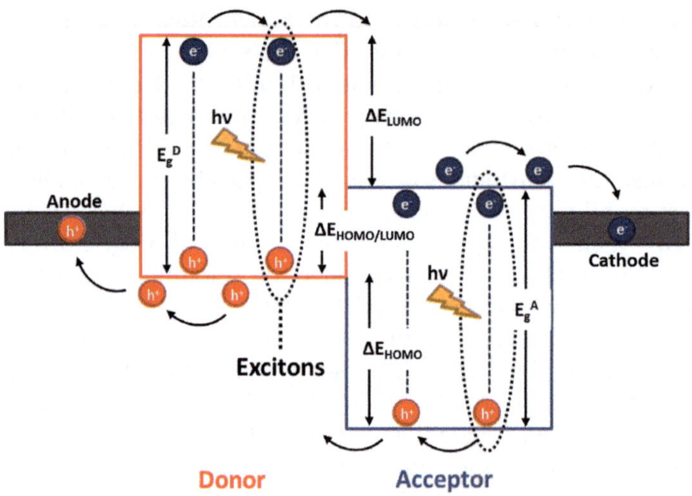

Figure 6.12 Schematic representation of a typical donor–acceptor interface in a bulk heterojunction (BHJ) solar cell with associated HOMO/LUMO energy level offsets. Formation of excitons and their transportation through electrons and holes to the corresponding electrode is also shown.

In these OPVCs, photons pass through the optically transparent indium tin oxide (ITO) upper layer and excite an electron from the HOMO to the LUMO of the donor, p-type layer. This excitation creates an electron–hole pair termed an exciton. In order to stop the electron relaxing and liberating a photon, which does not generate usable energy, the electron and hole must be spatially separated. This can be achieved by providing an acceptor layer that is composed of a material with a suitably positioned HOMO level that can accept the excited electron from the donor, and transport the electron to the cathode (Figure 6.12).

Although the chemistry of p-type polymers is highly developed with hundreds of structural variations in the literature, progress towards producing n-type acceptor materials has been much slower. Indeed the most widely studied OPVCs contain small molecule fullerene-containing residues as the acceptor residues, with progress over the past decade resulting[43,44] in power conversion efficiencies (PCE) approaching 10% in lab scale tests. The drawback of fullerene/polymer OPVCs is the challenge of processing the constituents with control over the final morphology of the cell proving hard to predict. Therefore, there is significant demand for a polymeric n-type polymer as the optimum all-polymer solar cell has remained relatively ineffective with efficiencies less than 5%.[45]

6.2.2.1 NDI–Thiophene Copolymers for OPV Applications

A variety of NDI–thiophene copolymer semiconductors have recently been reported by Jenekhe and co-workers (**3**) (Figure 6.3(a)),[46] Loi and co-workers (**4**) (Figure 6.3(b)),[47] amongst others,[3,48,49] and as n-type electron acceptors in solar cells that exhibit unipolar electron transport with high field-effect and bulk mobilities. These materials were blended with a known donor polymer such as the thiazolothiazole copolymer PSEHTT (Figure 6.13(a))[50,51] or P3HT (Figure 6.13(b))[52] to produce working cells for testing.

A NDI–thiophene copolymer **3** (Figure 6.3(a)) was synthesised using Stille coupling methodology by Jenekhe and co-workers.[46] To evaluate its use as an acceptor in bulk heterojunction (BHJ) organic solar cells, films were cast from a 1 : 1 (w/w) blend of the co-polymer acceptor **3** and the PSEHTT donor **18** (Figure 6.13(a)). It was shown that the film displayed a PCE of 1.3%. Power conversion efficiency is a measure of power out of the solar cell as a percentage of the total power in. Although the value is relatively poor, it is comparable to many other solar cells based on all-polymer systems such as polybenzothiadiazoles or polyphenylenevinylenes which are blended with a donor, demonstrating PCE values less than 2%, and often below 1%.[45] The short-circuit current density (J_{sc}) was measured (3.80 mA cm^{-2}) in the blend **3** + **18** (Figure 6.14(a)) and shown to be comparable to other all-polymer solar cells that do not use rylene diimides as acceptors where typical values of J_{sc} are less than 5 mA cm^{-2}.[45] An EQE of 28% (Figure 6.14(b)) was measured for

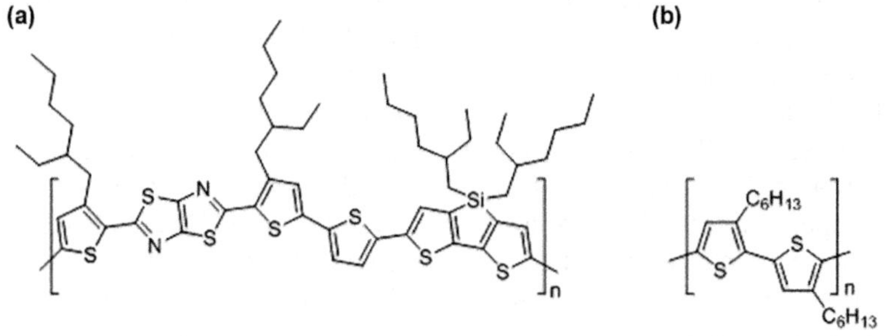

18 **19**

Figure 6.13 p-Type donors PSEHTT (**18**) and P3HT (**19**) used in photovoltaic devices.[46,47]

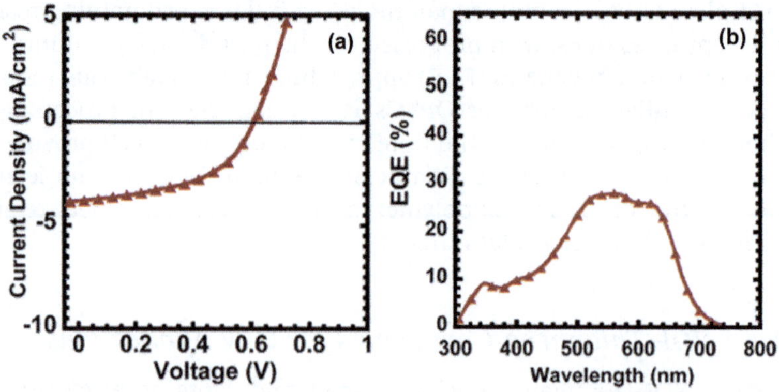

Figure 6.14 J_{sc} and EQE for the polymer blend of **3** + **18** reported by Jenekhe and co-workers.
Adapted with permission from ref. 46. Copyright 2013 American Chemical Society.

the blend **3** + **18**, which showed an improvement when compared to other all-polymer solar cells at the time.[45]

In 2011, Loi *et al.* studied a NDI–dithiophene copolymer **4** with an additional thiophene residue when compared to **3** (Figure 6.12(b)), blended with the donor P3HT **19** (Figure 6.14(b)) in a 2 : 1 ratio, respectively, for use in solar cells.[47] A high fill factor was demonstrated for the first time in the all-polymer solar cells cast from chlorobenzene or *o*-dichlorobenzene, with values in excess of 0.65 achieved, which compared well with typical values of less than 0.30 in all-polymer solar cells.[45] The increased fill factor suggested high charge separation efficiency and balanced carrier mobility within the blend. The high electron-transporting blend between 4 and 19 is comparable with those reported for fullerene-based devices.[42] However, the operating properties of the solar cell were modest when compared to the favourable charge transport conditions. Short-circuit charge density measurements of J_{sc} = 0.49 resulted in low power conversion efficiency (PCE = 0.16%). One possible explanation for the diminished properties was the photoactive blend morphology, which showed fibre-like structures in topography images by AFM with feature sizes in the order of 10–35 nm, demonstrating that optimising the chemical structure of the polymeric components in isolation is unlikely to solve the problem of commercialising OPVCs.

Further to this, McNeill and co-workers[53] reported in 2015 an analogous low band-gap, all-polymer photovoltaic blend where efficiencies of over 4% and an open circuit voltage of greater than 0.9 V are achieved. It was noted that whilst *high-efficiency* all-polymer systems exhibit similar fill-factors, the limitations observed in these systems are likely to be general across the field and further work will be required to realise long-lived charge-separated states in all-polymer blends akin to polymer/fullerene blends.

Polymers with a fused-thiophene moiety attached to an NDI core through the 2-position are also known.[54] Luscombe *et al.* have produced[54] a series of

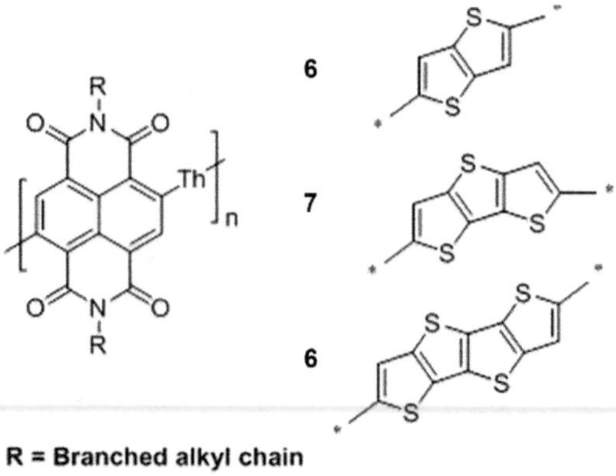

Figure 6.15 NDI-fused-thiophene copolymers as produced by Luscombe *et al.*[54]

high molecular weight polymers with between two and four fused-thiophene residues **6–8** (Figure 6.15), which demonstrate potential for use in solar cells. Increasing the number of fused-thiophene moieties within an NDI–copolymer backbone is shown to not only enable tuning of the electronic structure but also improve charge carrier mobilities within the active layer of organic field-effect transistors (7.0×10^{-3} $cm^2 V^{-1} s^{-1}$ to 1.2×10^{-3} $cm^2 V^{-1} s^{-1}$ with increasing number of fused thiophenes). Devices comprising blends of **19** with NDI-fused-tetrathiophene **8** yield the largest J_{sc} (0.57 mA cm^{-2}) and fill factor (0.55) in addition to the highest measured PCE (0.13%), although these values are modest when compared not only to other all polymer solar cells, but also alternative thiophene-functionalised NDI polymers.[45]

Whilst 1,4,5,8-NDI derivatives have been used widely in the production of all-polymer rylenyl diimide solar cells, the isomeric 1,2,5,6-NDI (iso-NDI) derivatives are less commonly studied. Zheng *et al.* have produced[55] a series of n-type polymers using an angular-shaped NDI derivative as the building block with thiophene chain lengths from 1 to 3 (**20–22**, Figure 6.16(a)) in an attempt to tune the energy levels and produce all-polymer solar cells when blended with P3HT **19** (Figure 6.14(b)). All polymers within the series exhibited good solubility in common organic solvents such as chloroform, THF and chlorobenzene, as a result of the octyldodecyl substituents located on the imide unit. In addition, high-lying LUMO levels led to large enough energy gaps between the HOMO and LUMO of the polymer iso-NDI-thiophenes (**20–22**) and P3HT (**19**) to guarantee efficient exciton dissociation.[56]

All polymer solar cells were fabricated by blending differing ratios of iso-NDI–thiophene copolymers (**20–22**) containing polymers (Figure 6.16(a)) with the p-type donor P3HT (**19**) (Figure 6.14(b)). The highest PCE of 0.32% was obtained by using **22** as the acceptor in a 1 : 1 w/w blend with **19** although this is significantly less than that of the 1,4,5,8-NDI analogue **5** (Figure 6.3(c)). iso-NDI-fused-dithiophene copolymers **23** have also been synthesised[57] (Figure 6.16(b)) in preparation for use in photovoltaic

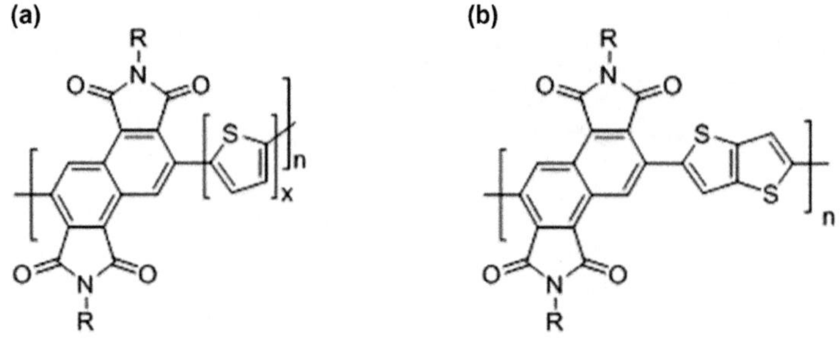

Figure 6.16 Copolymers of (a) iso-NDI with increasing thiophene chain lengths (**20–22**) as studied by Zheng *et al.*[55] and (b) co-polymer of iso-NDI with a fused dithiophene **23** reported by Gao and Di *et al.*[57]

applications by Gao and Di *et al.*, although further analysis is required to evaluate their use in flexible and large-scale optoelectronic applications.

These results indicate that the angular-shaped NDI is a promising building block for constructing all-polymer acceptors for solar cells although the efficiency of the material requires improvement before it can match those seen in fullerene/polymer solar cells.[42]

6.2.2.2 NDI–Selenophene Copolymers

Selenophene residues offer an alternative to copolymers produced with thiophenes. Further to their work on the NDI–thiophene copolymer **3** (Figure 6.3(a)), Earmme *et al.*,[46] amongst others,[58,59] have also produced a new semicrystalline NDI–selenophene copolymer acceptor (**24**) with hexyldecyl imide substituents (Figure 6.17), which was blended with a thiazolothiazole copolymer (PSEHTT) **18** (Figure 6.14(a)) donor (1 : 1 w/w) and evaluated in organic solar cells.

The blend exhibited a 3.3% PCE, an increase when compared to thiophene analogue **3** (Figure 6.3(a)). This efficiency is comparable to the performance of fullerene/PSEHTT devices.[60]

Side chain engineering was also investigated[61] by Jenekhe and co-workers by facile tuning of the composition and distribution of alkyl side chains on NDI–selenophene n-type conjugated polymers as a means of improving the photocurrent and performance. NDI–selenophene polymers with ratios of 10, 30 and 50 mol% of 2-butyloctyl side chains and a complementary portion of 2-decyltetradecyl-functionalised NDI monomers were synthesised by Stille coupling. The electronic structures (LUMO/HOMO levels) and optical absorption of the side-chain-modified NDI–selenophenes and PNDIS were

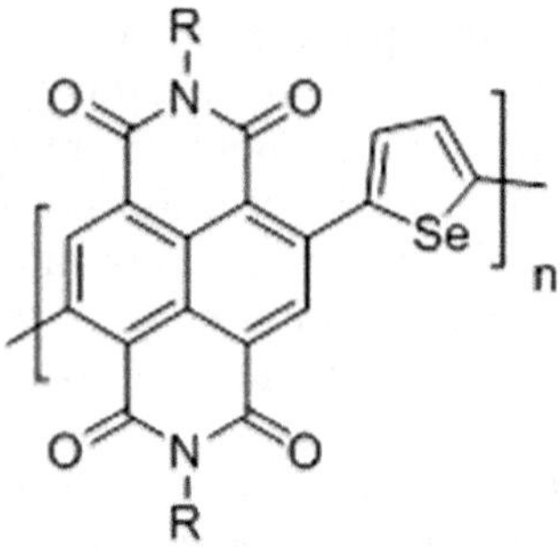

24

R = Branched alkyl chain

Figure 6.17 n-Type NDI–selenophene copolymer **24** synthesised by Jenekhe and co-workers.[46]

estimated from CV and UV-visible measurements of thin films. The LUMO energy level derived from the onset reduction potential in CVs was 3.9 eV for all polymer blends. The optical absorption spectra revealed two distinctive absorption bands at 300–400 nm and 500–750 nm, which are a result of π–π* transition and intramolecular charge transfer, respectively. Bulk heterojunction all-polymer solar cells were fabricated by blending the side-chain-functionalised NDI–selenophene polymer with a thiazolothiazole copolymer (PSEHTT) **18** (Figure 6.14(a)) in an inverted device structure. A device with 4.4% PCE, V_{oc} of 0.77 V and a photocurrent measured to be 10.4 mA cm^{-2} was achieved by using the optimum acceptor polymer composed of 2-butyloctyl-2-decyltetradecyl 30 : 70 mol%. Bulk electron and hole mobilities were also shown to increase compared to previous iterations to 6.5×10^{-5} and 2.1×10^{-4} cm^{2}V^{-1}s^{-1}, respectively, in the blended system. These increases are attributed to the increased crystallinity and electron mobility of the polymer in conjunction with the optimum polymer/polymer blend morphology.

Subsequently, Jenekhe and co-workers reported[62] a high-performance all-polymer solar cell with a record 7.7% power conversion efficiency, record short-circuit current density (18.8 mA cm^{-2}) and 85% peak external quantum efficiency by further exploring the use of NDI–selenophene copolymer **24** (Figure 6.17). Blending with a modified benzodithiophene–thieno[3,4-*b*]thiophene copolymer donor led to a bulk electron mobility of 7.25×10^{-3} cm^2 V^{-1} s^{-1} in the room-temperature-annealed donor polymer blend. A two-fold enhancement in efficiency compared to the thermally (175 °C) annealed devices was also noted. This result could be explained by the slower solvent evaporation rate at room temperature, enabled improvement in electron mobility and the favourable bulk morphology of the polymer blend afforded.

Furthermore, Jenekhe and co-workers[59] have reported a series of new semiconducting naphthalene diimide–selenophene/perylene diimide (PDI)–selenophene random copolymers. By controlling the bulk crystallinity of the acceptor polymer, the optical properties were tuned in the random co-polymer. The bulk hole (μ_h) and electron (μ_e) mobilities and carrier asymmetry (μ_h/μ_e) in the polymer/polymer blends were shown to vary dramatically with crystallinity of the acceptor polymer component, reaching their optimal values at the optimum crystallinity found with 30% incorporation of the PDI–selenophene monomer. The NDI/PDI acceptor was blended with a PBDTTT-CT donor and the resulting material was investigated as an electron acceptor in BHJ solar cells. Analysis demonstrated enhanced performance (PCE = 6.3%, I_{sc} = 18.6 mA cm^{-2}, EQE = 91%) owing to the pairing of crystalline domain size (L_c) between the new NDI/PDI copolymer (L_c = 5.11 nm) with the PBDTTT-CT donor to yield compatible blends for all-polymer solar cells. It was suggested that the bulk crystallinity, described with respect to the average crystalline domain size, L_c, is a material property that can be used as an important principle for selecting donor–acceptor pairs in polymer/polymer blend solar cells.

6.2.2.3 Conjugated Donor–Acceptor Block Copolymers for OPVC

The first example of a fully conjugated donor–acceptor block copolymer incorporating NDI–dithiophene acceptor segments and P3HT donor segments in a single copolymer for use in the fabrication of all-polymer solar cells using fully conjugated donor–acceptor block copolymers as acceptor materials was presented by Nakabayashi *et al.* (Figure 6.18).[63] The conjugated donor–acceptor copolymer **25** was synthesised successfully in good yield (85%) by a combination of quasi-living Grignard metathesis polymerisation and Yamamoto coupling reactions. The optical band gap of 1.38 eV was much smaller than those of previously reported donor–acceptor block copolymers (1.6–1.8 eV),[45] with the HOMO and LUMO energy levels in the ranges of 5.57–5.60 eV and 4.22–4.27 eV, respectively. The all-polymer solar

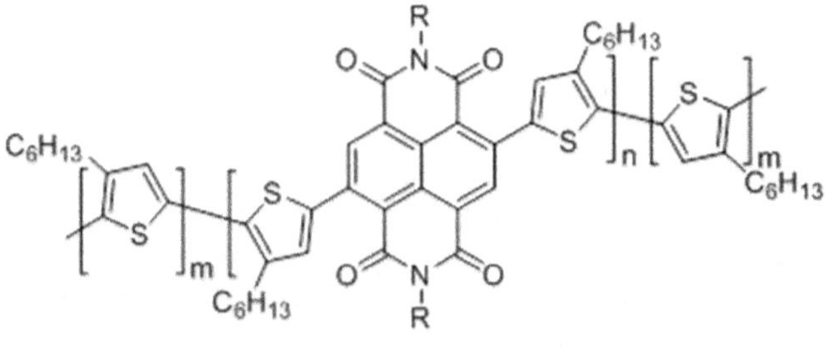

25

R = Branched alkyl chain

Figure 6.18 A fully conjugated donor–acceptor block copolymer (**25**), which consists of regioregular P3HT-based donor and NDI–dithiophene-based acceptor segments, as reported by Nakabayashi and co-workers.[63]

cells were fabricated using films cast from a 1:1 (w/w) blend of donor material P3HT (**19**) and the conjugated donor–acceptor copolymer (**25**) in chlorobenzene, which was subsequently annealed at 200 °C. The best performance was recorded in the conjugated donor–acceptor copolymer **25** that was synthesised from P3HT units of $M_n = 6.3$ kDa and an overall M_n of 26 kDa, which gave a PCE of 1.28%. These results demonstrate that fully conjugated donor–acceptor block-copolymers are potentially applicable as non-fullerene acceptor materials, although further optimisation will be required as the efficiency of the photovoltaic materials is lower than that of other NDI-derived materials previously reported.

6.2.3 Conclusions for NDI-based Polymers for Organic Electronic Devices

The incorporation of NDI residues within the main chains of polymers through functionalisation across the naphthalene ring has resulted in polymers that show excellent potential for producing OFETs. To produce components that can compete with silicon semiconductors requires further interplay between synthetic chemists and materials scientists and engineers to optimise all aspects of device design and fabrication. Furthermore, the use of NDI-containing polymers has progressed rapidly over the previous 10 years. The electron deficient nature of NDI combined with the highly developed chemistry surrounding its derivatisation has resulted in the production of a range of conjugated polymeric materials that have shown valuable electron accepting properties. Their use as n-type materials in OPVCs is particularly important as there are few other structural motifs that encompass the stability and electronic properties of NDI-containing polymers.

6.3 Chemosensors Containing NDI Residues

6.3.1 NDI Polymers for pH Sensors

In the past 20 years, the development and application of a wide variety of chemical sensors and biosensors have grown rapidly.[64] The detection and control of pH and ions has an important role in a wide variety of fields such as chemistry, biochemistry, cellular biology and drug delivery with two solutions most prominent. One solution uses materials that swell as a function of pH value while the alternative solution is based on the use of optical probes (indicators).

NDIs are well-known as fluorophores for use in optic-electronic functional materials owing to their long absorption band, light-fastness and high fluorescence quantum yield. Moreover, the facile tuning of their photo- and electrochemical properties by chemical modification provides scope for a means of visualising a wide variety of sensor systems.[65] One such core modification is the addition of alkyl amines, which allows the emission wavelength to be tuned from 387 nm to *ca.* 630 nm.[9,66]

To this end, Zhu *et al.* have successfully incorporated a core-substituted NDI chemosensor into a polymeric backbone by a covalent linker to yield a long-wavelength fluorescent hydrophilic copolymer.[67] The sensor can be fixed onto microbioreactors and is free from concentration quenching of the chromophore.

To produce the sensor, piperidine groups were introduced to the NDI core in a stepwise manner before a methacrylate group was added to facilitate polymerisation (Scheme 6.3). Initially, *N*-methylpiperidine **27** was linked to the 2-position **28** through nucleophilic substitution of the bromo group of the NDI core, before the addition of piperidine **29** at the bromo-substituted 6-position **30** and subsequent introduction of methylacryloyl chloride **31** to yield a monomer **32** that could be easily co-polymerised with other monomers. Hydroxyethylmethacrylate (HEMA) **33** was chosen as the co-monomer as the main polymeric matrix of poly(hydroxyethylmethacrylate) P(HEMA) is hydrophilic and able to form uniform films. This polymer design avoids the fluorescence self-quenching induced by aggregation by means of controlling the monomer feed ratio. The NDI–HEMA copolymer chemosensor **34** was synthesised by radical polymerisation in a 1 : 1100 ratio of NDI–HEMA using 2,2′-azodiisobutyronitrile (AIBN) as an initiator (Scheme 6.3).

Copolymer films were produced by spin-coating onto a quartz substrate from a 5 wt% solution of the NDI–HEMA copolymer chemosensor **34** in methanol and the optical responses (both absorption and emission) were measured in deionised water with aqueous NaOH and HCl used to adjust the pH values of the solution. Absorption peaks for the film were red-shifted slightly from 561 to 569 nm as pH values were decreased from 8.0 to 4.6 and a clear isosbestic point at 572 nm was identified (Figure 6.19(a)). Analogous results were also observed then using phosphate buffers as the medium for adjustment of the pH. A three-fold increase in emission intensity was

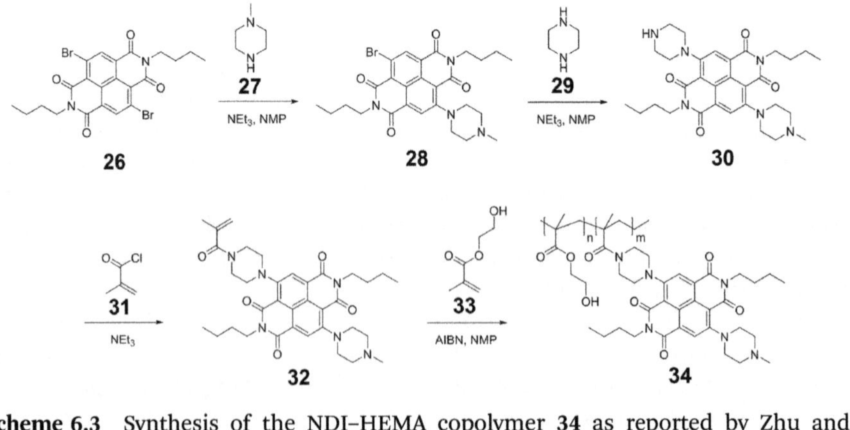

Scheme 6.3 Synthesis of the NDI–HEMA copolymer **34** as reported by Zhu and co-workers.[67]

observed between pH 8.0 and 4.6 (Figure 6.19(b)), demonstrating excellent sensitivity for this sensor over a wide pH range. Furthermore, the sensor exhibited a linear response to pH with respect to absorption intensity between pH 8.0 and 4.6, which satisfies the physiological pH range for most biosystems.

An alternative method of sensing pH has been reported by functionalisation of silica-gel partials with NDI chromophores.[68] Mallouk *et al.* have shown that α-zirconium phosphonate sheets can work as inorganic polymer spacers for polyelectrolyte chromophore multilayer films in flat supports and in silica suspensions (**36**, Scheme 6.4). In addition, Baptista *et al.*[69] have

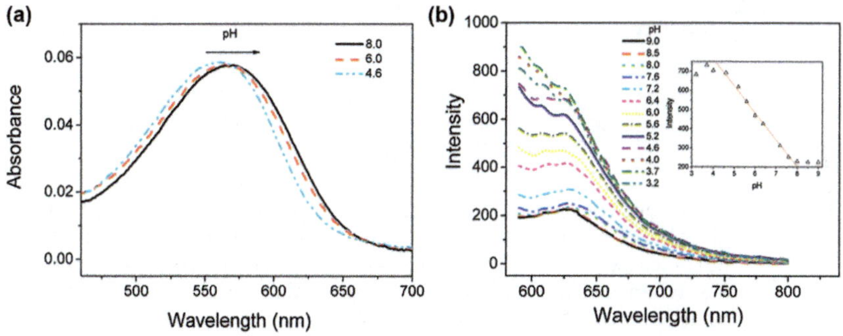

Figure 6.19 (a) UV-vis spectra of the NDI–HEMA copolymer chemosensor film **34** showing slight red-shift in absorption maxima at decreasing pH. The isosbestic point was observed at 572 nm. (b) Fluorescence spectral of the NDI–HEMA copolymer film **34** excited at the isosbestic point (572 nm) in an aqueous system showing emission intensity over a wide pH range. Inset: fitting curve of pH-dependent response to fluorescence intensity of the NDI–HEMA copolymer film **34** at 628 nm with linear response and relative coefficient of 0.998 in the pH range of 4.6–8.0. Reprinted with permission from ref. 67. Copyright 2011 American Chemical Society.

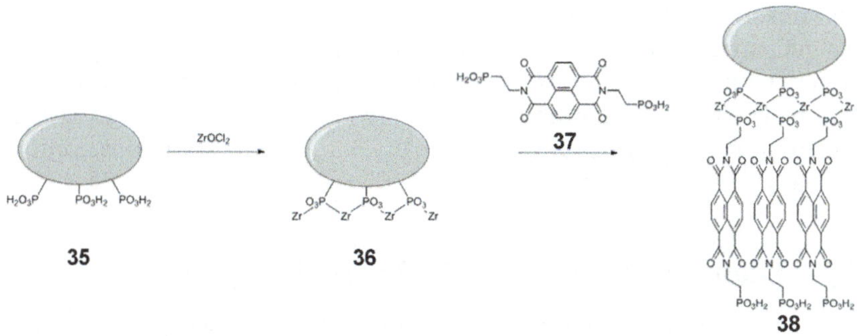

Scheme 6.4 Representation of the silica-gel particles modified with imide-functionalised NDI photoactive molecules *via* phosphonate zirconium chemistry.[68] Note: the scheme is not to scale.

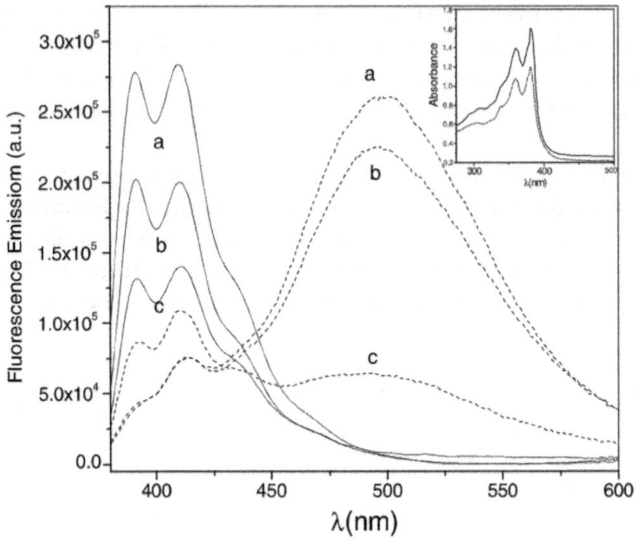

Figure 6.20 Emission spectra of unattached NDI chromophore **37** in ethylene glycol–buffer solution (—) 9 : 1 v/v, and of NDI chromophore–silica-gel particles **38** (—) at different pH values: (a) pH 1.0; (b) pH 6.8, and (c) pH 9.3 when excited at 350 nm. Inset: absorption spectra (pH 1) of unattached NDI chromophore **37** in 9 : 1 v/v ethylene glycol–buffer solution (—) and bound to the silica-gel particles **38** (—).
Reproduced from M. A. Rodrigues *et al.* Photophysical and photochemical properties of porphyrin and naphthalene-diimide-modified silica-gel particles, *J. Non-Cryst. Solids*, **69**, Copyright 2002 with permission from Elsevier.

shown that by using silica-gel particles modified by phosphonate-zirconium chemistry, photoactive *N,N′*-bis(2-phosphonoethyl)-1,4,5,8-naphthalene diimide chromophores **37** can be affixed to produce a novel pH sensor (**38**, Scheme 6.4).

The photophysical and photochemical properties of the silica-gel-bound NDI chromophore **38** was shown to differ from their unbound **37** solution properties. In the fluorescence spectrum of the NDI chromophore bound to silica-gel particles **38**, a band centred at 500 nm was evident, arising from the excimer, which was not observed with the unbound chromophore **37** in solution. The ratio of monomer (415 nm) to excimer emission was shown to correlate to solution pH, ranging from 1.0–9.3, providing the possibility for use as a pH sensor (Figure 6.20).

6.3.2 Fluoride Sensors from NDI-containing Polymers

Anions are able to interact with NDIs though anion–π-acceptor,[70,71] charge transfer,[72,73] and electron transfer[74,75] interactions as a result of the electron deficient nature of NDIs, mimicking π-donor–π-acceptor interactions.[76–78] Lewis basic ions such as F^- and AcO^- are able to transfer electrons to π-acidic NDIs,[74,75] whilst electron transfer events are not triggered by basic anions. Electrospray ionisation mass spectrometry[14] (ESIMS) and theoretical models[79] have revealed several NDI–anion complexes, however, the very first crystallographic evidence of DNI–anion complexes was reported by Saha *et al.*[80] and the data revealed an association between the electron deficient fused aromatic system of the NDIs with the lone-pair electrons of an anion or neutral molecules through anion–π-acceptor or lone-pair–π-acceptor interactions.

The treatment of *N*,*N*′-di(4-pyridyl)NDI **39** and Pd(dppe)$(TfO)_2$ **40** in a 1 : 1 ratio at room temperature yielded a Pd(II) coordination polymer **41** (Scheme 6.5) with a 'zigzag' architecture in which Pd(II) metal centres are bridged by linear DPNDI through the pyridine rings.[81]

The crystal structure of the zig-zag polymer **41** was attained from crystals obtained from THF vapour diffusion into MeCN revealing a distorted square planar geometry (N–Pd(II)–*N* = 88.2°), which was attributed to the bulky bidentate dppe ligand (Figure 6.21). The *zigzag* coordination polymer **41** displayed evidence of non-covalent interactions between the π-acidic NDI and the oxygen lone-pair electrons of the TfO^- counterion and THF.

Fortuitously, in a manner similar to the *N*,*N*′-di(4-pyridyl)NDI monomer/ligand **39**,[16] the colourless coordination polymer (Pd(II)/*N*,*N*′-di(4-pyridyl)NDI)$_n$ **41** can optically detect ions such as the toxic[82] F^- in aqueous DMSO solutions (15 : 85, H_2O–DMSO) through a colour change from colourless to orange and finally pink with increasing concentration of F^-. UV-vis spectroscopy was utilised to demonstrate this capability, with absorptions at $\lambda_{max} = 475$ nm (orange) and $\lambda_{max} = 542$ nm (pink), respectively, and an isosbestic point at 394 nm (Figure 6.22).

The electrochemically generated *N*,*N*′-di(4-pyridyl)NDI^- radical anion and *N*,*N*′-di(4-pyridyl)NDI^{2-} dianion confirmed that the F^- induced two-step reduction of *N*,*N*′-di(4-pyridyl)NDI is responsible for its colorimetric changes over two distinct intervals. CV analysis revealed two fully reversible one-electron reduction peaks at −391 and −778 mV corresponding to the formation of the *N*,*N*′-di(4-pyridyl)NDI^- radical anion and

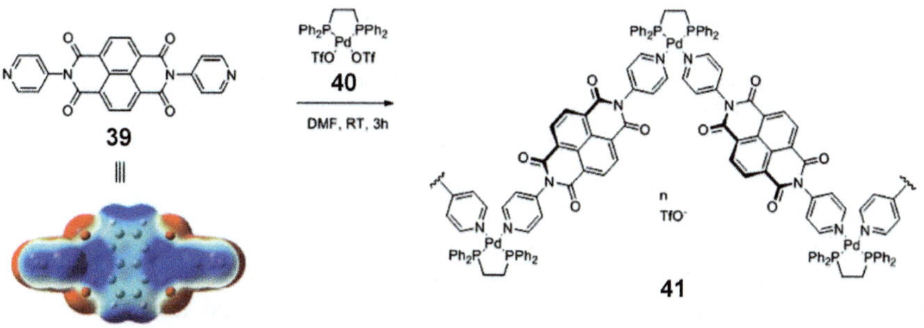

Scheme 6.5 Synthesis of a Pd(II) and DPNDI-based zig-zag coordination polymer. The ESP map of the DPNDI ligand shows electron rich (red) and deficient (blue) domains. Reproduced from ref. 80 with permission from the Royal Society of Chemistry.

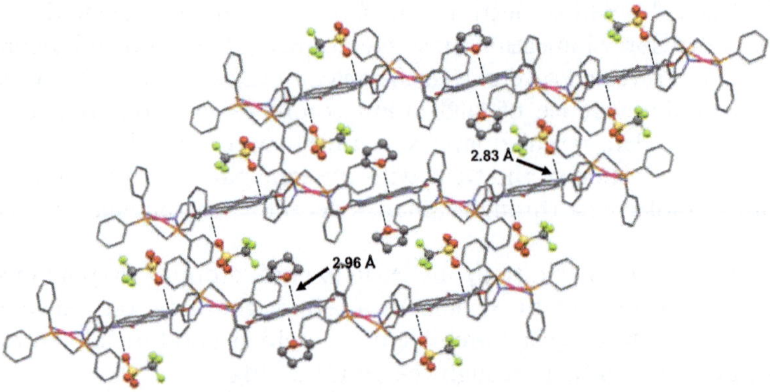

Figure 6.21 Crystal structure of $[Pd(dppe)DPNDI]_n(TfO)_{2n}$ coordination polymer (pink: Pd(II), blue: N, red: O, orange: P, yellow: S, green: F, grey: C, and white: H) showing the packing of the polymer backbones, TfO^- anions, and THF molecules.
Reproduced from ref. 80 with permission from the Royal Society of Chemistry.

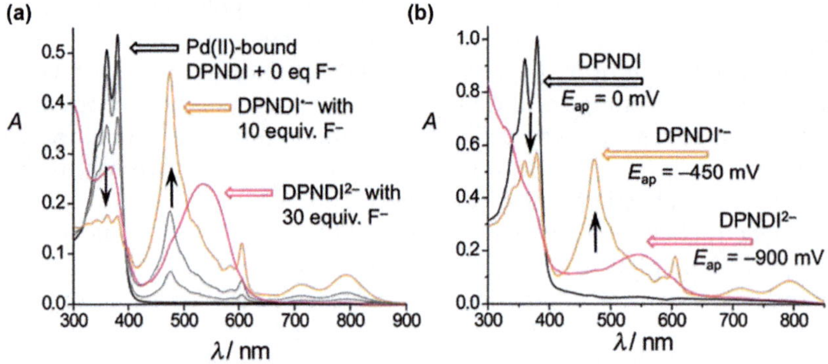

Figure 6.22 UV-vis spectroscopic changes of (a) $(Pd(II)N,N'$-di(4-pyridyl)NDI$)_n$ coordination polymer **41** (*ca.* 2.5 μM in DMSO) with an increasing concentration of F^- showing neutral *N,N′*-di(4-pyridyl)NDI (black), *N,N′*-di(4-pyridyl)NDI^- radical anion (orange), and *N,N′*-di(4-pyridyl)NDI^{2-} dianion (pink), and (b) free *N,N′*-di(4-pyridyl)NDI ligand (0.5 μM/0.1 M TBAPF6/DMF) upon direct electrochemical reductions (*vs.* Ag/AgCl).
Reproduced from ref. 80 with permission from the Royal Society of Chemistry.

N,N′-di(4-pyridyl)NDI^{2-} dianion, respectively. The same analysis also revealed a quasi-reversible $Pd^{2+/+}$ reduction peak at −896 mV, showing the dissociation of TfO^- from the electrochemically generated $DPNDI^-$ radical anions (Figure 6.23). It was proposed that once F^- reduces the *N,N′*-di(4-pyridyl)NDI ligands, electrostatic repulsion would prevent electron-rich TfO^- and THF from interacting with the reduced *N,N′*-di(4-pyridyl)NDI species any further. Saha and co-workers had demonstrated previously that a strongly

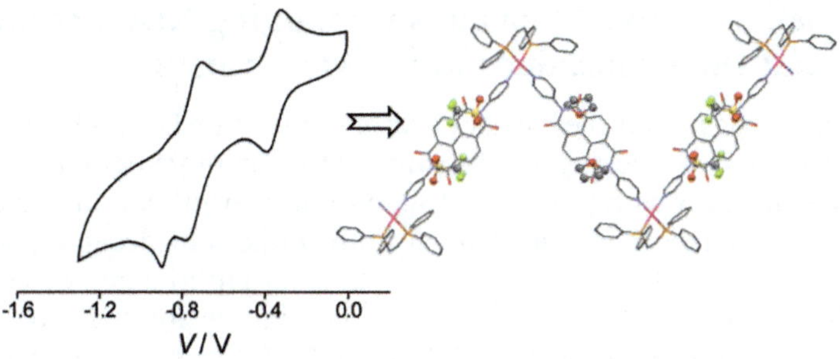

Figure 6.23 Cyclic voltammetry (CV) of (Pd(II)*N*,*N*′-di(4-pyridyl)NDI)$_n$ coordination polymer (*vs.* Ag/AgCl using 0.1 M TBAPF6/DMF).
Reproduced from ref. 80 with permission from the Royal Society of Chemistry.

Lewis basic F^- anion reduces *N*,*N*′-di(4-pyridyl)NDI to *N*,*N*′-di(4-pyridyl)NDI$^-$ and *N*,*N*′-di(4-pyridyl)NDI^{2-} and prohibits nucleophilic attack of F^- on *N*,*N*′-di(4-pyridyl)NDI, resulting in a covalently bound NDI-F species forming.[16] Owing to the lack of anion-induced electron transfer to *N*,*N*′-di(4-pyridyl)NDI, other anions such as Cl^-, Br^-, I^-, NO_2^-, NO_3^-, AcO^-, $H_2PO_4^-$ and PF_6^- did not cause any spectroscopic changes in the coordination polymer, revealing a highly selective sensor towards F^- anions.

6.3.3 Conclusions for NDI-containing Chemosensors

The electronic and film-forming properties of NDI-based materials make them strong candidates as the responsive components in a multitude of sensing arrays. However, work is still ongoing to move these materials from the laboratory into the commercial sector.

6.4 Self-assembled Poly-NDI Structures through Non-covalent Interactions

Exploiting the predictable and directional interactions between π-electron-deficient (acceptor) and π-electron-rich (donor) molecules has led to the realisation of some of the most exquisite structures in supramolecular chemistry.[83–88] The origin of these donor–acceptor interactions can be found in chapters elsewhere in this book (Chapter 1).

The reversible nature of donor–acceptor interactions has resulted in a considerable body of work investigating their introduction into polymeric systems with the aim of producing stimuli-responsible materials. The architecture of these materials is highly varied but can be broadly segregated into two themes: (i) both the donor and acceptor residues are contained either within the same polymer, or (ii) on two different polymers. Both classes of materials are covered in sections 6.4 and 6.5, respectively, of this chapter.

6.4.1 Self-assembled Polymers Containing NDI (Acceptor) and Donor Residues in the Main Chain

One of the earliest examples of a π-electron-deficient NDI residue (**42**), which demonstrated π–π stacking interactions with a complementary π-electron-rich molecule **44**, was reported by Iverson and co-workers during efforts to produce oligomeric molecules with a predictable and designed secondary structure.[76] Readily functionalisable 1,5-dialkoxynaphthalene was chosen as the donor owing to the stability of the complexes formed with NDIs in aqueous solutions (Figure 6.24) (**42** + **44**) (130 M^{-1} as measured by ^{1}H NMR and UV-vis spectroscopy).

The co-crystals formed between the structurally related monomer units (**43** + **45**) revealed an alignment of the aromatic rings, which was assumed to be the energy minimised conformation in solution (Figure 6.25).

Molecular mechanics simulations suggested that separating the electronically complementary motifs by L-aspartic acid would allow the resulting

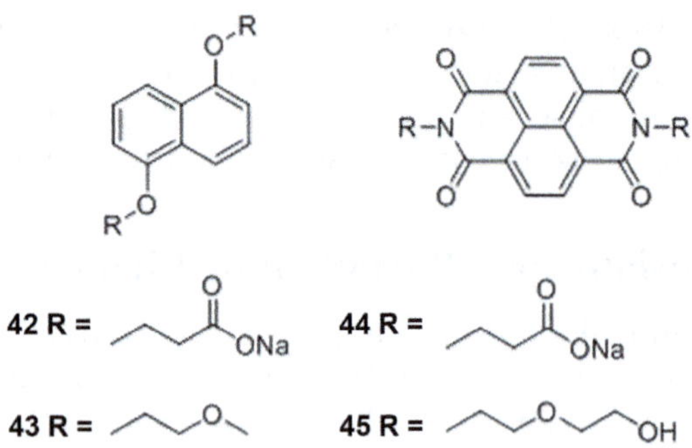

Figure 6.24 Model compounds of π-electron-deficient NDI and π-electron-rich naphthalene used for aqueous binding titration (**42** and **44**) and X-ray diffraction (**43** and **45**).[76,89]

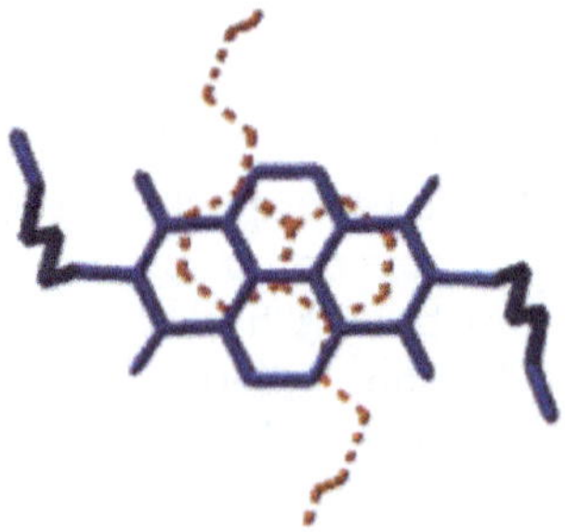

Figure 6.25 X-Ray structure of the co-crystals of **43** and **45**, which exhibits an infinite stack of alternating donor and acceptor residues.
Adapted with permission from ref. 89. Copyright 2001 American Chemical Society.

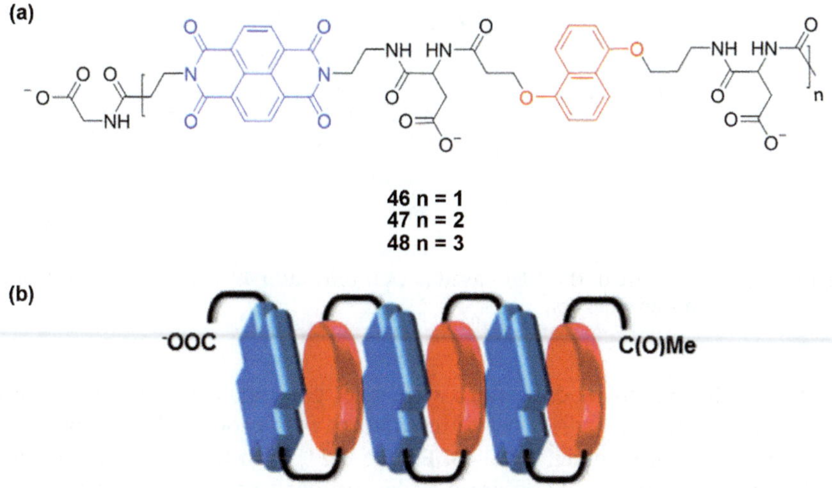

Figure 6.26 (a) Foldamer synthesised by Iverson *et al.*,[90] $n = 1$ (**46**), 2 (**47**), and 3 (**48**). (b) Schematic of complementary motifs in **48** forming π–π stacking interactions.

oligomer to fold into a designed secondary structure driven by maximising the number of retaining π–π stacking interactions. Thus, the ter-oligomer **48** was synthesised by solid-phase methods (Figure 6.26).

The aqueous solutions of the oligomer **46** had a deep red colour as a result of a broad charge transfer band in the visible region ($\lambda_{max} = 522$ nm), identical to that of the monomers used for spectroscopic studies (**42** and **44**, Figure 6.24), whilst a red shift of 18 nm was observed in oligomers with increased repeat units (**46**–**48**).[91] ^{1}H NMR spectroscopy studies[92] of the folding oligomers demonstrated an upfield shift of between 0.20 ppm (**46**) and 0.7 ppm (**48**), indicating that the oligomer **46** was only partially folded. COSY ^{1}H NMR spectroscopy also revealed that many of the methylene hydrogens on the backbone of the oligomers were diastereotopic in nature, indicating restricted rotation, whilst nuclear Overhauser effect enhancements were observed between protons in adjacent aromatic rings, suggesting that the oligomers formed well defined pleated structures despite the inherent flexibility of the amino acid backbone.

Further studies conducted by Iverson *et al.*[93] resulted in a series of polyintercalators based solely on NDI residues (**49**) (Figure 6.27), which was shown to intercalate with double stranded DNA, by insertion of one or more aromatic groups between the base pairs of the double helix.[94] Molecular dynamics simulations using a flexible peptide backbone on a bisintercalated DNA template[95] were used to predict that four amino acid residues between diimide moieties gave the optimum combination of length and flexibility within the polymer backbone to bridge two base pairs, resulting in

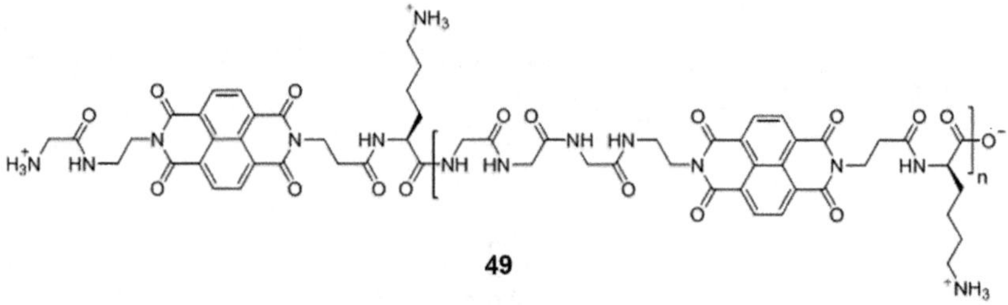

49

Figure 6.27 Structure of the NDI-based DNA intercalator **49** synthesised[93] by Iverson and co-workers.

polyintercalation in accordance with the general principle of nearest neighbour exclusion.[96] The polyintercalators (**49**) produced were based on those reported previously[76] (**42–45**, Figure 6.24) with between one and four NDI residues, and a lysine residue added on each segment to provide electrostatic attraction to DNA. A solid-phase peptide synthesis methodology[97] was used to produce the oligomers with between 0 and 3 repeat units (Figure 6.27).

Binding studies using calf thymus DNA (CT-DNA) were only possible for the monomeric NDI residue because the free ligand could not be detected in a concentration regime low enough to evaluate ($>10^7$). However, unwinding studies showed the helix extension parameter (lengthening of the linear fragment of double-stranded CT-DNA) increased with the number of NDI moieties showing that these systems exhibit a powerful and tuneable ability to interact with DNA.[98–101]

In 2015, Hart *et al.*[102,103] used low molecular weight polymers, which are able to self-assemble through tuneable π–π stacking interactions between π-electron-rich pyrene (**50** and **51**) or perylene (**52**) polymers and chain-folding of a π-electron-deficient NDI-PEG copolymer (**53**), to form extended supramolecular polymer networks for inkjet printing applications (Figure 6.28).

The electronically complementary polymers were successfully deposited using three different printing techniques (piezoelectric inkjet, thermal inkjet and continuous inkjet). Sequential overprinting of the components individually (Scheme 6.6(a)) resulted in supramolecular network formation. The complementary π–π stacked polymer blends (**50** + **53**, **51** + **53** and **52** + **53**) generated strongly coloured materials as a result of charge-transfer absorption bands in the visible spectrum, potentially negating the need for pigments or dyes in the ink formulation and also revealed fluorescence quenching of the polycyclic aromatics (Scheme 6.6(b)). Indeed, the final colour of the deposited material was tailored by varying the end-groups of the π-electron rich polymer component, with pyrenyl complexes producing red images and perylene generating green images.

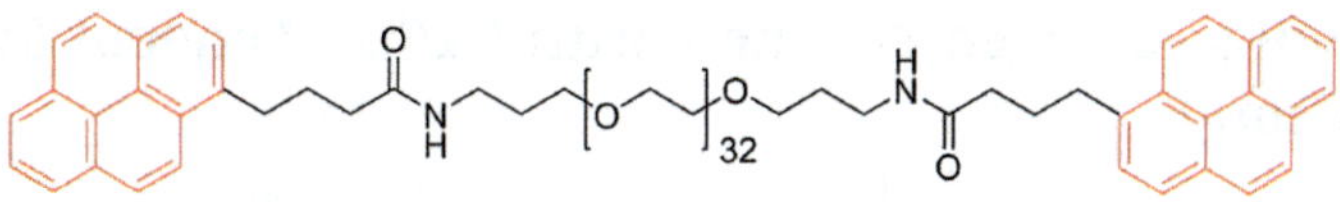

50

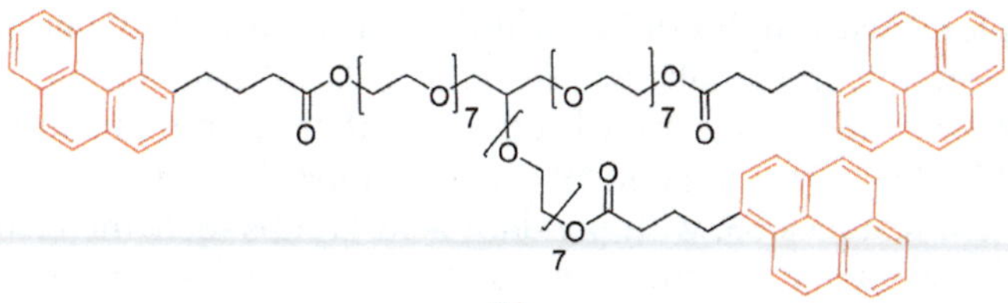

51

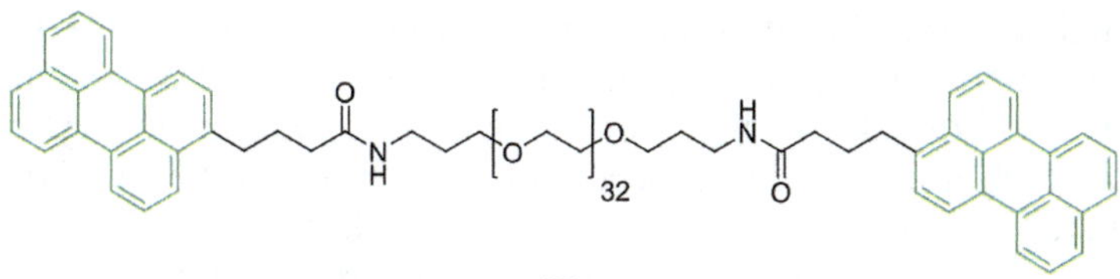

52

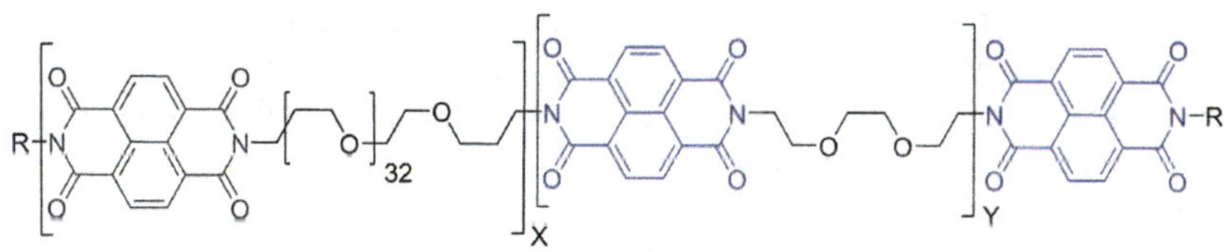

53

R = PEG (Mn = 1500)

Figure 6.28 Structures of π-electron-rich pyrenyl (**50**, **51**) and perylene (**52**) terminated polyethers, as well as a chain-folding π-electron-deficient NDI-PEG copolymer (**53**) developed for inkjet printing by Hart *et al.*[102]

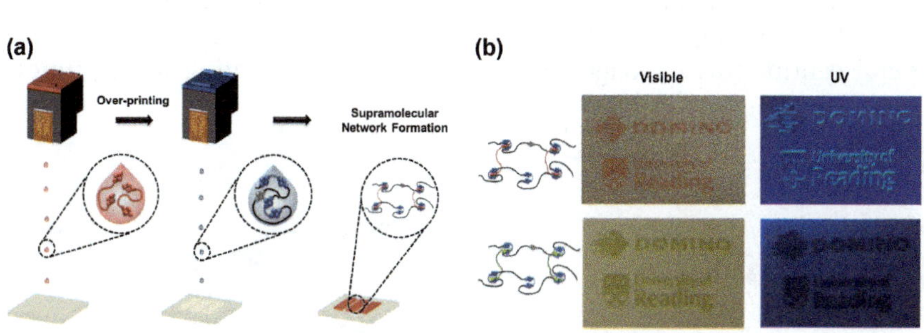

Scheme 6.6 (a) Sequential overprinting of the π-electron-rich polymers (red **50, 51** and **52**) with the chain-folding π-electron-deficient NDI-PEG copolymer **53**. (b) Photographs of the inkjet printed materials under visible and UV light, showing tuning of colour and fluorescence quenching. Adapted with permission from ref. 102. Copyright 2015 American Chemical Society.

6.4.2 Self-assembled Polymers with NDI as Pendent Side Groups

In 2006, the Iverson group[100] produced a conceptually distinct water-soluble supramolecular complex, exploiting interactions between complementary 1,5-dialkylnaphthalene (DAN) and NDI oligomers, which are able to form hetero duplexes with high binding affinities in water.[104] Polymer analogues were produced by functionalising high molecular weight commercially available polyethylene-*alt*-maleic anhydride with either amino-functionalised NDI **54** ($M_n = 173$ kD) with a pendent acid residue (Scheme 6.7(a)) or DAN ($M_n = 146$ kD) residues **55**, with pendant acid groups (Scheme 6.7(b)), which were able to self-assemble into the three-dimensional structures **54** and **55** (Scheme 6.7(c)).

Although the diimide functionalised polymer **54** was soluble in 0.2 M NaOH solution 3% (w/v), the naphthyl derivatised polymer **55** was insoluble at this pH without the addition of a surfactant (sodium dodecyl sulfate). However, an equimolar (with respect to binding motifs) solution of these two polymers (3% w/v) was soluble in 0.2 M NaOH solution generating a purple, homogeneous solution. The viscosity of the blend (11.9 cP) was more than four times greater than that of a solution of individual components (**54** = 1.7 cP, **55** = 2.5 cP) at the same concentration, demonstrating the high degree of interchain interactions exhibited by mixing these two polymers.

Thin films of the individual polymers (**54** and **55**), as well as an equimolar polymer blend (**54** + **55**) spun onto glass supports from solution and washed with HCl and water, were produced for AFM imaging in tapping mode. The DAN-appended polymer **55** revealed roughly circular features with average widths and heights of 500 nm and 75 nm, respectively, characteristic of micelles deposited on a surface.[105] A smooth, uniform film was observed for the NDI-functionalised polymer **54** with a thickness of *ca.* 100 nm, whilst the polymer blend between NDI and DAN polymers (**54** + **55**) revealed structures approximately 700 nm in height, 1.6 μm across, and averaging 14 μm in length. In an attempt to align the polymer chains, long, delicate fibres up to several centimetres in length were formed by precipitating the equimolar

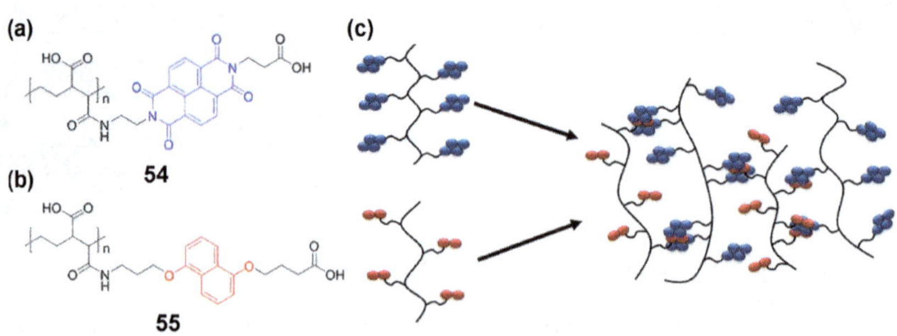

Scheme 6.7 Structures of the (a) NDI **54** and (b) 1,5-dialkylnaphthalene **55** functionalised polymers produced by Iverson and co-workers, which produce a supramolecular polymer complex (c).[100,104]

blended polymer solution through a small aperture into 1 M HCl. The individual components (**54** and **55**) formed only fine precipitates, demonstrating the importance of the supramolecular interaction in the formation of substantive materials. SEM analysis of the polymer blend fibre demonstrated densely packed threads, orientated in a uniform direction from an elongated fibre.

In 2015, supramolecular nanotubes containing a polyurethane scaffold with pendant NDI chromophores were reported by Ghosh and co-workers.[106] NDI chromophores were attached as pendant functional groups to a novel polyurethane (**56**) (Figure 6.29). The resulting polymer is able to adopt a folded conformation in aliphatic hydrocarbon solvents such as methylcyclohexane with high thermal stability through intrachain hydrogen bonding interactions between the urethane groups.

The helical organisation of NDI moieties formed by intrachain folding resulted in spatial organisation of the pendant groups, resulting in face-to-face π–π interactions (Scheme 6.8). In the folded state, the self-assembled

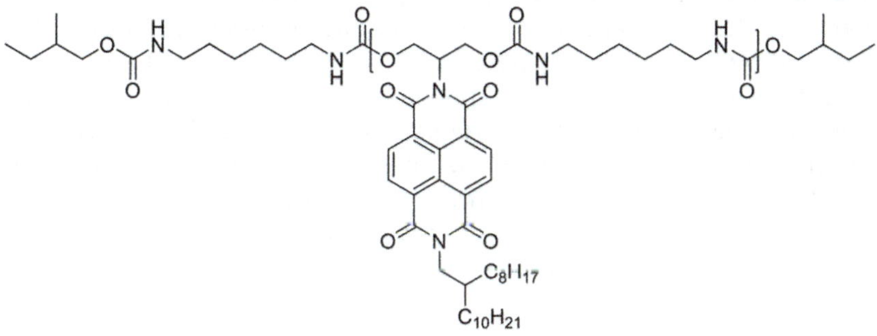

56

Figure 6.29 A polyurethane derivative containing pendant naphthalene diimide chromophores and branched alkyl chains (**56**) to provide steric insulation against interchain H-bonding reported by Ghosh and co-workers.[106]

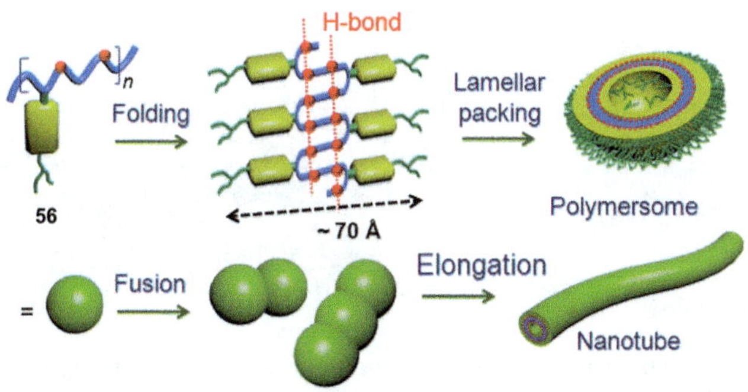

Scheme 6.8 Proposed model for self-assembly of NDI-decorated polyurethane **56**. Reprinted with permission from ref. 106. Copyright 2015 American Chemical Society.

polymer adopted a vesicle-like spherical structure, which upon aging, yielded micrometre-long multiwall nanotubes through fusion of these discrete spherical polymersomes (Scheme 6.8). Fluorescence microscopy studies were conducted to examine the morphologies of the solution over time. Initially, green emitting spherical objects were observed, which were converted to an entangled 1D network structure over time, resulting in gelation. High resolution transmission electron microscopy images revealed spherical objects in the freshly prepared solution whilst nanotubular structures were observed in the aged sample. The inner diameter and wall thickness of the nanotubes were measured to be 85 ± 5 nm and 26 ± 2 nm, respectively, and the overall length of each nanotube extended over 5–9 μM. To explore the molecular interactions, a variety of spectroscopic techniques were employed. UV-vis spectroscopy revealed π–π stacking interactions through diminishing absorption at 382 and 361 nm, respectively, which were shown to be stable to 85 °C. The green nature of the gels was explained by fluorescence spectroscopy, which showed the appearance of a broad emission band at *ca.* 500 nm, corresponding to a NDI excimer produced by stacking of NDI chromophores. Distinct adsorptions at 3360 (N – H stretching), 1665 (carbonyl stretching), and 1537 cm^{-1} (N – H bending), as measured by FT-IR spectroscopy, provided further evidence of hydrogen bonding amongst the urethane groups. Circular dichroism spectroscopy experiments were also conducted to investigate the packing and folding within the polymersome-like structure and nanotubes. The poly-NDI-derived scaffolds were shown to possess electronic properties that could have uses in photocatalysis and light harvesting materials.

6.4.3 Self-assembled Polymers Utilising Electronically Complementary NDI-containing End Groups

In 2013, Colquhoun and Greenland *et al.* designed a single supramolecular motif that contained both NDI and pyrenyl residues[107] (**57**) (Figure 6.30(a)). Solution state studies (UV-vis and ^1H NMR spectroscopies) conducted on

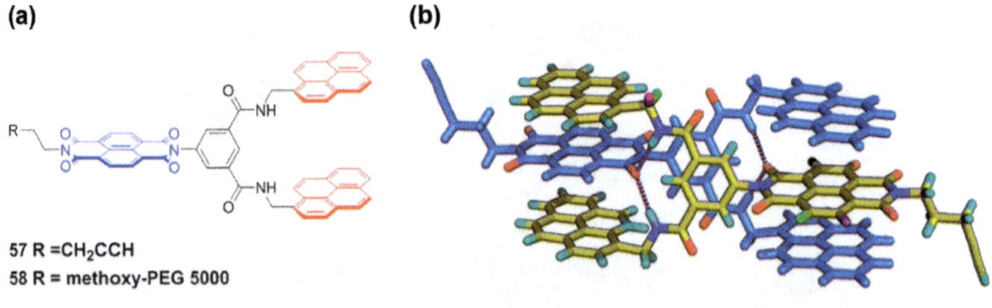

Figure 6.30 (a) Structure of the self-complementary NDI-containing small molecule **57** and polymer **58** end group synthesised by Colquhoun, Greenland and co-workers. (b) Solid state structure of the dimeric complex of **57**. Reproduced from ref. 107 with permission from the Royal Society of Chemistry.

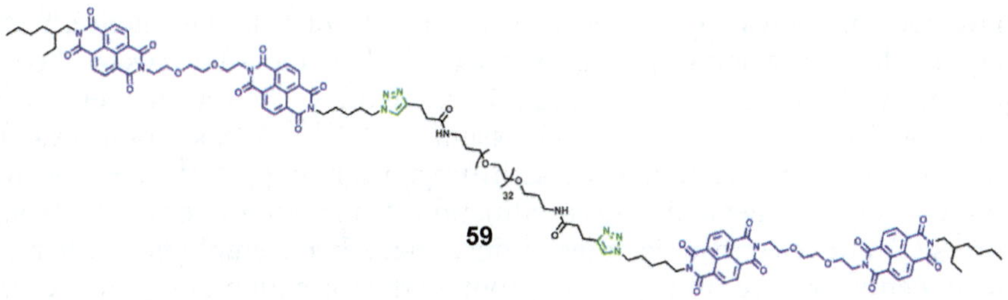

Figure 6.31 Structure of π-electron-deficient polyimide **59** with a well-defined architecture synthesised by 'click' chemistry by Hart *et al.*[108]

discrete molecules of this compound suggested that it formed and extremely high affinity 1 : 1 supramolecular complex ($K_a \approx 10^5$ M^{-1}). The structure of **57** was determined by X-ray crystallography, which confirmed that the molecule dimerised in the solid state to form a self-complementary complex supported by multiple π–π stacking and hydrogen bonding interactions (Figure 6.30(b)). Despite the hydrophilic nature of this supramolecular motif, it could be solubilised in water by attaching it to a polyethylene glycol oligomer ($M_n = 5000$ Da) (**58**) (Figure 6.30(a)). The resulting supramolecular polymer showed a doubling of molecular weight by GPC analysis in comparison to the starting PEG materials in addition to exhibiting self-assembled micelles of approximately 20 nm diameter as observed by cryo-TEM analysis.

Hayes, Colquhoun and co-works have also developed[108] a discrete NDI-containing polymer architecture for improved inkjet printing (Scheme 6.6). Highly efficient cycloaddition or 'click' chemistry was employed to afforded a well-defined poly(ethylene glycol) featuring chain-folding diimide end groups (**59**) (Figure 6.31), which possesses greatly enhanced solubility in THF relative to previous iterations[102] of polymers featuring random diimide sequences. Blending the new polyimide with a complementary poly(ethylene glycol) system bearing pyrene end groups (**50**) (which bind to the chain-folding diimide units) overcomes the limited solubility of chain-folding polyimides in inkjet printing applications. The solution state properties of the resulting polymer blend were assessed *via* viscometry to confirm the presence of a supramolecular polymer before depositing the two electronically complementary polymers by inkjet printing techniques. This resulted in a deeper red image being produced as quantified by colour space analysis of the deposited materials. The novel materials so produced also offer an insight into ways of controlling the properties of printed materials through tuning the structure of the polymer at the molecular and supramolecular level.

6.5 Healable Supramolecular Polymers Utilising NDI

The design and synthesis of novel polymeric materials whose physical characteristics are governed by highly directional non-covalent interactions have become increasingly important in the field of material science.[109–112] Supramolecular chemistry utilises relatively low molecular weight components with complementary recognition motifs[113–115] within their structures which can reversibly self-assemble.[116–119] 'Weak' non-covalent intermolecular interactions such as hydrogen bonding,[120–122] π–π stacking interactions,[10–12] metal–ligand coordination,[123] in addition to van der Waals forces,[121] and electrostatic interactions,[1] have been employed, either independently or in combination, to promote the formation of highly ordered three-dimensional networks.

Polymers with the ability to repair themselves after damage have the potential to extend the lifetime of materials in many applications.[125,126] Most notably, these materials demonstrate the ability to heal autonomically[127] or upon the application of external stimuli such as heat,[128,129] light[130] or pressure.[131] Supramolecular polymer blends which can assemble *via* designed, specific non-covalent interactions, most notably by hydrogen bonding[131–135] metal–ligand interactions[130,136] or π–π interactions[113,121–125] between complementary moieties have been shown recently as effective healable materials.

The association between electronically-complementary aromatic π-systems has found widespread applications in supramolecular chemistry[142] since the first examples were reported in the early 1980s.[83,143] More recently, work by Colquhoun, Hayes and co-workers has demonstrated that such interactions ('complementary π–π stacking') can also be exploited in the design of healable supramolecular polymer blends.[137] Computational and spectroscopic studies on model compounds showed that a readily accessible 'chain-folding' motif[144–146] comprising two π-electron deficient 1,4,5,8-NDI residues separated by a triethylenedioxy linking unit (**60**) provides an ideal stereoelectronic centre for binding to complementary π-electron-rich aromatic residues, namely pyrene **61** (Figure 6.32(a)). Indeed, studies reported by Colquhoun, Hayes and co-workers[137] revealed that this designed motif

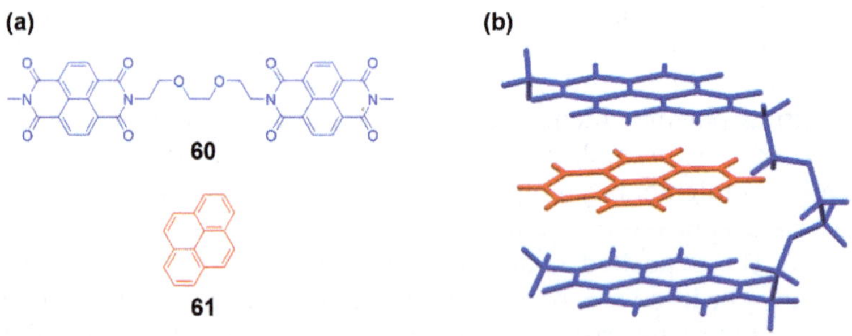

Figure 6.32 (a) Structures of model compounds **60** and **61** used for computational modelling, with π-electron-deficient naphthalene diimide units spaced by a 2,2-(ethylenedioxy)bis(ethylamine)-derived linker and π-electron-rich pyrenyl residue. (b) An energy minimised model of the chain-folding tweezer complex between complementary π-electron-rich and π-electron-deficient species showing a close fit with an interplanar spacing of 3.5 Å between the π-systems.

was effective, with a pyrenyl end-group intercalating into the designed chain-folds to produce a well-defined π-stacked complex (Figure 6.32(b)) with a series of model supramolecular complexes synthesised, which exhibited binding constants of 150–11 000 M^{-1} relative to the number of face-to-face π-stacking interactions.

6.5.1 Chain-folding NDI Homopolymers

The key structural features of these supramolecular chain-folded complexes were introduced into polymeric systems and thus, a series of blend-type materials were produced[138] by Colquhoun, Hayes and co-workers. A polydiimide (**62**) (Figure 6.33(a)) containing multiple NDI chain-folding units was synthesised by condensation polymerisation between the diamine, 2,2′-(ethylenedioxy)-bis(ethylamine) and 1,4,5,8-naphthalenetetracarboxylic dianhydride in the presence of a small quantity of end-capping monoamine to control the molecular weight, which yielded the homopolymer ($M_n = 3200$, $Đ = 2.30$). The NDI polymer was found to bind an electronically complementary, pyrene end-capped poly(siloxane) (**63**) (Figure 6.33(b)) to form a π–π stacked network, which could be broken and healed.

A film was cast from an equimolar blend (with respect to binding motifs) of chain-folding NDI homopolymer **62** and pyrenyl-end-capped poly(siloxane) **63** in 2,2,2-trichloroethanol to yield a visually homogeneous polymer film (**62** + **63**); in contrast, the individual polymers did not form a film, but instead produced a tan powder or waxy solid. The initial polymer films produced were cut with a scalpel and healed at temperatures above 90 °C. At elevated temperatures, the supramolecular π–π stacking interactions are able to dissociate, resulting in a decrease in viscosity, allowing

(a)

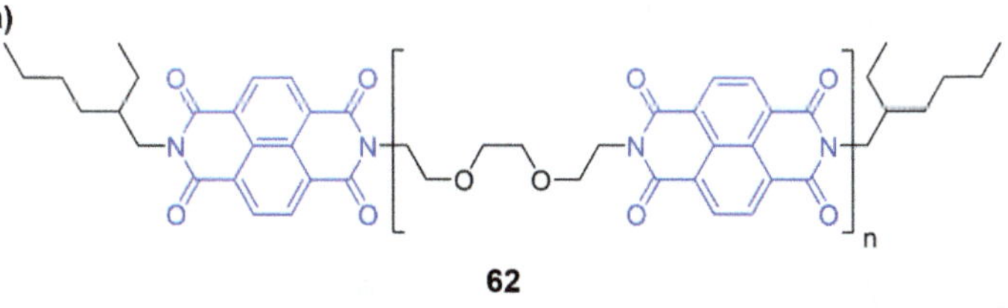

62

(b)

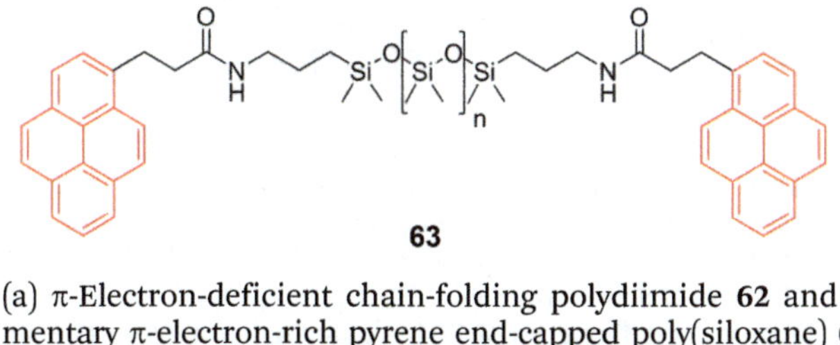

63

Figure 6.33 (a) π-Electron-deficient chain-folding polydiimide **62** and (b) complementary π-electron-rich pyrene end-capped poly(siloxane) **63** produced by Colquhoun, Hayes and co-workers.[138]

the polymers to flow into the void. Upon cooling, the material is healed and the initial physical properties retained as the non-covalent π–π stacking interactions are progressively re-established. A control π-electron-rich polymer with benzyl terminal groups was also studied, which was unable to form π–π stacking interactions. The polymer blend did not form a homogeneous film and did not heal when subjected to the same conditions as the pyrenyl analogue, demonstrating the importance of the π–π stacking interaction. Although the mechanical properties of the healable polymer blend were relatively poor, variable temperature environmental scanning electron microscopy (ESEM) images demonstrated healing at temperatures in excess of 90 °C (Figure 6.34).

It was proposed that the thermal energy resulting from the increase in temperature initiates the healing process of the supramolecular network by disrupting the π–π stacking interactions. The free components can then diffuse, refilling the fracture void and re-establishing the complementary π–π stacking interactions on cooling (Scheme 6.9).

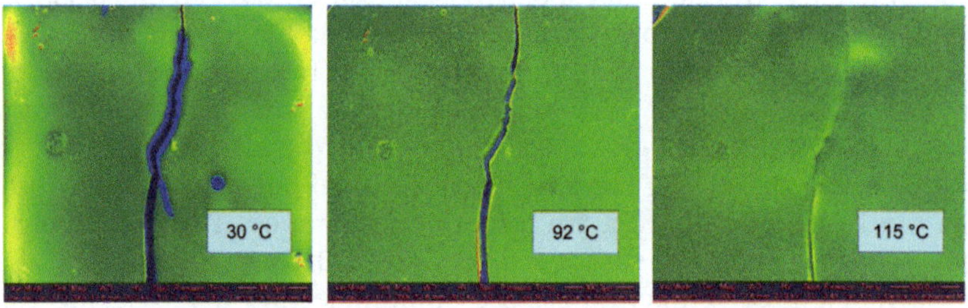

Figure 6.34 ESEM images (magnification ×200) of the fractured film of the blend between a chain-folding NDI homopolymer (**62**) and pyrenyl terminated poly(siloxane) (**63**) at increasing temperatures.
Reproduced from ref. 138 with permission from the Royal Society of Chemistry.

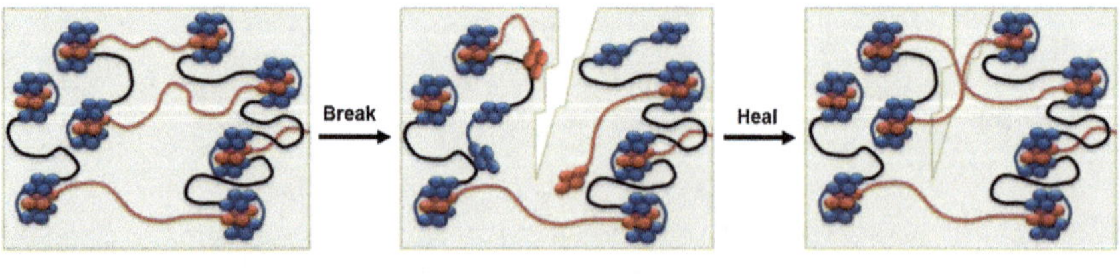

Scheme 6.9 Proposed healing mechanism for the aromatic π–π stacking supramolecular polymer blend comprising an 'electron-deficient' oligomeric chain-folding NDI copolymer (in blue), which is able to host an 'electron-rich' bis-pyrenyl end-capped polysiloxane guest (in red).
Reproduced from ref. 147 with permission from the Royal Society of Chemistry.

6.5.2 Chain-folding NDI Copolymers

While maintaining the π-electron-deficient and π-electron-rich residues, the polymeric midblocks of the material were next modified to produce a second generation of materials with greatly improved physical properties (Figure 6.35(a) and (b)).[129] This was achieved by introducing an oligomeric diamine (Jeffamine®) into the polymer backbone to produce a π-electron-deficient NDI–Jeffamine® copolymer **64**, and π-electron-rich pyrenyl end-capped-Jeffamine® copolymer **65**, which, when blended, could be cast into a flexible, free-standing film. Healing of this second-generation material (Figure 6.35(c) and (d)) was visualised by variable temperature ESEM, which demonstrated that fractures up to 70 μM wide could be rapidly eradicated (in less than a minute) by heating the material above 80 °C.

Quantitative data on the physical properties of the healed material (**64** + **65**) were obtained by cutting the samples, overlapping the cut edges

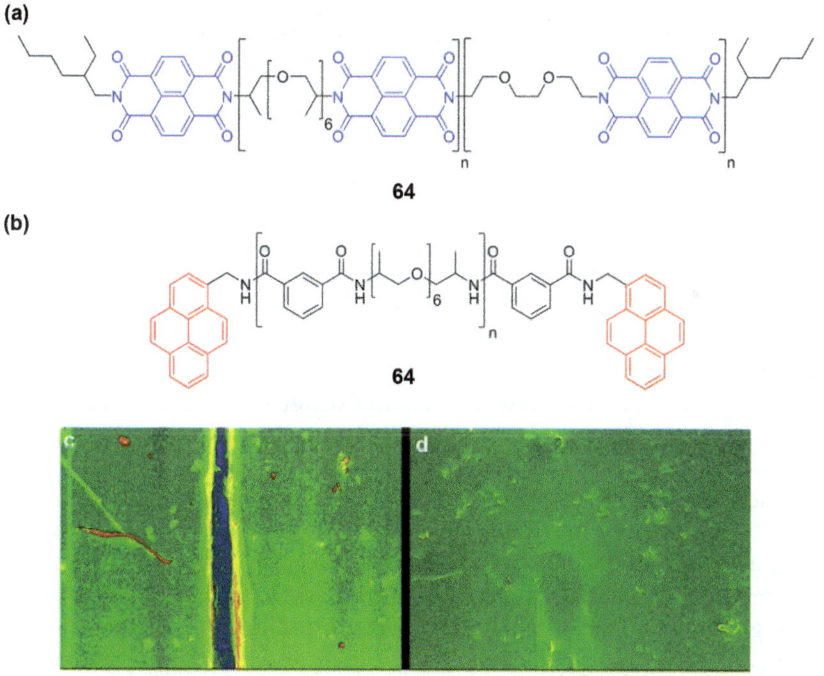

Figure 6.35 Components of a supramolecular polymer blend produced by Colquhoun, Hayes and co-workers[129] showing polyimide-containing multiple π–π electron-deficient chain-folds (**64**) (a) and the telechelic polyamides end-capped with electronically-complementary pyrenyl residues (**65**) (b). False colour ESEM images of a fracture zone of the supramolecular healable material (**64** + **65**) at ambient temperature (c) and 87 °C (d). The fracture is approximately 70 μM in width.[129] Adapted from ref. 129 with permission from the Royal Society of Chemistry.

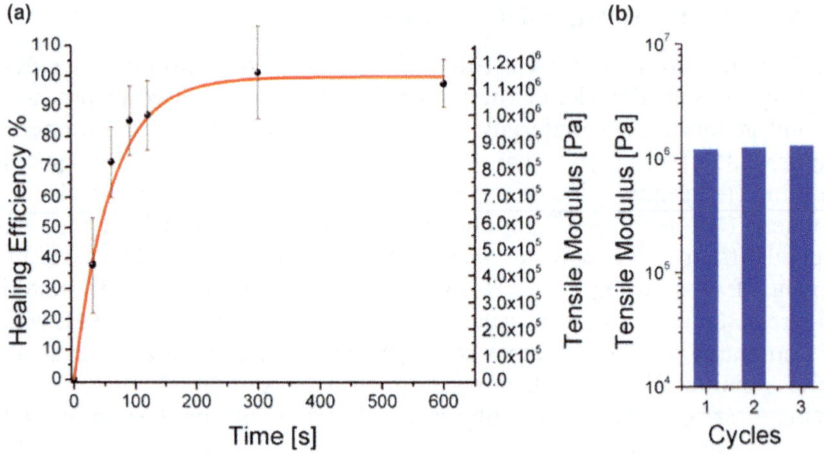

Figure 6.36 (a) Recovery of tensile modulus as a function of healing time at 50 °C for supramolecular blend (**64** + **65**) and (b) final tensile modulus after each of three break–heal cycles of this type.[125,129]
Adapted from ref. 125 and 129 with permission from the Royal Society of Chemistry.

and keeping them in contact for varying periods of time (up to 10 minutes) at 50 °C prior to measuring the strength of the healed sample. Comparing the tensile modulus of the pristine sample (1 MPa at 30 °C) with the values measured after healing demonstrated that the material exhibited a healing efficiency (η_{eff}) of 100% after approximately five minutes (Figure 6.36(a)). A similarly high η_{eff} was recorded over three break–heal cycles (Figure 6.36(b)), and no loss of efficiency was observed up to at least 24 hours between breaking and healing.

Colquhoun, Hayes and co-workers also demonstrated that it was possible to produce healable materials containing two distinct types of supramolecular interaction.[78] By blending a previously reported[129] chain-folding polyimide **64** (Figure 6.35(a)) with a low molecular weight polyurethane having pyrenyl end-groups and a poly(butadiene) polymer backbone, a tough, elastomeric material was produced. This material incorporated both hydrogen bonding moieties (ureas and urethanes) and complementary π–π-stacking residues. Blends of this hydrogen bonding polymer with a chain-folding NDI–Jeffamine copolymer were tough and elastomeric showing up to 170% elongation (2.7 times the initial length) before breaking (Figure 6.37(a)). When this material was broken and re-healed, it repeatedly regained 77% of its modulus of toughness (Figure 6.37(b)), 95% of its tensile modulus (Figure 6.37(c)), and 91% of its elongation to break after annealing for 240 minutes at 100 °C.

A further series of healable materials, containing *tweezer-type* dipyrenyl end-groups blended with a chain-folding NDI–Jeffamine® copolymer was also reported.[139] Such materials display similar tensile moduli (1.5 MPa) to

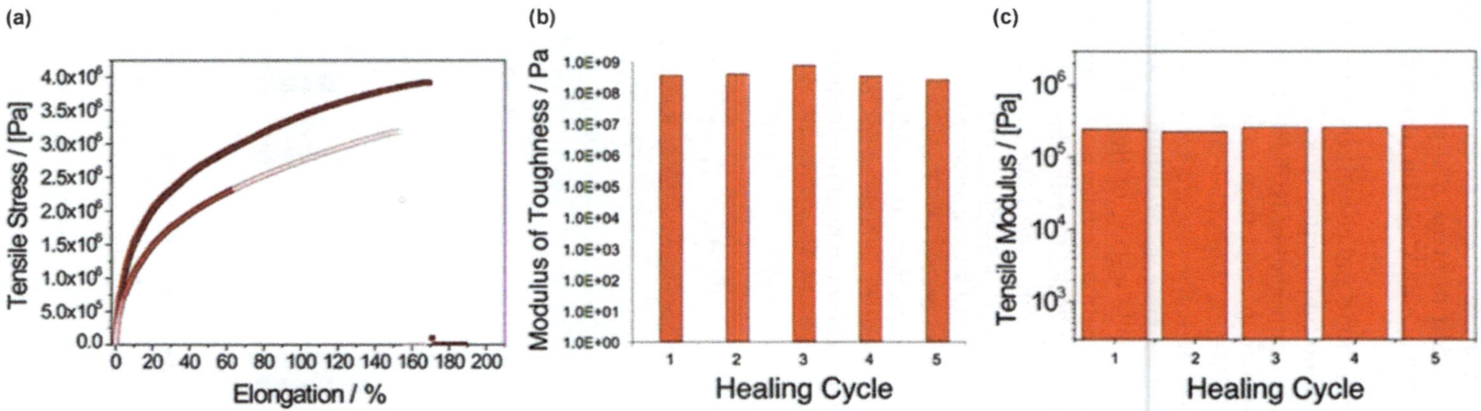

Figure 6.37 (a) Stress–strain curves for the pristine network (•), and the healed network after annealing at 100 °C for four hours (◦). (b) Recovery of the modulus of toughness and of the tensile modulus (c) for the supramolecular blend over five break–heal cycles.[78]
Adapted with permission from ref. 78. Copyright 2010 American Chemical Society.

the analogous materials with single pyrenyl termini (**59**) (Figure 6.34(b)),[129] but exhibit far higher elongation to break and greatly increased ultimate tensile strength. When healed at 140 °C for up to 160 minutes, the tensile modulus was fully restored.

More recently, Hayes, Colquhoun and co-workers have investigated[147] the relationship between supramolecular cross-link density and mechanical strength. To achieve this, a chain-folding NDI–Jeffamine® copolymer (**64**) (Figure 6.35) was blended with pyrenyl terminated poly(ethylene glycol)s (**50** or **51**, Figure 6.28) of differing supramolecular valency (Figure 6.28). Healing was first examined optically, *via* electron microscopy, which revealed that the systems were healable, as exemplified by the trivalent blend (**51** + **64**) in Figure 6.38(a). Tensile testing of the pristine sample demonstrated that increasing the cross-link density of the material not only increased the tensile modulus by 70% (49 MPa) but also yielded a 16-fold improvement in the modulus of toughness (22.6 MPa, Figure 6.38(b)) when comparing trivalent (**51** + **64**) and divalent (**50** + **64**) healable systems. Furthermore, the physical properties were retained over three break–heal cycles with a healing efficiency in excess of 95%.

Alternative electronically complementary residues have been investigated[148] by Hart *et al.*, substituting electron-rich pyrenyl moieties with perylene residues. Perylene was shown to bind to the chain-folding NDI motif

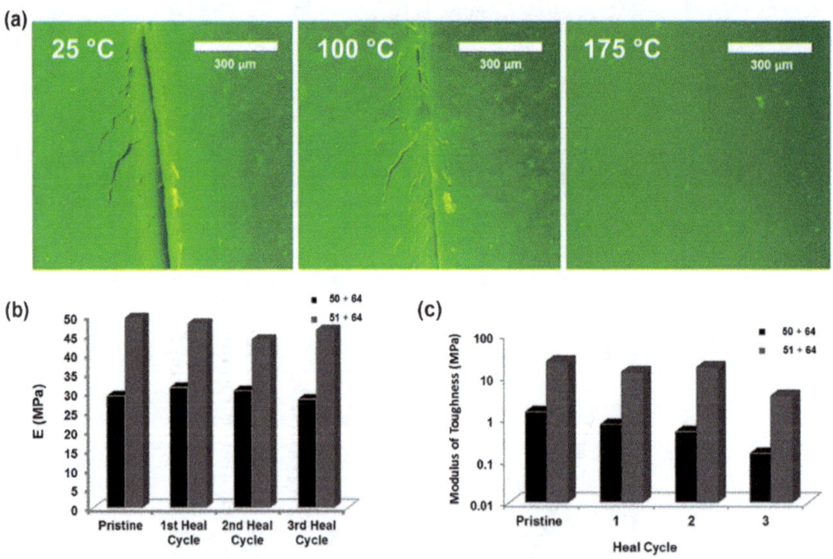

Figure 6.38 (a) False colour ESEM images of the healable polymer blend (**51** + **64**). (b) Tensile moduli and (c) modulus of toughness of the blends containing the divalent pyrenyl-polymer (**50** + **64**) and trivalent pyrenyl-polymer (**51** + **64**).
Adapted from ref. 147 with permission from the Royal Society of Chemistry.

five times as strongly ($K_a = 227\ M^{-1}$) as the pyrenyl analogue ($K_a = 50\ M^{-1}$). It was hypothesised that the stronger association constant would result in enhanced mechanical performance in polymer blends between a chain-folding NDI–Jeffamine® (**64**) (Figure 6.35) and perylene-terminated polyethers (**52**) (Figure 6.28). The new polymer blend (**52** + **64**) demonstrated a viscosity increase when compared to the individual polymers alone, confirming the presence of supramolecular interactions. Furthermore, films of the blended material demonstrated healability *via* ESEM microscopy as well as by mechanical analysis, which revealed 100% recovery of tensile modulus over three break–heal cycles. A decrease in tensile modulus (10 MPa) was noted when compared to the tensile modulus of the analogous pyrenyl polymer blend (**50** + **64**, 29 MPa). Although the association constant for perylene is greater than that of pyrene, it was suggested that the 'on–off' rate has a greater influence on the tensile properties of the material than the association constant, resulting in a weaker material.

6.5.3 Nanoparticle-reinforcing Chain-folding NDI Copolymers

6.5.3.1 Cellulous Nanocrystal NDI-containing Supramolecular Materials

The healable supramolecular polymers described thus far have all been elastomeric in nature. A potential route to achieving high modulus materials whilst still retaining the attractive reversible characteristics offered by supramolecular non-covalent interactions between the components is to generate polymer composites by addition of appropriate fillers to the polymer matrix. In these materials, high-strength fillers such as carbon fibres or carbon nanotubes are incorporated into the polymer to produce a composite material with far better mechanical properties than either of its constituent components.

In 2012, Fox *et al.* approached[140] this problem by reinforcing the relatively weak but thermally responsive supramolecular polymer blend containing a chain-folding NDI copolymer (**64**) and a pyrenyl end-capped Jeffamine® polymer (**65**) (Figure 6.35) with rigid, bio-sourced cellulose nanocrystals (CNCs; tensile strength *ca.* 140 GPa), which are well studied.[149] This procedure afforded healable nanocomposite materials containing various proportions of CNCs, up to 20.0 wt%, within the healable supramolecular polymer matrix. It was found that the tensile modulus of the material increased from 8 MPa to 261 MPa as the proportion of filler increased from 0 to 10 wt% (Figure 6.39(a)). Above this loading level, the materials became weaker, as a consequence of phase separation of the filler and matrix polymer producing a non-homogeneous dispersion of CNCs. Films were cut with a razor blade and found to heal upon exposure to elevated temperatures of 85 °C over various periods of time (Figure 6.39(b)).

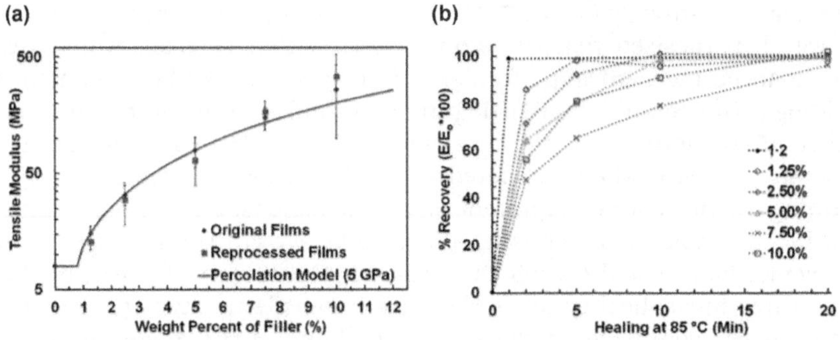

Figure 6.39 (a) Variation of the tensile modulus of pristine and healed supramolecular nanocomposite materials as a function of wt% cellulose nanocrystal filler. The blue line shows the predicted moduli of the materials as calculated by the percolation model and the red squares show experimental values for the corresponding composite films. (b) Healing efficiencies and timescales for composites as a function of the loading of CNCs.
Adapted with permission from ref. 140. Copyright 2010 American Chemical Society.

Comparison of the moduli of the healed samples to those measured in pristine samples revealed healing efficiencies of at least 90% for CNC loading levels up to 10 wt% (Figure 6.39(a)). In addition, it was found that all samples, both pristine and healed, exhibited tensile moduli in close agreement with values predicted from the percolation model,[150] suggesting that in all cases the CNCs remained homogeneously dispersed, and in contact with one another, throughout the supramolecular nanocomposite. The optimum healing efficiency and mechanical properties were obtained from the nanocomposite containing 7.5 wt% CNCs. This exhibited an enhanced tensile modulus over 20 times that of the pristine matrix material and could be fully healed within 30 minutes.

6.5.3.2 Gold-nanoparticle-reinforced NDI-containing Supramolecular Materials

Further work from the Hayes and Colquhoun groups[141,151] investigated the effect of introducing gold nanoparticles (GNPs) into supramolecular materials. Unlike in their previous work concerning cellulous nanocrystals, the GNPs were functionalised with multiple π-electron-rich pyrenyl residues. Addition of these functionalised GNPs to the supramolecular polymer blend of **64** and **65** (Figure 6.40) resulted in materials with a greater tensile modulus and modulus of toughness than those that did not contain the pyrenyl-functionalised GNPs. Importantly, the introduction of the filler component did not inhibit healing of these nanocomposites and all exhibited healing efficiencies of *ca.* 100% over 10 minutes at 50 °C.

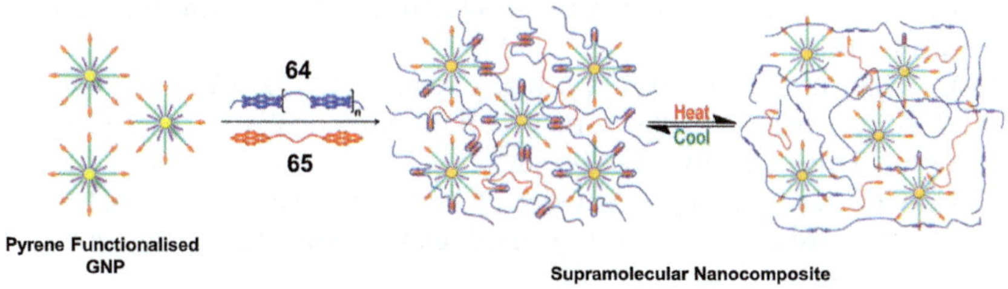

Figure 6.40 Schematic production of a healable supramolecular nanocomposite containing GNPs, imide-containing polymer **64** (blue) and pyrene-containing polymer **65** (red), as exemplified in Figure 6.35. Reprinted with permission from ref. 141 and 151. Copyright 2012 American Chemical Society and the Royal Society of Chemistry.

6.5.4 Conclusions for Healable Supramolecular Polymers

Over recent years, a multitude of supramolecular interactions have been harnessed to produce healing materials including hydrogen bonding, ionic and metal–ligand interactions. However, the structural performance of NDI-based materials, combined with the inert nature of charge-transfer interactions to environmental contaminants (including water), make them ideally suited for use in healable materials.

6.6 Conclusions

Polymers containing NDIs have become an important field of research over the past decade. By exploiting the rich chemistry of the NDI residue, it is possible to produce materials with a range of architectures (Figure 6.1). By design it is possible to produce materials that exhibit semiconducting properties for use in FETs and OPVCs, chemo sensing, self-assembled structures or healable materials. In each of these areas, the laboratory-scale preliminary studies have been completed yet it remains a future challenge to bring these materials to market.

References

1. X. Guo, D. Tu and X. Liu, *J. Energy Chem.*, 2015, **24**, 675–685.
2. Y. Lin and X. Zhan, *Mater. Horiz.*, 2014, **1**, 470–488.
3. J. Choi, H. Song, N. Kim and F. S. Kim, *Semicond. Sci. Technol.*, 2015, **30**, 064002.
4. H. E. Katz and J. Huang, *Annu. Rev. Mater. Res.*, 2009, **39**, 71–92.
5. H. Usta, A. Facchetti and T. J. Marks, *Acc. Chem. Res.*, 2011, **44**, 501–510.
6. A. Facchetti, *Mater. Lett.*, 2007, **10**, 28–37.
7. J. Cornil, J.-L. Brédas, J. Zaumseil and H. Sirringhaus, *Adv. Mater.*, 2007, **19**, 1791–1799.

8. C. Thalacker, C. Röger and F. Würthner, *J. Org. Chem.*, 2006, **71**, 8098–8105.
9. C. Röger and F. Würthner, *J. Org. Chem.*, 2007, **72**, 8070–8075.
10. C. A. Hunter, K. R. Lawson, J. Perkins and C. J. Urch, *J. Chem. Soc., Perkin Trans. 2*, 2001, **2**, 651–669.
11. C. G. Claessens and J. F. Stoddart, *J. Phys. Org. Chem.*, 1997, **10**, 254–272.
12. M. E. Østergaard and P. J. Hrdlicka, *Chem. Soc. Rev.*, 2011, **40**, 5771–5788.
13. J. Steed and J. L. Atwood, *Supramolecular Chemistry*, John Wiley & Sons, 2nd edn, 2009.
14. L. M. Salonen, M. Ellermann and F. Diederich, *Angew. Chem., Int. Ed.*, 2011, **50**, 4808–4842.
15. S. R. Forrest, *Nature*, 2004, **428**, 911–918.
16. R. Hajlaoui and F. Garnier, *Science*, 2013, **265**, 1684–1686.
17. S. Allard, M. Forster, B. Souharce, H. Thiem and U. Scherf, *Angew. Chem., Int. Ed.*, 2008, **47**, 4070–4098.
18. M. Mas-Torrent and C. Rovira, *Chem. Soc. Rev.*, 2008, **37**, 827–838.
19. A. R. Murphy and J. M. J. Fréchet, *Chem. Rev.*, 2007, **107**, 1066–1096.
20. M. Sommer, *J. Mater. Chem. C*, 2014, **2**, 3088–3098.
21. M. M. Durban, P. D. Kazarinoff and C. K. Luscombe, *Macromolecules*, 2010, **43**, 6348–6352.
22. T. B. Singh, S. Erten, S. Günes, C. Zafer, G. Turkmen, B. Kuban, Y. Teoman, N. S. Sariciftci and S. Icli, *Org. Electron.*, 2006, **7**, 480–489.
23. B. Crone, A. Dodabalapur, Y. Lin, R. Filas, Z. Bao, A. LaDuca, R. Sarpeshkar, H. Katz and W. Li, *Nature*, 2000, **403**, 521–523.
24. Z. Chen, Y. Zheng, H. Yan and A. Facchetti, *J. Am. Chem. Soc.*, 2009, **131**, 8–9.
25. H. Yan, Z. Chen, Y. Zheng, C. Newman, J. R. Quinn, F. Dötz, M. Kastler and A. Facchetti, *Nature*, 2009, **457**, 679–686.
26. X. Guo, F. S. Kim, M. J. Seger, S. A. Jenekhe and M. D. Watson, *Chem. Mater.*, 2012, **24**, 1434–1442.
27. R. Matsidik, H. Komber, A. Luzio, M. Caironi and M. Sommer, *J. Am. Chem. Soc.*, 2015, **137**, 6705–6711.
28. B. A. Jones, A. Facchetti, M. R. Wasielewski and T. J. Marks, *J. Am. Chem. Soc.*, 2007, **129**, 15259–15278.
29. G. H. Gelinck, H. E. A. Huitema, E. van Veenendaal, E. Cantatore, L. Schrijnemakers, J. B. P. H. van der Putten, T. C. T. Geuns, M. Beenhakkers, J. B. Giesbers, B.-H. Huisman, E. J. Meijer, E. M. Benito, F. J. Touwslager, A. W. Marsman, B. J. E. van Rens and D. M. de Leeuw, *Nat. Mater.*, 2004, **3**, 106–110.
30. Y. Kim, D. X. Long, J. Lee, G. Kim, T. J. Shin, K.-W. Nam, Y.-Y. Noh and C. Yang, *Macromolecules*, 2015, **48**, 5179–5187.
31. R. Kim, P. S. K. Amegadze, I. Kang, H.-J. Yun, Y.-Y. Noh, S.-K. Kwon and Y.-H. Kim, *Adv. Funct. Mater.*, 2013, **23**, 5719–5727.
32. D. S. Chung, S. J. Lee, J. W. Park, D. B. Choi, D. H. Lee, J. Park, S. C. Shin, Y. Kim, S. Kwon and C. E. Park, *Chem. Mater.*, 2008, **20**, 3450–3456.

33. Y.-J. Hwang, N. M. Murari and S. A. Jenekhe, *Polym. Chem.*, 2013, **4**, 3187–3195.
34. Y. Kim, J. Hong, J. H. Oh and C. Yang, *Chem. Mater.*, 2013, **25**, 3251–3259.
35. W. Zhou, Y. Wen, L. Ma, Y. Liu and X. Zhan, *Macromolecules*, 2012, **45**, 4115–4121.
36. A. Babel and S. A. Jenekhe, *J. Am. Chem. Soc.*, 2003, **125**, 13656–13657.
37. M. M. Durban, P. D. Kazarino, Y. Segawa and C. K. Luscombe, *Macromolecules*, 2011, **44**, 4721–4728.
38. Y. Li, G. Zhang, Z. Liu, X. Chen, J. Wang and D. Zhang, *Macromolecules*, 2013, **46**, 5504–5511.
39. G. C. Berry, *Discuss. Faraday Soc.*, 1970, **49**, 121–136.
40. S. A. Jenekhe and P. Johnson, *Macromolecules*, 1990, **23**, 4419–4429.
41. S. Günes, H. Neugebauer and N. S. Sariciftci, *Chem. Rev.*, 2007, **107**, 1324–1338.
42. B. C. Thompson and J. M. J. Fréchet, *Angew. Chem., Int. Ed.*, 2008, **47**, 58–77.
43. J. You, L. Dou, K. Yoshimura, T. Kato, K. Ohya, T. Moriarty, K. Emery, C.-C. Chen, J. Gao, G. Li and Y. Yang, *Nat. Commun.*, 2013, **4**, 1446.
44. Z. He, C. Zhong, S. Su, M. Xu, H. Wu and Y. Cao, *Nat. Photonics*, 2012, **6**, 591–595.
45. A. Facchetti, *Mater. Today*, 2013, **16**, 123–132.
46. T. Earmme, Y.-J. Hwang, N. M. Murari, S. Subramaniyan and S. A. Jenekhe, *J. Am. Chem. Soc.*, 2013, **135**, 14960–14963.
47. S. Fabiano, Z. Chen, S. Vahedi, A. Facchetti, B. Pignataro and M. A. Loi, *J. Mater. Chem.*, 2011, **21**, 5891–5896.
48. C. Lee, H. Kang, W. Lee, T. Kim, K.-H. Kim, H. Y. Woo, C. Wang and B. J. Kim, *Adv. Mater.*, 2015, **27**, 2466–2471.
49. J. M. Salazar-Rios, W. Gomulya, V. Derenskyi, J. Yang, S. Z. Bisri, Z. Chen, A. Facchetti and M. A. Loi, *Adv. Electron. Mater.*, 2015, **1**, 1500074.
50. H. Xin, S. Subramaniyan, T. Kwon, S. Shoaee, J. R. Durrant and S. A. Jenekhe, *Chem. Mater.*, 2012, **24**, 1995–2001.
51. S. Subramaniyan, H. Xin, F. S. Kim, S. Shoaee, J. R. Durrant and S. A. Jenekhe, *Adv. Energy Mater.*, 2011, **1**, 854–860.
52. H. Sirringhaus, P. J. Brown, R. H. Friend, M. M. Nielsen, B. M. W. Langeveld-Voss, K. Bechgaard, A. J. H. Spiering, R. A. J. Janssen, E. W. Meijer, P. Herwig and D. M. de Leeuw, *Nature*, 1999, **401**, 685–688.
53. K. D. Deshmukh, T. Qin, J. K. Gallaher, A. C. Y. Liu, E. Gann, K. O'Donnell, L. Thomsen, J. M. Hodgkiss, S. E. Watkins and C. R. McNeill, *Energy Environ. Sci.*, 2015, **8**, 332–342.
54. M. Yuan, M. M. Durban, P. D. Kazarinoff, D. F. Zeigler, A. H. Rice, Y. Segawa and C. K. Luscombe, *J. Polym. Sci., Part A: Polym. Chem.*, 2013, **51**, 4061–4069.
55. S. Chen, Q. Zheng, Q. Zhang, D. Cai, J. Wang, Z. Yin and C. Tang, *J. Polym. Sci., Part A: Polym. Chem.*, 2013, **51**, 1999–2005.
56. G. Dennler, M. C. Scharber and C. J. Brabec, *Adv. Mater.*, 2009, **21**, 1323–1338.

57. Z. Zhao, F. Zhang, X. Zhang, X. Yang, H. Li, X. Gao, C. Di and D. Zhu, *Macromolecules*, 2013, **46**, 7705–7714.
58. T. Earmme, Y.-J. Hwang, S. Subramaniyan and S. A. Jenekhe, *Adv. Mater.*, 2014, **26**, 6080–6085.
59. Y.-J. Hwang, T. Earmme, B. A. E. Courtright, F. N. Eberle and S. A. Jenekhe, *J. Am. Chem. Soc.*, 2015, **137**, 4424–4434.
60. M. C. Scharber, D. Mühlbacher, M. Koppe, P. Denk, C. Waldauf, A. J. Heeger and C. J. Brabec, *Adv. Mater.*, 2006, **18**, 789–794.
61. Y.-J. Hwang, T. Earmme, S. Subramaniyan and S. A. Jenekhe, *Chem. Commun.*, 2014, **50**, 10801–10804.
62. Y.-J. Hwang, B. A. E. Courtright, A. S. Ferreira, S. H. Tolbert and S. A. Jenekhe, *Adv. Mater.*, 2015, **27**, 4578–4584.
63. K. Nakabayashi and H. Mori, *Macromolecules*, 2012, **45**, 9618–9625.
64. X.-D. Wang and O. S. Wolfbeis, *Anal. Chem.*, 2013, **85**, 487–508.
65. S. V. Bhosale, C. H. Jani and S. J. Langford, *Chem. Soc. Rev.*, 2008, **37**, 331–342.
66. S. V. Bhosale, A. L. Sisson, P. Talukdar, A. Fürstenberg, N. Banerji, E. Vauthey, G. Bollot, J. Mareda, C. Röger, N. Sakai, S. Matile, A. Furstenberg, C. Roger and F. Wurthner, *Science*, 2006, **313**, 84–86.
67. L. Shen, X. Lu, H. Tian and W. Zhu, *Macromolecules*, 2011, **44**, 5612–5618.
68. D. M. Kaschak, J. T. Lean, C. C. Waraksa, G. B. Saupe, H. Usami and T. E. Mallouk, *J. Am. Chem. Soc.*, 1999, **121**, 3435–3445.
69. M. A. Rodrigues, D. B. Tada, M. J. Politi, S. Brochsztain and M. S. Baptista, *J. Non-Cryst. Solids*, 2002, **304**, 116–125.
70. M. Mascal, I. Yakovlev, E. B. Nikitin and J. C. Fettinger, *Angew. Chem., Int. Ed.*, 2007, **46**, 8782–8784.
71. O. B. Berryman and D. W. Johnson, *Chem. Commun.*, 2009, 3143–3153.
72. R. E. Dawson, A. Hennig, D. P. Weimann, D. Emery, V. Ravikumar, J. Montenegro, T. Takeuchi, S. Gabutti, M. Mayor, J. Mareda, C. A. Schalley and S. Matile, *Nat. Chem.*, 2010, **2**, 533–538.
73. H. T. Chifotides, B. L. Schottel and K. R. Dunbar, *Angew. Chem., Int. Ed.*, 2010, **49**, 7202–7207.
74. S. Guha and S. Saha, *J. Am. Chem. Soc.*, 2010, **132**, 17674–17677.
75. S. Guha, F. S. Goodson, S. Roy, L. J. Corson, C. A. Gravenmier and S. Saha, *J. Am. Chem. Soc.*, 2011, **133**, 15256–15259.
76. B. L. Iverson and R. S. Lokey, *Nature*, 1995, **375**, 303–305.
77. C. A. Hunter and J. K. M. Sanders, *J. Am. Chem. Soc.*, 1990, **112**, 5525–5534.
78. S. Burattini, B. W. Greenland, D. Hermida Merino, W. Weng, J. Seppala, H. M. Colquhoun, W. Hayes, M. E. Mackay, I. W. Hamley and S. J. Rowan, *J. Am. Chem. Soc.*, 2010, **132**, 12051–12058.
79. B. L. Schottel, H. T. Chifotides and K. R. Dunbar, *Chem. Soc. Rev.*, 2008, **37**, 68–83.
80. S. Guha, F. S. Goodson, R. J. Clark and S. Saha, *CrystEngComm*, 2012, **14**, 1213–1215.

81. P. J. Stang and D. H. Cao, *J. Am. Chem. Soc.*, 1994, **116**, 4981–4982.
82. M. Cametti and K. Rissanen, *Chem. Commun.*, 2009, 2809–2829.
83. H. M. Colquhoun, J. F. Stoddart, D. J. Williams, J. B. Wolstenholme and R. Zarzycki, *Angew. Chem., Int. Ed. Engl.*, 1981, **120**, 1051–1053.
84. M. C. T. Fyfe and J. F. Stoddart, *Acc. Chem. Res.*, 1997, **30**, 393–401.
85. D. B. Amabilino and J. F. Stoddart, *Chem. Rev.*, 1995, **95**, 2725–2828.
86. H. M. Colquhoun, Z. Zhu, J. S. Shaw, C. J. Cardin, S. Burattini, J. M. Elliott, S. Basu, T. B. Gasa and J. F. Stoddart, *Org. Lett.*, 2009, **11**, 5238–5241.
87. H. Y. Au-Yeung, F. B. L. Cougnon, S. Otto, G. D. Pantoş and J. K. M. Sanders, *Chem. Sci.*, 2010, **1**, 567–574.
88. E. Tamanini, N. Ponnuswamy, G. D. Pantoş and J. K. M. Sanders, *Faraday Discuss.*, 2010, **145**, 205–218.
89. M. S. Cubberley and B. L. Iverson, *J. Am. Chem. Soc.*, 2001, **123**, 7560–7563.
90. C. Peebles, R. Piland and B. L. Iverson, *Chem. - Eur. J.*, 2013, **19**, 11598–11602.
91. A. C. Benniston, A. Harriman and V. M. Lynch, *Tetrahedron Lett.*, 1994, **35**, 1473–1476.
92. J. A. Chudek, R. Foster and D. R. Twiselton, *J. Chem. Soc. Perkin Trans. 2*, 1983, 1385–1389.
93. R. S. Lokey, Y. Kwok, V. Guelev, C. J. Pursell, L. H. Hurley and B. L. Iverson, *J. Am. Chem. Soc.*, 1997, **119**, 7202–7210.
94. S. F. Yen, E. J. Gabbay and W. D. Wilson, *Biochemistry*, 1982, **21**, 2070–2076.
95. M. E. Peck, L. A. Lipscomb, J. A. Bertrand, Q. Gao, B. P. Roques, C. Garbay-Jaureguiberry and L. D. Williams, *Biochemistry*, 1994, **33**, 3794–3800.
96. J. SantaLucia Jr., *Proc. Natl. Acad. Sci. U. S. A.*, 1998, **95**, 1460–1465.
97. E. Atherton and R. C. Sheppard, *Solid Phase Peptide Synthesis: A Practical Approach*, Oxford University Press, 1989.
98. J. Q. Nguyen and B. L. Iverson, *J. Am. Chem. Soc.*, 1999, **121**, 2639–2640.
99. M. M. Murr, M. T. Harting, V. Guelev, J. Ren, J. B. Chaires and B. L. Iverson, *Bioorg. Med. Chem.*, 2001, **9**, 1141–1148.
100. J. J. Reczek and B. L. Iverson, *Macromolecules*, 2006, **39**, 5601–5603.
101. P. M. P. Alvey, R. J. R. Ono, C. C. W. Bielawski and B. L. Iverson, *Macromolecules*, 2013, **46**, 718–726.
102. L. R. Hart, J. L. Harries, B. W. Greenland, H. M. Colquhoun and W. Hayes, *ACS Appl. Mater. Interfaces*, 2015, **7**, 8906–8914.
103. L. R. Hart, J. L. Harries, A. Clifton, H. M. Colquhoun and W. Hayes, WO 2014111722 A1, 2014.
104. G. J. Gabriel and B. L. Iverson, *J. Am. Chem. Soc.*, 2002, **124**, 15174–15175.
105. I. W. Hamley, S. D. Connell and S. Collins, *Macromolecules*, 2004, **37**, 5337–5351.
106. T. Mondal, T. Sakurai, S. Yoneda, S. Seki and S. Ghosh, *Macromolecules*, 2015, **48**, 879–888.

107. B. W. Greenland, M. B. Bird, S. Burattini, R. Cramer, R. K. O'Reilly, J. P. Patterson, W. Hayes, C. J. Cardin and H. M. Colquhoun, *Chem. Commun.*, 2013, **49**, 454–456.
108. L. R. Hart, J. L. Harries, B. W. Greenland, H. M. Colquhoun and W. Hayes, *Polym. Chem.*, 2015, **6**, 7342–7352.
109. R. P. Sijbesma, F. H. Beijer, L. Brunsveld, B. J. Folmer, J. H. Hirschberg, R. F. Lange, J. K. Lowe and E. W. Meijer, *Science*, 1997, **278**, 1601–1604.
110. J.-M. Lehn, *Science*, 2002, **295**, 2400–2403.
111. L. Brunsveld, B. J. Folmer, E. W. Meijer and R. P. Sijbesma, *Chem. Rev.*, 2001, **101**, 4071–4098.
112. J.-M. Lehn, *Polym. Int.*, 2002, **51**, 825–839.
113. D. J. Cram and J. M. Cram, *Science*, 1974, **183**, 803–809.
114. H.-J. Schneider, *Angew. Chem., Int. Ed. Engl.*, 1991, **30**, 1417–1437.
115. E. P. Kyba, R. C. Helgeson, K. Madan, G. W. Gokel, T. L. Tarnowski, S. S. Moore and D. J. Cram, *J. Am. Chem. Soc.*, 1977, **99**, 2564–2571.
116. G. M. Whitesides and B. Grzybowski, *Science*, 2002, **295**, 2418–2421.
117. J. N. Israelachvili, D. J. Mitchell and B. W. Ninham, *J. Chem. Soc., Faraday Trans. 2*, 1976, **72**, 1525–1568.
118. G. M. Whitesides, J. P. Mathias and C. T. Seto, *Science*, 1991, **254**, 1312–1319.
119. F. J. M. Hoeben, P. Jonkheijm, E. W. Meijer and A. P. H. J. Schenning, *Chem. Rev.*, 2005, **105**, 1491–1546.
120. J. Emsley, *Prog. Inorg. Chem.*, 1980, **9**, 91–124.
121. T. F. A. De Greef, M. M. J. Smulders, M. Wolffs, A. P. H. J. Schenning, R. P. Sijbesma, E. W. Meijer, T. F. A. De Greef and C. Counterpart, *Chem. Rev.*, 2009, **109**, 5687–5754.
122. J. Watson and F. Crick, *Nature*, 1953, **171**, 737–738.
123. R. Quesada and P. A. Gale, *Annu. Rep. Prog. Chem., Sect. B: Org. Chem.*, 2005, **101**, 148–170.
124. X. Zhang and C. Wang, *Chem. Soc. Rev.*, 2010, **40**, 94–101.
125. S. Burattini, B. W. Greenland, D. Chappell, H. M. Colquhoun and W. Hayes, *Chem. Soc. Rev.*, 2010, **39**, 1973–1985.
126. L. R. Hart, J. L. Harries, B. W. Greenland, H. M. Colquhoun and W. Hayes, *Polym. Chem.*, 2013, **4**, 4860–4870.
127. S. R. White, N. R. Sottos, P. H. Geubelle, J. S. Moore, M. R. Kessler, S. R. Sriram, E. N. Brown and S. Viswanathan, *Nature*, 2001, **409**, 794–797.
128. X. Chen, M. A. Dam, K. Ono, A. Mal, H. Shen, S. R. Nutt, K. Sheran and F. Wudl, *Science*, 2002, **295**, 1698–1702.
129. S. Burattini, H. M. Colquhoun, J. D. Fox, D. Friedmann, B. W. Greenland, P. J. F. Harris, W. Hayes, M. E. Mackay and S. J. Rowan, *Chem. Commun.*, 2009, 6717–6719.
130. M. Burnworth, L. Tang, J. R. Kumpfer, A. J. Duncan, F. L. Beyer, G. L. Fiore, S. J. Rowan and C. Weder, *Nature*, 2011, **472**, 334–338.
131. P. Cordier, F. Tournilhac, C. Soulié-Ziakovic, L. Leibler and C. Soulie, *Nature*, 2008, **451**, 977–980.

132. Y. Chen, A. M. Kushner, G. A. Williams and Z. Guan, *Nat. Chem.*, 2012, **4**, 467–472.
133. A. W. Bosman, R. P. Sijbesma and E. W. Meijer, *Mater. Today*, 2004, **7**, 34–39.
134. A. W. Bosman, in *Proceedings of the First International Conference on Self Healing Materials*, 2007, pp. 1–6.
135. B. Folmer, R. P. Sijbesma, R. M. Versteegen, J. A. J. van der Rijt and E. W. Meijer, *Adv. Mater.*, 2000, **12**, 874–878.
136. G. L. Fiore, S. J. Rowan and C. Weder, *Chim. Int. J. Chem.*, 2011, **65**, 745.
137. B. W. Greenland, S. Burattini, W. Hayes and H. M. Colquhoun, *Tetrahedron*, 2008, **64**, 8346–8354.
138. S. Burattini, H. M. Colquhoun, B. W. Greenland and W. Hayes, *Faraday Discuss.*, 2009, **143**, 251–264.
139. S. Burattini, B. W. Greenland, W. Hayes, M. E. Mackay, S. J. Rowan and H. M. Colquhoun, *Chem. Mater.*, 2011, **23**, 6–8.
140. J. Fox, J. J. Wie, B. W. Greenland, S. Burattini, W. Hayes, H. M. Colquhoun, M. E. Mackay and S. J. Rowan, *J. Am. Chem. Soc.*, 2012, **134**, 5362–5368.
141. R. Vaiyapuri, B. W. Greenland, H. M. Colquhoun, J. M. Elliott and W. Hayes, *Polym. Chem.*, 2013, **4**, 4902–4909.
142. L. Fang, M. A. Olson, D. Benítez, E. Tkatchouk, W. A. Goddard and J. F. Stoddart, *Chem. Soc. Rev.*, 2010, **39**, 17–29.
143. H. M. Colquhoun, E. P. Goodings, J. M. Maud, J. F. Stoddart, J. B. Wolstenholme and D. J. Williams, *J. Chem. Soc., Perkin Trans. 2*, 1985, 607.
144. Z. Zhu, C. J. Cardin, Y. Gan and H. M. Colquhoun, *Nat. Chem.*, 2010, **2**, 653–660.
145. H. M. Colquhoun, Z. Zhu, C. J. Cardin, Y. Gan and M. G. B. Drew, *J. Am. Chem. Soc.*, 2007, **129**, 16163–16174.
146. H. M. Colquhoun and Z. Zhu, *Angew. Chem., Int. Ed.*, 2004, **43**, 5040–5045.
147. L. R. Hart, J. H. Hunter, N. A. Nguyen, J. L. Harries, B. W. Greenland, M. E. Mackay, H. M. Colquhoun and W. Hayes, *Polym. Chem.*, 2014, **5**, 3680–3688.
148. L. R. Hart, N. A. Nguyen, J. L. Harries, M. E. Mackay, H. M. Colquhoun and W. Hayes, *Polymer*, 2015, **69**, 293–300.
149. S. J. Eichhorn, A. Dufresne, M. Aranguren, N. E. Marcovich, J. R. Capadona, S. J. Rowan, C. Weder, W. Thielemans, M. Roman, S. Renneckar, W. Gindl, S. Veigel, J. Keckes, H. Yano, K. Abe, M. Nogi, A. N. Nakagaito, A. Mangalam, J. Simonsen, A. S. Benight, A. Bismarck, L. A. Berglund and T. Peijs, *J. Mater. Sci.*, 2010, **45**, 1–33.
150. J. R. Capadona, O. van den Berg, L. A. Capadona, M. Schroeter, S. J. Rowan, D. J. Tyler and C. Weder, *Nat. Nanotechnol.*, 2007, **2**, 765–769.
151. R. Vaiyapuri, B. W. Greenland, S. J. Rowan, H. M. Colquhoun, J. M. Elliott and W. Hayes, *Macromolecules*, 2012, **45**, 5567–5574.

CHAPTER 7

Tunable Electronic Interactions between Aromatic Diimides and Anions

DILLIP K. PANDA AND SOURAV SAHA*

Department of Chemistry, Clemson University, Clemson SC 29634, USA
*Email: souravs19@gmail.com

7.1 Introduction

7.1.1 Electronic Properties and Utilities of Aromatic Diimides

Owing to the strong electron-withdrawing nature of imide rings fused to aromatic cores, electron-deficient aromatic diimides (Figure 7.1) constitute a fascinating family of n-type organic semiconductors.[1–3] Compared to all-carbon organic semiconductors, such as fullerenes, carbon nanotubes, and graphenes, aromatic diimides are much easier to synthesize and functionalize, and their electronic and optical properties can be easily manipulated by introducing different electron-donating and withdrawing groups at aromatic cores as well as on terminal imide rings.[1–3] The planar structures of aromatic diimides are replete with four carbonyl oxygen atoms, which allow them to participate in a range of noncovalent interactions, such as hydrogen bonding, π-stacking, and electron-donor/acceptor charge-transfer (CT) interactions. These structural, electronic, and molecular recognition properties render them promising building blocks for various functional supramolecular, polymeric, and hybrid materials.

Monographs in Supramolecular Chemistry No. 21
Naphthalenediimide and its Congeners: From Molecules to Materials
Edited by G. Dan Pantoş

Published by the Royal Society of Chemistry, www.rsc.org

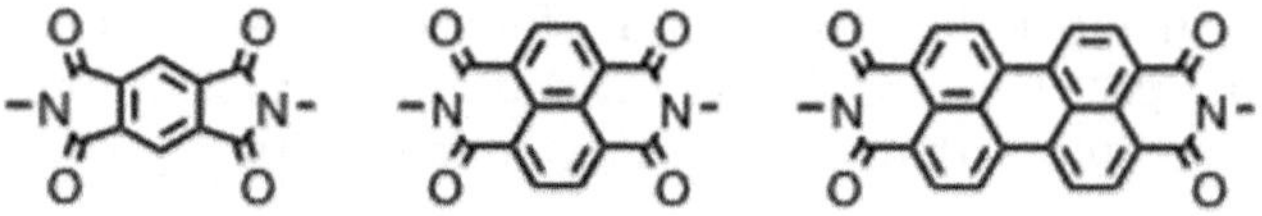

Figure 7.1 Chemical structures of pyromellitic diimide (PMI), naphthalenediimde (NDI), and perylene diimide (PDI).

Aromatic imides and diimides have long been used in many biological and medicinal applications. For example, thalidomide[4] was first introduced as an anti-nausea drug in the 1950s and it eventually played an important role in the evolution of stereochemistry, whereas other imides were found to act as DNA intercalators.[5] However, their electronic and optical properties remained largely unexplored and untapped until the 1990s. The situation changed dramatically once Iverson and Lokey[6] demonstrated that oligopeptide chains grafted with alternating electron-deficient naphthalenediimide (NDI) and electron-rich naphthalene units adapt pleated secondary structures and display interesting colorimetric changes indicative of donor–acceptor CT interactions. Arguably, this landmark discovery offered the first glimpse into the electronic and molecular recognition properties of aromatic diimides and initiated an onslaught from supramolecular chemists and materials scientists alike that led to their widespread applications in a variety of functional materials. These include organic field-effect transistors and semiconductors,[7] thin-films,[8] gelators,[9] nanotubes and nanorods,[10,11] light-harvesting and light-emitting materials,[12] and molecular switches in the form of mechanically interlocked rotaxanes and catenanes,[13] and metal–organic frameworks (MOFs)[14] to name a few. While much of these applications stemmed from π-donor/acceptor electron transfer (ET) and CT interactions of diimides, their noncovalent and electronic interactions with anions began to emerge much more recently.[15]

7.1.2 Importance of Anions and Anion Recognition

Anions are ubiquitous; many play critical roles in important biological, chemical, environmental, and industrial processes, while others are toxic and environmental pollutants. For example, in biology, anions undergo transmembrane passage—*i.e.*, symport and antiport of salts—to maintain ionic strength inside cells[16] and facilitate signal transduction through synapses,[17] hormone biosynthesis, and numerous enzymatic processes.[18] In chemistry, anions are used as bases, nucleophiles, reducing agents,[19] templating agents,[20] and cofactors.[21] In agriculture, NO_3^- and phosphates are used as fertilizers.[22] Radioactive $^{99m}TcO_4^-$ and $^{186}ReO_4^-$ ions are used in nuclear medicine and bio-imaging.[23] On the other hand, naturally occurring and anthropogenic CN^-, F^-, and chromates are highly toxic and environmental pollutants,[24] ClO_4^- is used as an explosive in rocket fuel,[25] and radioactive $^{99}TcO_4^-$ generated from nuclear power plants is a high-level nuclear waste.[26] Because of their diverse properties and functions, anion recognition chemistry has arisen to the forefront of supramolecular chemistry over the past two decades and it continues to inspire researchers to create new receptors that can regulate anions with high precision and selectivity.

7.1.3 Scope and Limitations of Traditional Anion Receptors

Most anion receptors developed thus far have relied primarily on H-bonding and electrostatic interactions.[27] While ion–dipole and ion–ion interactions can stabilize anions significantly and thereby allow receptors to overcome the dehydration energy of anions, they normalize the intrinsic electronic properties of anions. Therefore, the majority of the anion receptors cannot discriminate anions based on their Lewis basicity, charge density, and solvation energies; instead they rely heavily on size and shape matching for anion selectivity. Moreover, these noncovalent interactions are intrinsically nonchromogenic. Therefore, they do not usually engender any colorimetric and/or fluorimetric response in receptors unless the binding sites are coupled with separate signaling units covalently in a binding-site-signaling unit strategy or noncovalently in a signaling unit displacement strategy.[28] These factors require sophisticated receptor design and complicate the synthesis of anion sensors.

On the other hand, CT and ET interactions between π-donors and acceptors are known to produce characteristic optical responses, which can be easily exploited for visual detection and decoding of these species based on their electronic levels, *i.e.*, HOMO and LUMO energies, which match with the oxidation and reduction potentials, respectively. Although electron-deficient π-systems appear to be capable of forging electronic interactions with anions, surprisingly, such phenomena remained obscured and unexplored until the turn of this century.

7.1.4 Anion–π Interactions: A New Paradigm of Anion Recognition Chemistry

Conceived in theory[29] in the late 1990s and validated experimentally[30] throughout the past decade, anion–π interactions have added a new paradigm to anion recognition chemistry and revealed the possibilities of differentiating anions on the basis of their electronic properties. Mirroring cation–π interactions between cations and electron-rich π-systems,[31] anion–π interactions take place between anions and electron-deficient π-systems with positive quadrupole moments.[29,30] Despite their ostensible similarities, there remains a significant difference in the locations of guest cations and anions with respect to the respective π-systems. While virtually all cation–π complexes are centroid—*i.e.*, the cations are usually located above the center of the aromatic rings, which happen to be the most electron-rich regions of π-donors—the locations of anions in anion–π complexes vary with the structures of π-acceptors. For instance, centrosymmetric π-acceptors, such as six-fold symmetric hexafluorobenzene and three-fold symmetric cyanuric acid, are known to form perfectly centroid anion–π complexes,[29] as the centers of these π-acids are the most electropositive domains. In contrast, noncentrosymmetric π-acceptors tend to position guest anions in non-centroid locations, preferably in the vicinity of the strongest electron-withdrawing groups (EWGs) as they render that region more electropositive. While these differences are fully expected because of the fact that they are perfectly consistent with Coulombic interactions, they raised the question of whether anion–π complexes are actually held together by noncovalent interactions or if they involve some sort of covalent bonds, known as σ-complex or Meisenheimer complex formation, between the guest anion and π-acceptors. It is worth noting here that, although some computational studies projected the latter scenario,[32] no concrete experimental evidence—neither spectroscopic, nor crystallographic—could be found to support these theories. In fact, all experimental results, as well as advanced computational models, consistently refuted this viewpoint and showed that the interactions between anions and π-acceptors are primarily electronic in nature.

The early examples of anion–π interactions involved weak π-acceptors, such as C_6F_6, triazine, tetrazine, and cyanuric acid.[30] These interactions were so weak that they did not perturb the electronic properties of the receptors enough to elicit any optical response. However, the introduction of stronger π-acceptors, such as tetracyanopyrazine (TCP),[33] NDI[34–37] and PDI derivatives,[38] 1,4,5,8,9,12-hexaazatriphenylene hexacarbonitrile ($HAT(CN)_6$),[39,40] C_{60},[41] and tetracyanoquinodimethane (TCNQ)[41] revealed that anion–π interactions can be extended all the way to CT and ET events, which display distinct UV/Vis, fluorescence, NMR, and EPR spectroscopic changes.

While the nonchromogenic anion–π interactions between NDIs and PDIs have been thought to be operational for transmembrane anion and electron transport,[42,43] the recent discovery of chromogenic CT and ET events offers a completely new approach for optical detection and fingerprinting of anions on the basis of their Lewis basicity.[33–41] Herein, we will present an overview of different modes of interactions between anions and electron-deficient aromatic imides and their applications in (1) transmembrane anion transport, (2) colorimetric anion sensing and fingerprinting, and (3) radical reactions.

7.2 Transmembrane Anion Channels Based on Rigid NDI and PDI Rods

7.2.1 The Significance and Mechanism of Anion Channels

Transportation of ions across cell membranes is an important biological process that is responsible for maintaining proper cellular compositions and functions. Transmembrane proteins containing hydrophilic channels catalyze these events by compensating for the partial dehydration of ions as they traverse through hydrophobic membranes.[44] Ion transport through natural ion channels usually takes place through a multi-ion hopping mechanism:[44] the entry of an ion into a partially occupied ion channel pushes an already bound ion deeper into the channel, which in turn facilitates the ejection of another ion from the other end through tandem electrostatic repulsion. While cations are often actively transported through ATP-driven ion pumps, the passage of anions is either coupled with the movement of another anion in the opposite direction, known as antiport, or the movement of a cation in the same direction, namely symport. Transmembrane passage of ions is orchestrated by the interplay of two opposing factors: while strong interactions between anions and binding sites are necessary for high selectivity, tight binding impedes anion movement. Natural ion channels mitigate this quandary by offering multiple binding sites such that once anions enter into a channel they can slide through multisite interactions, while electrostatic repulsion between anions propels them forward under the influence of concentration gradients. Because of such intricacies, design, assembly, and investigation of artificial anion channels remain a formidable task. Nevertheless, it is an important and fascinating challenge, because malfunctions of natural ion channels are at the root of many physiological disorders. For example, gene-regulated malfunction of cystic fibrosis transmembrane conductance regulator (CFTR) chloride ion channels is responsible for a fatal disease.[45]

7.2.2 Anion Antiport through Rigid NDI Rod-based Anion Channels

The last decade has seen the proponents and practitioners of the anion–π field invest a lot of effort into unearthing and analyzing various anion–π complexes.[29–33] Anxious to deploy these weak noncovalent interactions for anion transport across lipid bilayer membranes, Matile and coworkers[42] constructed π-acidic rigid rod systems (**1**–**3**) by connecting three electron-deficient NDI units with two tetramethyl-*p*-phenylene linkers (Figure 7.2). The lengths of these rods were designed to match the membrane thickness (*ca.* 3.4 nm) of an egg yolk phosphatidylcholine large unilamellar vesicle (EYPCLUV) such that the resulting channels could bridge the intravesicular medium with the extravesicular one. To investigate the effects of cationic sites at one or both ends of these rods, neutral (**1**), monocationic (**2**), and dicationic (**3**) rods were prepared. In the hope of facilitating the formation of transmembrane channels and capturing cations simultaneously, two hairpin NDI rods (**4** and **5**) were also constructed by connecting two NDI rods with diethylene glycol chains.

The exchange of intravesicular Cl^- and Br^- ions with extravesicular F^-, Cl^-, Br^-, I^-, AcO^-, SCN^-, ClO_4^-, and SO_4^{2-} ions through the NDI-based anion channels made was investigated as a function of the pH-responsive fluorescence change of the 8-hydroxy-1,3,6-pyrenetrisulfonate (HPTS) fluorophore loaded inside the vesicles. No fluorescence change was observed

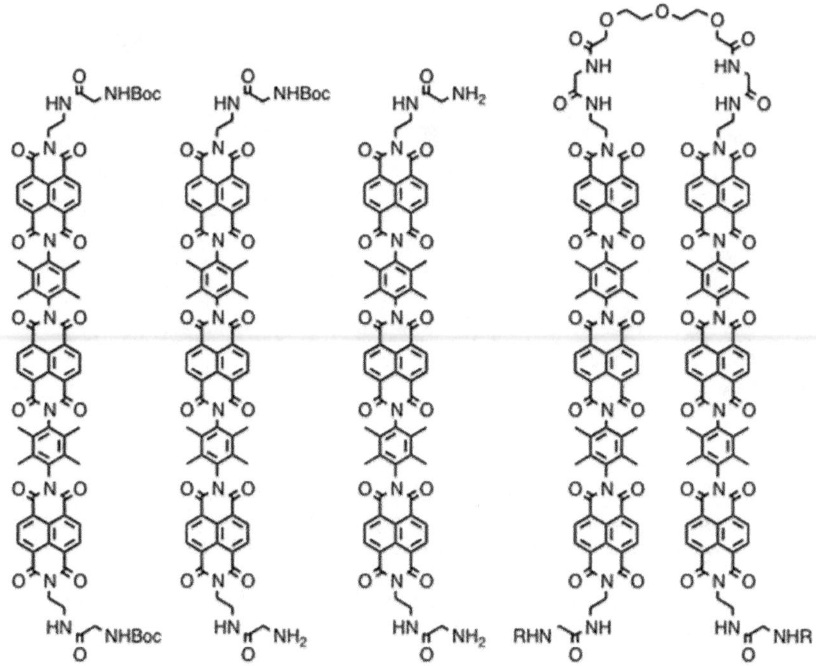

1 2 3 4 (R = Boc) 5 (R = H)

Figure 7.2 Rigid NDI rods used as transmembrane anion channels. Adapted from ref. 42a with permission from the Royal Society of Chemistry.

until a pH gradient was created across the membrane by adding NaOH to the external medium and the NDI rods were incorporated into membranes, which purportedly formed transmembrane channels and consequently turned ON anion antiport (Figure 7.3), a chain of events that triggered a fluorescence change. The influx of external OH^- ions into the vesicles engendered a dramatic pH-driven fluorescence change of HPTS. The rate of HPTS fluorescence change was dependent on the concentration of NDI rods. More efficient exchange of external anions X^- with internal Cl^- and Br^- ions also led to a faster and more dramatic fluorescence change. The rate of pH-induced fluorescence change was faster when the NDI rods were introduced prior to applying the base pulse than when they were introduced in the reverse order. In the former case, the NDI rods could already assemble into transmembrane channels before the change in pH and thereby facilitated anion transport more effectively. The anion transport activity of NDI rods remained unaffected by the presence of different cations, such as Na^+, K^+, Rb^+, Cs^+, indicating that these channels did not transport any cations.

When a base pulse was applied before the introduction of neutral NDI rod **1**, the ensuing ion channels displayed the following anion selectivity trend: $AcO^- \approx F^- > Cl^- > SO_4^{2-} > Br^- > I^- > SCN^- > ClO_4^-$, which contradicted with

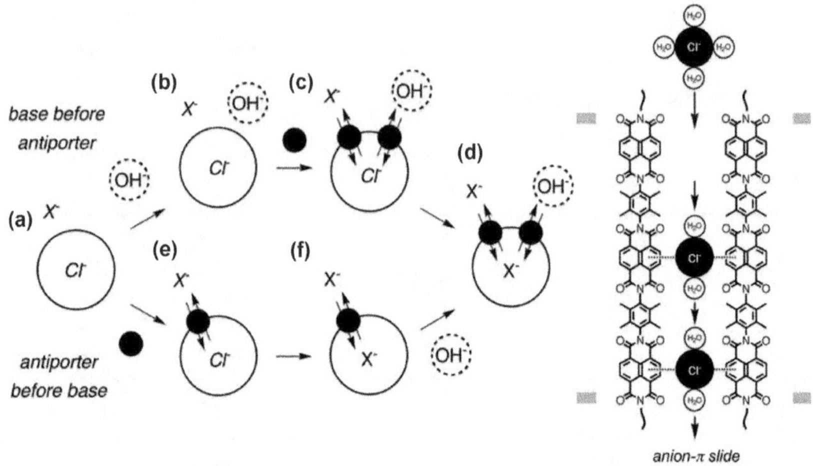

Figure 7.3 The proposed mechanism of anion antiport through artificial anion channels composed of π-acidic rigid NDI rods in a typical HPTS assay. In the presence of the NDI-based ion-channels, OH^- and a non-Lewis basic X^- anion enter into the EYPC vesicles in exchange of the release of Cl^- from inside the vesicles, as shown by a HPTS fluorescence change. Reproduced from ref. 42a with permission from the Royal Society of Chemistry.

the Hofmeister bias,[46] as the anions with greater charge density and hydration energies were surprisingly transported more effectively through the hydrophobic channels. The anion selectivity trend remained more or less the same, *i.e.*, the more hydrated external anions exchanged more effectively with the internal Cl^- and Br^- anions when the NDI rods were introduced before the base pulse. Essentially the same trend was observed in the presence of cationic rod **2**.

Among the different NDI rods (**1–5**), the singly charged one (**2**) was found to be the most effective for anion antiporter, presumably because of its better solubility in water. Highly water-soluble rod **3** containing cationic sites at both ends, however, turned out to be a poor antiporter. Surprisingly, the neutral and cationic hairpin NDI rods **4** and **5**, which were expected to form more effective anion channels because of intramolecular π–π-stacking interactions between the two tethered NDI rods, displayed exceptionally weak anion transport capacity. The unexpectedly poor anion transport activity of hairpin NDI rods was attributed to their possible aggregation into micelles or vesicles, although no evidence was presented in support of this speculation.

Anomalous mole fraction effect (AMFE) analysis suggested the manifestation of a multiion hopping mechanism.[44] In particular, the anion transport activity of rod **1** for an extravesicular Cl^- and I^- mixture was found to be less than the sum of activity of each anion, whereas transport of a Cl^-/ClO_4^- mixture showed a perfect additivity of each anion. These results indicated that while transport of Cl^- and I^- was competitive, rod **1** did not transport ClO_4^-. AMFE also suggested that binding of more than one Cl^- ion was needed for its fast transport, indicating the involvement of a multiion hopping mechanism.

7.2.3 Can Weak Anion–π Interactions Overcome Dehydration Penalties of Anions?

While dissipation of the pH gradient across EYPCLUV membranes in the presence of rigid NDI rods offers a strong indication of the exchange of internal Cl^- and Br^- anions with external anions (X^-) through an antiport mechanism, the nature of interactions between the anions and NDI rods is not obvious from these experiments. To decipher these interactions in an aqueous medium, Matile and coworkers invoked gas-phase computational studies.[42]

The electrostatic potential (ESP) maps of model NDI compounds show that the imide (pyridine dione) rings are the most electron-deficient parts of these π-acidic molecules. The quadrupole moments ($Q_{zz} \geq +14$ B) of NDIs far exceed that of the well-known π-acid hexafluorobenzene (+9.5 B) making them one of the best candidates for anion–π interactions.[42] Computational studies confirmed that in the gas phase, Cl^- and Br^- ions preferentially interact with the electron-deficient imide rings of the NDI molecules through anion–π interactions and the separation between the two entities depends on the size of the anion as well as the steric congestion above the imide rings. In the gas phase, the maximum energies of anion–π interactions between a model NDI compound and Cl^- and Br^- were estimated to be -18.2 and -15.7 kcal mol^{-1}, respectively, which are significantly dwarfed by the large hydration energies of these anions: *ca.* -86 and -80 kcal mol^{-1}, respectively.[47] The anion–π interactions of NDIs are further assisted by CH···anion H-bonding interactions between the guest anion and the N-aryl ring projected perpendicularly to the core NDI plane. On the basis of these gas-phase computational results, authors suggested that weak anion–π interactions between NDI rods and anions may be responsible for their transmembrane passage.[42] It was proposed that in the resting state of the channel, each terminal NDI unit binds an anion through an anion–π interaction. When a new anion enters the channel, it repels the anion bound to a terminal NDI unit deeper into the channel, which in turn expels the other NDI-bound anion due to Coulombic repulsion.

Despite these simplistic explanations, a closer look raises the question of how weak anion–π interactions worth less than 20 kcal mol^{-1} stabilization can overcome the large dehydration penalty of SO_4^{2-}, AcO^-, F^-, Cl^-, and Br^- anions as they travel through the hydrophobic channels made of NDI rods. In reality, it is thermodynamically impossible for weak anion–π interactions ($\leq$20 kcal mol^{-1}) to overcome the large dehydration penalties ($\geq$80 kcal mol^{-1}) of these anions. To address this discrepancy, authors have proposed[42] that perhaps the partially hydrated anions interacted with NDI rods *via* anion–π interactions avoiding a complete dehydration of guest anions. However, it is unlikely that the hydrated anions would experience any further stabilization upon interacting with the hydrophobic π-surfaces of NDI rods, when the anion–π interactions involving the naked anions are extremely weak in the first place. Therefore, the question remained whether or not anion–π interactions were involved at all for the transport of anions across EYPCLUV membranes replete with hydrophobic NDI-based channels. Recently, Matile's[32] and Saha's[34–36] groups have separately presented mass spectrometric (MS) evidence for anion–NDI complexes, however, it should be pointed out that MS data do not offer any insights into the nature of these complexes, nor do they confirm whether the same interactions would be taking place in aqueous media.

An alternate, and perhaps more logical scenario, could be H-bond formation between the carbonyl-O atoms of NDI rods and H_2O molecules surrounding the hydrated anions. Such a scenario would not require anions to be fully dehydrated before they enter into the hydrophobic NDI channels, yet it would provide enough stabilization to help them traverse through the NDI-based ion channels under concentration gradients. It is worth noting that H-bonding interactions are usually much stronger that weak anion–π interactions, especially when there is no contribution from charge transfer or electron transfer events between the guest anions and π-acidic receptors.

7.3 Photosynthetic Activity of PDI Rods

Encouraged by the anion transport activity of NDI rods, Matile and coworkers[42,43] attempted to couple anion transport with a photoinduced electron transfer (PET) process by using rigid rods made of core-substituted PDI (cPDI) chromophores that absorb visible light (Figure 7.4). One bis-PDI rod was composed of two electron-rich cPDI units, each carrying two pyrrolidine pendants, while another was composed of a pyrrolidine-substituted PDI unit and an aryl ether-substituted PDI unit connected by a *p*-phenylene ring. To facilitate the penetration of these PDI rods into EYPCLUV membranes, one end was decorated with a polar tri-glutamate chain and the other with a hydrophobic cyclohexyl group. Once again, the intravesicular chamber was loaded with a pH-sensitive HPTS fluorophore, NaCl, and a redox-active $[Co(bpy)_3]^{3+}$ complex and the extravesicular medium was enriched with SO_4^{2-}, NO_3^-, ClO_4^-, Cl^-, and Br^- anions as well as EDTA, which was used as a reducing agent.

Like NDI rods, PDI rods also displayed anion transport activity, which authors attributed to anion–π interactions with the core-substituted PDI units. However, these PDI rods, especially the dipyrrolidine-substituted PDI units, are reasonably electron-rich and therefore, less likely to participate in anion–π interactions with hydrated anions. The ability of electron-rich PDI

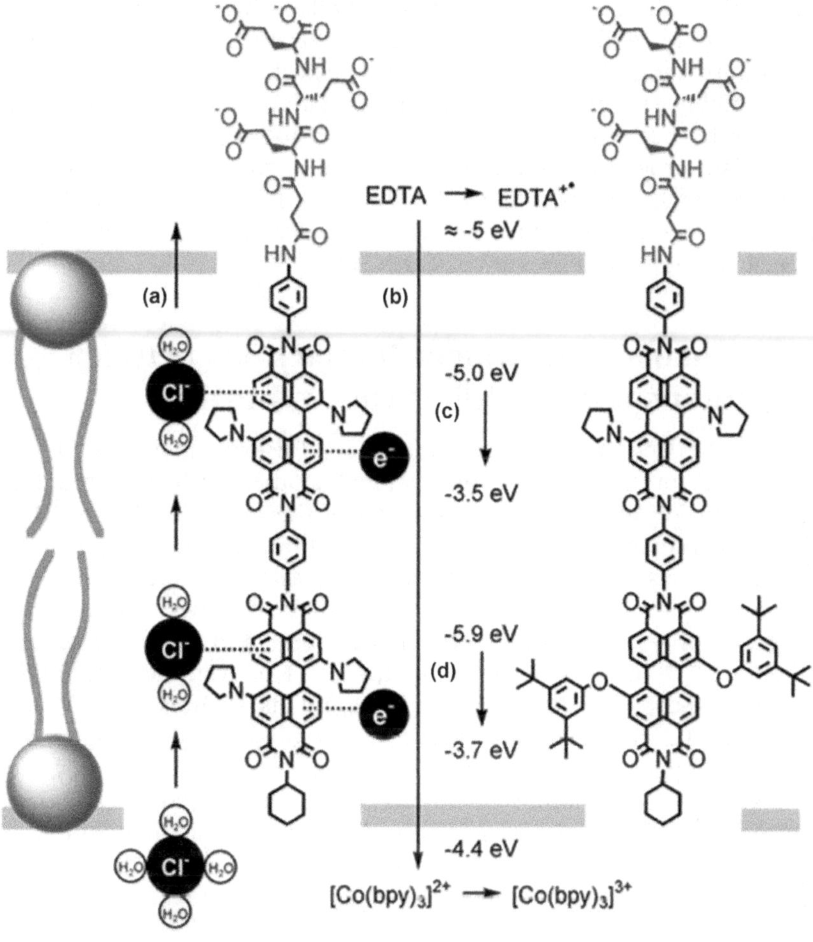

Figure 7.4 The proposed mechanism of transmembrane Cl^- passage coupled with photoinduced electron transfer involving PDI-based photosynthetic rods.
Reproduced with permission from ref. 43. Copyright © 2008 WILEY-VCH Verlag GmbH & Co. KGaA, Weinheim.

rods to transport anions in fact offers a strong support behind the potential involvement of H-bonding interactions between hydrated anions and the heteroatoms of PDI units instead of the perceived anion–π interaction between the hydrophobic π-surfaces of the electron-rich PDI units.

Sensitization of the membrane-bound PDI rods activated PET from the electron-rich dipyrrolidine-substituted PDI unit through the aryl ether-substituted PDI units to the intravesicular $[Co(bpy)_3]^{3+}$ complex, which was reduced to the corresponding Co^{2+} complex, as observed by its absorption change. The extravesicular EDTA acted as a reducing agent to reduce the photooxidized PDI units to reproduce their neutral forms.

While this initial work deserves to be extolled, there remained many discrepancies in the early explanations. Perhaps aware of these inconsistencies, the authors candidly declared:[42] "Many aspects of introduced concepts remain highly speculative. Several interpretations and conclusions made may thus be best considered as tentative, made with the only intension of stimulate progress in the field". Surely, this goal was soon realized, as several research groups,[34–41] delved into this field to glean a clear and accurate picture of electronic interactions between anions and π-acidic NDI and PDI receptors in different solvent media.

7.4 Tunable Noncovalent Interactions between Anions and π-acidic NDIs and PDIs

Although the lack of appropriate and insightful experimental studies and over-reliance on gas-phase computational studies early on have led to much confusion and mischaracterization of the nature of anion–NDI and anion–PDI interactions,[34–38] a clear and accurate understanding of these interactions has surfaced recently through extensive spectroscopic and crystallographic analyses. The advancement of the field was facilitated by the emergence of CT complexes between Cl^-, Br^-, and I^- ions and strong π-acidic TCP[33] and $HAT(CN)_6$ compounds.[39] However, the possibility of electron transfer from strong Lewis basic anions such as OH^-, F^-, CN^-, AcO^-, and $H_2PO_4{}^-$ to π-acids remained unexplored until recently. Nevertheless, analogous π-donor/acceptor CT and ET interactions[48] motivated us to investigate whether anion–π-acid interactions can be extended all the way to formal electron transfer by adjusting the electron donating abilities, *i.e.*, HOMO levels of anions and electron accepting abilities, *i.e.*, LUMO levels of the π-acids.

7.4.1 Electronic Constraints of Donor–Acceptor ET and CT Interactions

In donor–acceptor systems, $[D\cdot A]$ charge-transfer states and $[D^{\bullet+}\cdot A^{\bullet-}]$ electron-transfer (ET) states belong to the same energy continuum (Figure 7.5).[34–36,38] The manifestation of CT or ET depends on the relative energies of the HOMO and LUMO levels of electron donors and acceptors, respectively, and their spectroscopic signals are quite distinctive. For instance, if the HOMO of an electron-donor is located well above the LUMO of an electron-acceptor (Figure 7.5(a)), then thermal ET ($\Delta G^\circ_{ET} < 0$) should generate paramagnetic $D^{\bullet+}$ and $A^{\bullet-}$ radical ions.[36] In contrast, if the HOMO of an electron-donor is located below the LUMO of an electron-acceptor, then thermal ET should be turned off ($\Delta G^\circ_{ET} > 0$, forbidden). However, if the donor-HOMO is located above the photogenerated SOMO-1 of the electron-acceptor, a photoinduced electron transfer (PET) pathway could be activated, generating the same $A^{\bullet-}$ radical anion. Regardless of the electron source

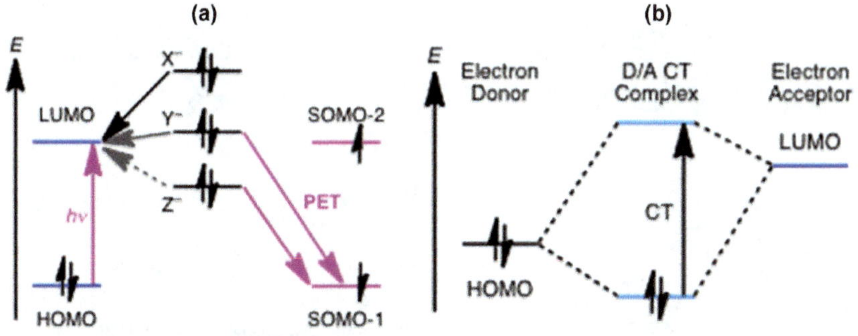

Figure 7.5 The HOMO and LUMO levels of electron-donating anions and π-acidic acceptors illustrating the thermodynamic requirements for thermal ET, PET, and CT interactions.
Adapted with permission from ref. 36. Copyright 2012 American Chemical Society.

(electrochemical *vs.* chemical reduction, thermal ET *vs.* PET), a given $A^{\bullet-}$ radical anion always displays the same characteristic signals (UV/Vis and EPR). The rate and extent of $A^{\bullet-}$ formation depend on the electron-transfer driving force (ΔG°_{ET} and ΔG°_{PET}). When both thermal and photo-induced ET are turned OFF because of a significantly low-lying donor-HOMO level, neutral D–A CT complexes could be formed *via* orbital mixing (Figure 7.5). CT complexes show broad UV/Vis absorption bands, the locations of which (λ_{CT}) depend on the donor and acceptor strengths, which is known as Mulliken dependence.[49]

7.4.2 UV/Vis Spectroscopic Evidence of Anion-induced ET and CT Events

UV/Vis, NMR, and EPR spectroscopies revealed that, depending on the (1) Lewis basicity of anions, (2) π-acidity of receptors, and (3) solvents (protic *vs.* aprotic), anions interact with π-acidic NDI and PDI receptors through discrete anion–π, CT, and ET interactions.[34–38] Precisely, in aprotic solvents, such as DMSO, DMF, MeCN, THF, and *o*-dichlorobenzene, where anions are less solvated, (1) strongly Lewis basic anions ($X^- = OH^-$ and F^-) trigger thermal ET to π-acidic NDIs, generating orange-colored $NDI^{\bullet-}$ radical anions and pink-colored NDI^{2-} dianions; (2) less Lewis basic anions ($Y^- = AcO^-$ and $H_2PO_4^-$) that cannot trigger thermal ET, can initiate photoinduced ET to the photogenerated SOMO-1 of excited NDIs, forming the $NDI^{\bullet-}$ radical anions; (3) poor Lewis basic anions ($Z^- = Br^-$ and I^-), which do not produce any $NDI^{\bullet-}$ *via* ET, can form charge CT complexes; and (4) charge-diffuse anions (TfO^- and ClO_4^-) can bind NDIs through nonchromogenic anion–π and CH···anion interactions.[50–52]

UV/Vis titration of unsubstituted *N,N′*-dipyridyl-NDI (DPNDI) and core-substituted dicyano-NDI (DCNDI) derivatives with strong Lewis basic anions,

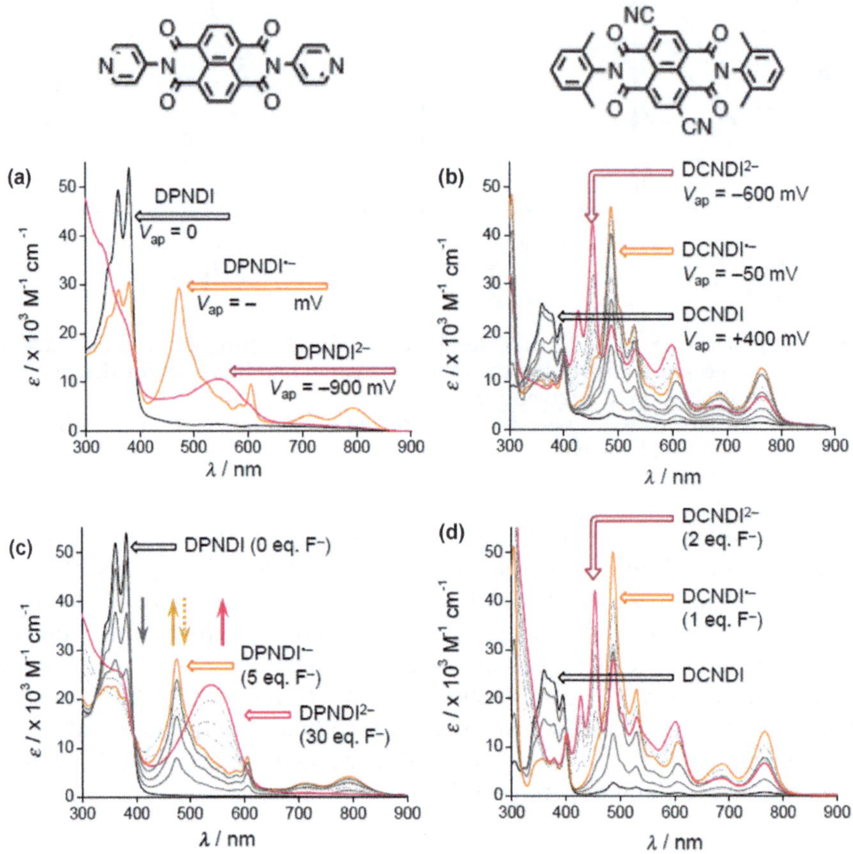

Figure 7.6 The absorption spectra of the electrochemically generated radical anion and dianion of (a) DPNDI and (b) DCNDI. The radical anion and dianion spectra of (c) DPNDI and (d) DCNDI generated *via* thermal ET from F^- anion.
Reproduced with permission from ref. 36. Copyright 2012 American Chemical Society.

such as OH^- and F^- in aprotic solvents produced spectroscopic features (Figure 7.6) that are identical to those displayed by electrochemically generated corresponding radical anions ($DPNDI^{\bullet -}$ and $DCNDI^{\bullet}$) and dianions ($DPNDI^{2-}$ and $DCNDI^{2-}$).[36] Not only do the final UV/Vis spectra of the anion-generated radical anions and dianions match perfectly with the corresponding electrochemically generated spectra, but also the isosbestic points observed during spectroscopic transitions engendered by both reduction processes are identical. These results clearly indicate that thermal ET from Lewis basic anions is responsible for the stepwise formation of the radical anions and dianions of these π-acids and rule out the formation of any covalent intermediates, namely Meisenheimer complexes, *via* nucleophilic attack of anions on π-acids. The latter scenario would have rendered anion-induced spectral change and isosbestic points different from those produced by electrochemical reduction. This is a significant revelation, because prior to in-depth spectroscopic analysis conducted by Saha and coworkers[34–36,38] and backed-up by others,[37,40,41] the F^--generated $DCNDI^{\bullet-}$ radical anion spectrum was incorrectly ascribed to a Meisenheimer complex by Matile and coworkers.[32] These experimental results also exposed the limitations of gas-phase computational analysis in explaining supramolecular interactions in solutions.[32]

Using a library of NDI derivatives with tunable electron accepting ability, Saha and coworkers[35,36] also demonstrated that the extent of the corresponding $NDI^{\bullet-}$ radical anion formation diminishes systematically with the gradually decreasing π-acidity of the receptors and Lewis basicity of anions. This trend confirmed that the efficacy of ET from anions to π-acids in aprotic solvents strictly depends on the energy gap between the donor HOMO and acceptor LUMO, *i.e.*, ΔG°_{ET}. When strong π-acidic receptors and Lewis basic anions are involved, the ΔG°_{ET} value is negative, which allows for facile thermal ET from anions. But it becomes smaller and smaller with the diminishing π-acidity of NDIs and Lewis basicity of anions, which eventually turns off thermal ET. Therefore, in the absence of light, less Lewis basic AcO^- and Cl^- anions can produce negligible amounts of $NDI^{\bullet-}$ radical anions *via* thermal ET. However, excitation of NDIs in the presence of these anions turns on PET from the anions' HOMO to the photogenerated SOMO-1 of NDIs, producing significant amounts of $NDI^{\bullet-}$ radical anions. The UV/Vis spectra of $NDI^{\bullet-}$ radical anions generated *via* thermal or photoinduced processes are identical.

Analogous to π-acidic NDI compounds, PDIs are also reduced to $PDI^{\bullet-}$ radical anions by OH^- and F^-, and to PDI^{2-} dianions in the presence of OH^- (Figure 7.7).[38] The reduction of these π-acids to the corresponding radical anions can also be effected by other Lewis basic anions, such as AcO^-, $H_2PO_4{}^{2-}$, and CN^-, as well as various amines and hydrazine.[34–38,53] The fact that strongly Lewis basic OH^- was able to produce both a $PDI^{\bullet-}$ radical anion and PDI^{2-} dianion, whereas F^- could produce only the $PDI^{\bullet-}$ radical anion, confirmed that more Lewis basic OH^- is indeed a stronger reducing anion than F^-. Similarly, when more electron-deficient DCNDI was introduced, both OH^- and F^- anions produced the corresponding radical anion and dianion, but less Lewis basic I^- could only lead to a partial formation of the radical anion but did not generate any dianion.[36] These results clearly suggest that in aprotic solvents, more Lewis basic OH^- and F^- anions indeed act as much stronger reducing agents than I^-.

Solvents play very important roles in the electron-donating abilities of anions. For example, while AcO^- can reduce more π-acidic NDI derivatives to the corresponding $NDI^{\bullet-}$ radical anions in DMSO and *o*-dichlorobenzene (ODCB) *via* thermal ET, this process is turned OFF in MeCN, which stabilizes the anion *via* CH···anion H-bond formation. However, upon irradiation of NDI, PET is turned ON, generating the characteristic $NDI^{\bullet-}$ radical anion spectra.[35,36]

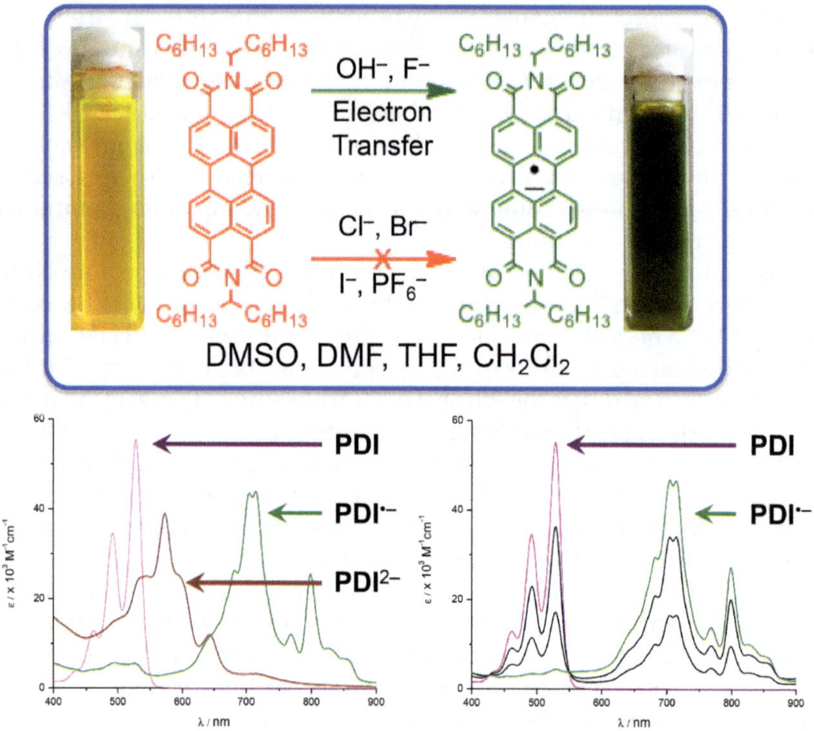

Figure 7.7 The radical anion and dianion spectra of PDI generated *via* thermal ET from OH^- and F^- anions.
Reproduced from ref. 38 with permission from the Royal Society of Chemistry.

7.4.3 Electron-donating Abilities of Anions in Aprotic Solvents

These results are fully consistent with the fundamental concepts of acid–base chemistry: weaker acids afford stronger conjugate bases, the electron-donating abilities of which follow their Lewis basicity in aprotic solvents.[47] The acid strength increases with increasing electronegativity of elements along the rows of the periodic table. Therefore, OH^- (pK_a of $H_2O = 32$ in DMSO) acts as a stronger reducing agent than F^- (pK_a of HF = 15 in DMSO), which explains why the former can reduce PDI all the way to its dianion while the latter can only produce the radical anion. The acid strength does not follow the electronegativity trend along the Group; instead it increases (HF <HCl <HBr <HI) with the increasing ionic radii of the corresponding conjugate bases ($I^- > Br^- > Cl^- > F^-$). Consequently, the Lewis basicity of halides follows the following trend: $F^- > Cl^- > Br^- > I^-$. This is because larger halide anions can accommodate the negative charge in a bigger sphere much more easily—*i.e.*, they experience much lower electronic reorganization energy—than a much smaller F^- anion. As a result, in aprotic solvents, such as DMSO, DMF, MeCN, THF, and ODCB, sparsely solvated F^- acts as a much better reducing agent than I^-.

Interestingly, the situation changes dramatically in protic solvents, which solvate small F^- anions containing a large charge density much more extensively than larger Br^- and I^- ions. Therefore, smaller anions have a much larger hydration energy than the larger ones. In protic solvents, heavily solvated F^- becomes so stable that its HOMO level drops below the LUMO of π-acids and consequently, ET from F^- is turned OFF. Therefore, in water, hydrated F^- ions act as much weaker electron-donors than less solvated I^- ions.

The stabilization of F^- and subsequent loss of its reducing capacity upon forming a H-bonded complex was further demonstrated by Ballester and coworkers.[40] They demonstrated that while the "naked" F^- ions in MeCN can easily reduce strongly π-acidic $HAT(CN)_6$ to its radical anion and dianion, a preformed $F^- \supset$calix[4]pyrrole complex could not reduce it at all because of a dramatic stabilization of F^- in the H-bonded complex. These experimental observations were further validated by advanced computational analysis, which revealed that in MeCN, the HOMO levels (eV) of anions closely follow their Lewis basicity trend: OH^- (−3.27) > F^- (−4.33) > Cl^- (−5.25) ≈ Br^- (−5.31) ≈ I^- (−5.14) > HF_2^- (−6.40). These values are fully consistent with the experimentally observed reducing ability of anions.[40]

7.4.4 NMR and EPR Evidence of Anion-induced Reduction of π-Acids

The anion-induced reduction of NDI and PDI derivatives to paramagnetic radical anion forms was further confirmed by NMR and EPR spectroscopies.[34,36,38] Upon titration of these neutral π-acids with Lewis basic anions, their characteristic NMR signals became broad and finally disappeared indicating paramagnetic radical anion formation. Addition of an oxidizing agent, $NOBF_4$, regenerated the neutral species, as their characteristic NMR signals returned to full glory demonstrating the complete reversibility of the chemical reduction process (Figure 7.8). Consistent with the NMR titration data, characteristic EPR signals of paramagnetic $NDI^{\bullet -}$ and $PDI^{\bullet -}$ radical anions emerged in the presence of Lewis basic anions, such as F^-, confirming anion-induced ET to π-acids.

^{19}F NMR spectroscopy revealed the oxidation of a NMR-active diamagnetic F^- anion to a NMR-silent $F^{\bullet}$ radical upon ET to π-acids.[34,36,38] In DMSO-d_6, $Bu_4NF \cdot 3H_2O$ displays a characteristic F^- signal at −102 ppm and HF_2^- peak at −142 ppm. Upon addition of NDI or PDI compounds into the fluoride solution, the F^- signal becomes broad and eventually disappears at a 1 : 1 molar ratio, indicating the oxidation of the anion to a paramagnetic radical. This extremely high-energy radical species immediately acts as a sacrificial agent, the product of which remains elusive to date. On the other hand, the HF_2^- signal remained essentially unaffected, indicating that it was the F^-

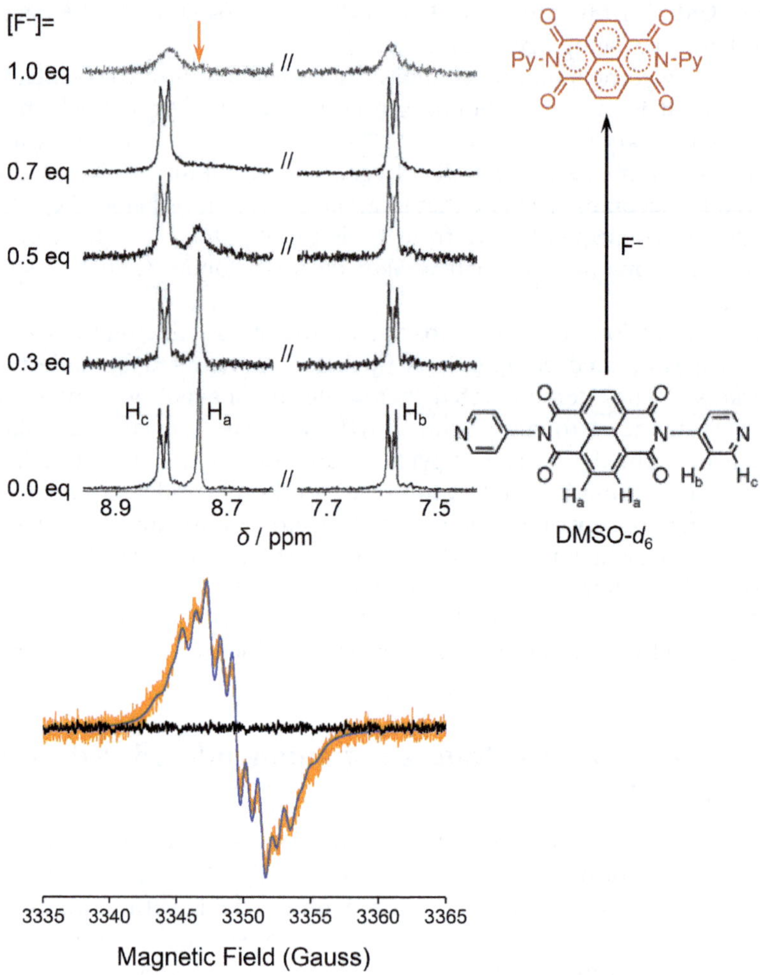

Figure 7.8 (a) 1H NMR titration of DPNDI with F^- and (b) EPR spectroscopy confirm the formation of paramagnetic $DPNDI^{\bullet -}$ radical anion at a 1:1 molar ratio.
Reproduced with permission from ref. 36. Copyright 2012 American Chemical Society.

anion that was responsible for the reduction of the π-acids to radical anions *via* thermal ET.

The NMR and EPR data not only confirmed the anion-induced ET phenomena, but also ruled out alternate scenarios, such as a Meisenheimer complex formation[32] or deprotonation of any C–H bond. Had the anion attacked the π-acids as a nucleophile producing covalent Meisenheimer complexes or acted as a base to deprotonate them, these products would have been diamagnetic species that should have displayed new NMR signals instead of the disappearance of the original NMR signals. Also, covalent bond formation between the anion and π-acid would have been an irreversible process and regeneration of neutral receptors by $NOBF_4$ would be highly improbable, if not impossible. However, none of these scenarios evolved, thereby eliminating these possibilities.

7.4.5 Anion-induced Reduction and Deprotonation of NDI Derivatives

Langford[54] and Mukhopadhyay[55] have separately shown that NDI derivatives containing acidic protons at their cores or at imide positions can be deprotonated by basic anions. The N–H protons of a core-substituted NDI containing sulfonamide groups are easily deprotonated by F^- ions, which changes the color from blue to green.[54] Similarly, the N–H group of a hydrazimide derivative of NDI also undergoes deprotonation in the presence of excess F^- leading to the formation of an N-centered dianion, $[M-2H]^{2-}$, that shows completely different spectroscopic signals from the NDI^{2-} dianion formed *via* thermal ET from Lewis basic anions (Figure 7.9).[55] In the NDI^{2-} dianion, the electron density is delocalized through the NDI core and the highest charge density is located on the imide-O atoms. In contrast, the deprotonated dianion $[M-2H]^{2-}$ contained the negative charges on the distal N atoms of hydrazimide, which are delocalized through the NDI core, but could participate in intramolecular CT within the adjacent electron-deficient aryl groups. The solvent and concentration-dependent formation of two discrete dianions was exploited for the development of molecular logic gates that display distinct UV/Vis spectra at particular concentrations of the F^- anion.

7.4.6 Cation-induced Oxidation of Electron-rich Amino-NDI Derivatives

In contrast to π-acidic NDI and PDI derivatives that interact with anions through tunable noncovalent interactions and can be reduced by strong

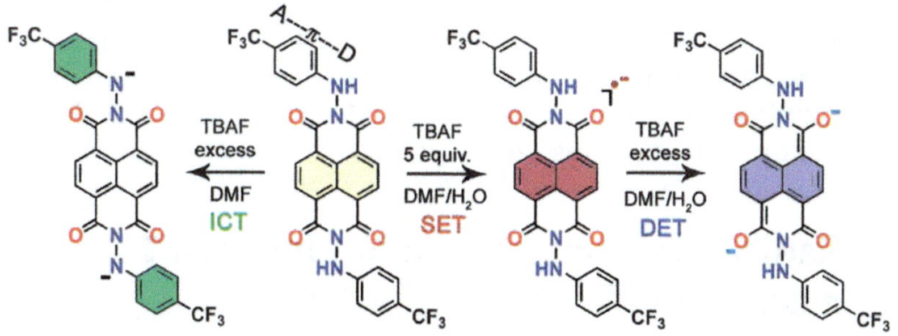

Figure 7.9 Fluoride-induced ET and deprotonation of a NDI derivative leading to the formation of dramatically different dianions.
Reproduced from ref. 55 with permission from the Royal Society of Chemistry.

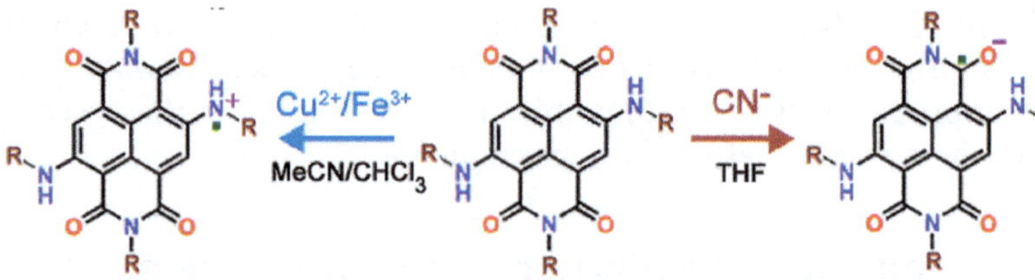

Figure 7.10 Cu(II) and Fe(III)-mediated oxidation and CN^--mediated reduction of an electron-rich diamino-NDI derivative.
Reproduced with permission from ref. 56. Copyright 2012 American Chemical Society.

Lewis basic anions, core amino-substituted diimides are electron-rich and can be oxidized to radical cations by Lewis acids, such as Cu(II) and Fe(III), as shown by Mukhopadhyay and coworkers (Figure 7.10).[56] Interestingly, the same diamino-NDI derivative can be reduced to its radical anion by a strongly Lewis basic CN^- anion. Steric crowding and intramolecular H-bonding between the core-NH group and imide-O render the radical cation fairly stable under ambient conditions. Electron spin densities of the radical anion and radical cation of diamino-NDI are located at different parts of the molecule. While the electron spin density in the radical anion is located along the molecular axis connecting two imide-N atoms, the charge density of the radical cation is located along the axis connecting two amine N-atoms on the NDI core. As a result, their electronic absorption spectra are distinctly different, which can be used for visual detection of oxidizing cations as well as reducing anions.

The above discussions illustrate how tunable electronic interactions, especially chromogenic ET and CT events, can be utilized for differentiating and detecting anions based on their intrinsic electronic and solvation properties, tasks that seemed impossible to achieve with traditional anion receptors because of the normalizing effects of H-bonding and electrostatic interactions. In addition to their immediate and obvious application in anion recognition and sensing, anion-induced ET events show great promise in the realms of functional electronic materials as well as radical chemistry. For example, the electron-donating abilities of OH^- and F^- anions have been exploited recently in high-performance polymer solar cells based on n-type organic semiconductors, such as TCNQ and C_{60} compounds,[41] which are reduced to the corresponding radical anions by these reducing anions.

7.5 Radical Reactions *via* Anion-induced ET to π-acidic Imides and Diimides

Long before the anion recognition properties of NDI and PDI derivatives were studied systematically, Kelly and coworkers[57] explored their ability to undergo PET from carboxylate groups and electron-rich aromatic side-chains of amino acids, such as tryptophan, tyrosine, and histidine, as well as nucleobases of DNA in aqueous buffers. Photoinduced reduction of diimides to radical anions and oxidation of electron-donating aromatic groups were confirmed by transient absorption spectroscopy. Interestingly, intra- and intermolecular PET from carboxylate anions led to radical decarboxylation followed by H-abstraction and radical coupling reactions. Griesbeck and coworkers[58] have also demonstrated that α-amino acids attached to π-acidic phthalimides undergo similar PET-driven radical decarboxylation, leading to the formation of the corresponding amines after hydrolysis of the imide derivatives. These results exemplify the utility of anion-induced ET processes to imides and diimides in synthetic radical chemistry.

7.6 Crystal Structures of Anion–NDI Complexes

The ultimate evidence of anion–π interactions of NDI receptors emerged from single crystal X-ray diffraction (SXRD) analysis. Saha and coworkers[50] were the first to report the crystal structure of a TfO^-/DPNDI anion–π interaction in a zigzag $[\{Pd(Ph_3PCH_2CH_2PPh_3)\}(DPNDI)\cdot(TfO)_2]_n$ coordination polymer (Figure 7.11). In this complex, every other DPNDI ligand formed anion–π complexes with two TfO^- anions—one located 2.83 Å above and the other below the two highly electropositive imide rings. Each of the other DPNDI ligands carried two THF molecules in the same fashion through lone-pair–π interactions. The separation between the TfO^- anions and Pd(II) centers is far greater than that between the anion and π-acidic DPNDI ligands, indicating that in this coordination polymer, soft anion–π interactions between charge-diffuse anions and π-acidic DPNDI ligands are favored over hard electrostatic interactions with the Pd(II) centers.

Further extending this work, Saha and coworkers[51] have demonstrated that NDI's ability to bind charge-diffuse anions noncovalently can be employed for template-directed synthesis and structural regulation of metal–organic frameworks (MOFs). A square-grid MOF was constructed from a Zn(II) ion and DPNDI ligands, each cavity (*ca.* 20×20 Å) of which was filled with two ClO_4^- anions (Figure 7.12). By occupying the cavities through CH···anion interactions with DPNDI ligands, the anions not only prevented network interpenetration but also bestowed significant thermostability to the porous architecture. This strategy also offers a unique opportunity to

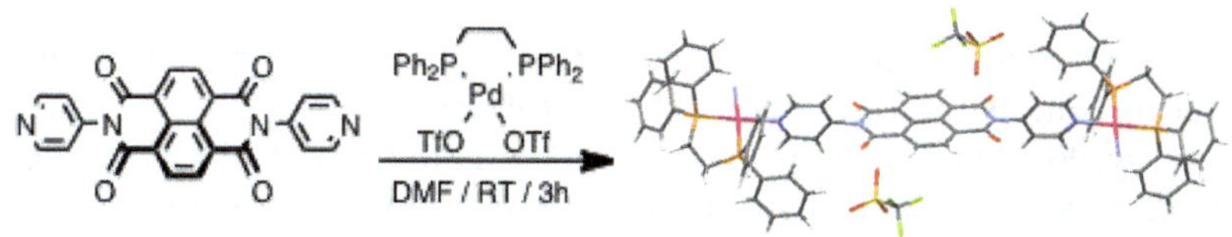

Figure 7.11 Crystal structure of a DPNDI-based Pd(II) coordination polymer showing an anion–π interaction between the TfO^- anions and imide rings. Reproduced from ref. 50 with permission from the Royal Society of Chemistry.

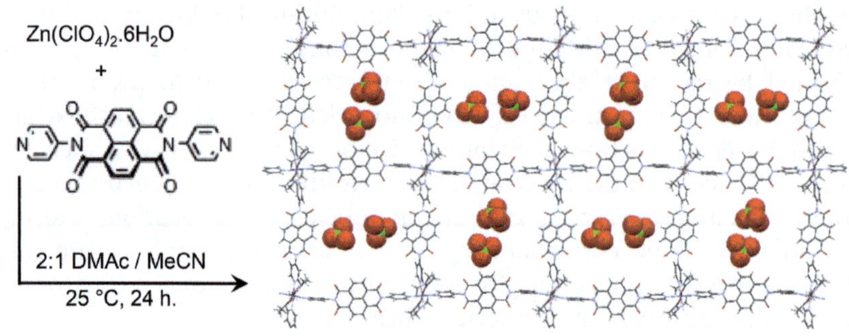

Figure 7.12 Crystal structure of a DPNDI-based square-grid MOF, each cavity of which is occupied by two ClO_4^- anions involved in CH···anion interactions.
Reproduced from ref. 51 with permission from the Royal Society of Chemistry.

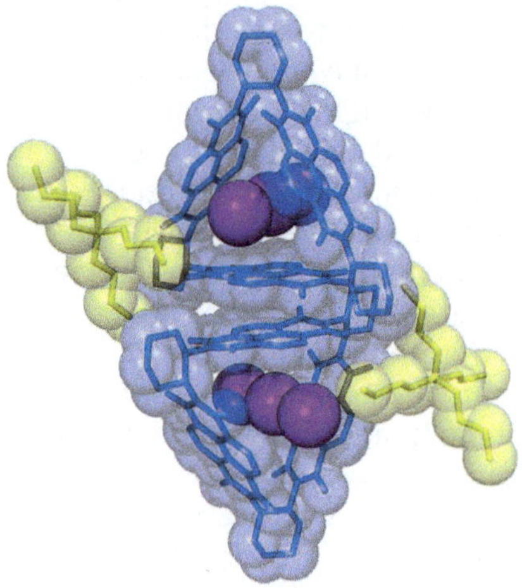

Figure 7.13 Crystal structure of a molecular triangular prism comprised of three NDI units that adapts helical superstructures upon encapsulating I_3^- anions *via* anion–π interactions.
Reproduced with permission from ref. 52. Copyright © 2013 WILEY-VCH Verlag GmbH & Co. KGaA, Weinheim.

sequester charge-diffuse anions into MOFs, which could be beneficial for separation and extraction of toxic charge diffuse anions.

Recently, Stoddart and coworkers[52] have constructed chiral molecular triangular prisms containing a tubular cavity encircled by three NDI sides. This architecture can encapsulate linear I_3^- anions through anion–π interactions, which in turn promotes stacking of these chiral cage molecules into helical superstructures with either right- or left-handed helicity (Figure 7.13).

7.7 Conclusions and Outlook

The foregoing discussions provide a complete and accurate depiction of the entire spectrum of tunable noncovalent interactions between anions and π-acidic aromatic diimides that range from weak anion–π and CH···anion interactions that do not perturb the electronic and optical properties of receptors all the way to charge-transfer and electron-transfer events that significantly modify the electronic and optical properties of the π-acids. In aprotic solvents, the intrinsic electronic properties of anions, *i.e.*, their Lewis basicity and electron-donating properties, are better preserved than in protic solvents, which exert normalizing effects *via* extensive solvation. For the same reasons, most traditional anion receptors that rely on H-bonding and electrostatic interactions for anion binding, are completely dependent on size- and shape-complementarity for selectivity. This impasse can be overcome by using π-acidic receptors that can decode, differentiate, and optically detect anions based on their intrinsic electronic properties. Although anion–π interactions involving aromatic diimides were first recognized and exploited in NDI and PDI-based transmembrane anion channels, the actual modes of these electronic interactions in solutions could not be always elucidated correctly by gas phase computational studies. However, in-depth spectroscopic and crystallographic analysis by many research groups in recent years has helped to eradicate these issues and exposed the actual nature of these interactions. These advancements in the past five years have expanded the scope of these electronic interactions beyond the realm of anion recognition chemistry, as they have already been implemented in high-performance polymer solar cells and molecular logic gates and radical reactions. Equipped with tunable electronic, optical, and molecular recognition properties, aromatic diimides and their congeners enjoy boundless potential in supramolecular and materials chemistry, much of which is still to be explored and exploited.

Acknowledgements

The authors acknowledge the National Science Foundation (Awards NSF-1507529 and NSF-1660329) for financial support.

References

1. B. A. Jones, A. Facchetti, M. R. Wasielewski and T. J. Marks, *J. Am. Chem. Soc.*, 2007, **129**, 15259.
2. S. V. Bhosale, C. H. Janiab and S. J. Langford, *Chem. Soc. Rev.*, 2008, **37**, 331.
3. N. Sakai, J. Mareda, E. Vauthey and S. Matile, *Chem. Commun.*, 2010, **46**, 4225.
4. M. E. Franke, G. R. Macpherson and W. D. Figg, *Lancet*, 2004, **363**, 1802.

5. V. Guelev, S. Sorey, D. W. Hoffman and B. L. Iverson, *J. Am. Chem. Soc.*, 2002, **124**, 2865.
6. R. S. Lokey and B. L. Iverson, *Nature*, 1995, **375**, 303.
7. H. E. Katz, A. J. Lovinger, J. Johnson, C. Kloc, T. Siegrist, W. Li, Y.-Y. Lin and A. Dodabalapur, *Nature*, 2000, **404**, 478.
8. M. A. Rodrigues, G. J. F. Demets and M. J. Ploiti, *J. Mater. Chem.*, 2002, **12**, 1250.
9. P. Mukhopadhyay, Y. Iwashita, M. Shirakawa, S.-I. Kawano, N. Fujita and S. Shinkai, *Angew. Chem., Int. Ed.*, 2006, **45**, 1592.
10. N. Ashkenasy, W. S. Horne and M. R. Ghadiri, *Small*, 2006, **2**, 99.
11. G. D. Pantoş, P. Pengo and J. K. M. Sanders, *Angew. Chem., Int. Ed.*, 2007, **46**, 194.
12. S. J. Langford, M. J. Latter and C. P. Woodward, *Photochem. Photobiol.*, 2006, **82**, 1530.
13. (a) T. Iijima, S. A. Vignon, H.-R. Tseng, T. Jarrosson, J. K. M. Sanders, F. Marchioni, M. Venturi, E. Apostoli, V. Balzani and J. F. Stoddart, *Chem. – Eur. J.*, 2004, **10**, 6375; (b) S. Burattini, B. W. Greenland, D. H. Merino, W. Weng, J. Seppala, H. M. Colquhoun, W. Hayes, M. E. H. Mackay and S. J. I. W. Rowan, *J. Am. Chem. Soc.*, 2010, **132**, 12051.
14. (a) K. L. Mulfort and J. T. Hupp, *J. Am. Chem. Soc.*, 2007, **129**, 9604; (b) K. L. Mulfort and J. T. Hupp, *Inorg. Chem.*, 2008, **47**, 7936; (c) O. K. Farha, C. D. Malliakas, M. G. Kanatzidis and J. T. Hupp, *J. Am. Chem. Soc.*, 2010, **132**, 950.
15. V. Gorteau, G. Bollot, J. Mareda, A. Perez-Velasco and S. Matile, *J. Am. Chem. Soc.*, 2006, **128**, 14788.
16. (a) R. Dutzler, E. B. Campbell, M. Cadene, B. T. Chait and R. MacKinnon, *Nature*, 2002, **415**, 287; (b) A. Accardi and C. Miller, *Nature*, 2004, **427**, 803.
17. D. Yamamoto, Y. Kiyozuka, Y. Uemura, C. Yamamoto, H. Takemoto, H. Hirata, K. Hioki, K. Tanaka and A. Tsubura, *J. Cancer Res. Clin. Oncol.*, 2000, **126**, 191.
18. (a) D. W. Christianson and W. N. Lipscomb, *Acc. Chem. Res.*, 1989, **22**, 62; (b) S. R. Davis-Kaplan, C. C. Askwith, A. C. Bengtzen, D. Radisky and J. Kaplan, *Proc. Natl. Acad. Sci. U. S. A.*, 1998, **95**, 13641; (c) J. J. Heymann, K. D. Weaver, T. A. Mietzner and A. L. Crumbliss, *J. Am. Chem. Soc.*, 007, **129**, 9704.
19. V. V. Rostovtsev, L. G. Green, V. V. Fokin and K. B. Sharpless, *Angew. Chem., Int. Ed.*, 2002, **41**, 2596.
20. (a) J.-F. Ayme, J. E. Beves, C. J. Campbell and D. A. Leigh, *Chem. Soc. Rev.*, 2013, **42**, 1700; (b) I. A. Riddell, Y. R. Hristova, J. K. Clegg, C. S. Wood, B. Beeiner and J. R. Nitschke, *J. Am. Chem. Soc.*, 2013, **135**, 2723; (c) M. S. Vickers and P. D. Beer, *Chem. Soc. Rev.*, 2007, **36**, 211; (d) M. D. Lankshear and P. D. Beer, *Coord. Chem. Rev.*, 2006, **250**, 3142; (e) R. Vilar and N. Gimeno, *Coord. Chem. Rev.*, 2006, **250**, 3161; (f) R. L. Paul, Z. R. Bell, J. C. Jeffrey, J. A. McCleverty and M. D. Ward,

Proc. Natl. Acad. Sci. U. S. A., 2002, **99**, 4883; (g) B. Hasenknop, J.-M. Lehn, N. Boumediene, A. Dupont-Gervais, A. Van Dorsselaer, B. Kneisel and D. Fenske, *J. Am. Chem. Soc.*, 1997, **119**, 10956.
21. J. S. Park, E. Karnas, K. Ohkubo, P. Chen, K. M. Kadish, S. Fukuzumi, C. W. Bielawski, T. W. Hundall, V. M. Lynch and J. L. Sessler, *Science*, 2010, **329**, 1324.
22. (a) B. Moss, *Chem. Ind.*, 1996, 407; (b) C. Glidwell, *Chem. Br.*, 1990, **26**, 137.
23. (a) K. Schwochau, *Angew. Chem., Int. Ed.*, 1994, **33**, 2258; (b) U. Y. Ryo, P. V. Vaidya, A. B. Schneider, C. Bekerman and S. M. Pinsky, *Radiology*, 1983, **148**, 819; (c) M. Eisenhut, *Int. J. Appl. Radiat. Isot.*, 1982, **33**, 99.
24. (a) S. Langard and T. Norseth, *Handbook on the Toxicity of Metals*, Elsevier, New York, 1979; (b) D. Browne, H. Whelton and D. O'Mullene, *J. Dent.*, 005, **133**, 177; (c) E. B. Bassin, D. Wypij, R. B. Davis and M. A. Mittleman, *Cancer, Causes Control, Pap. Symp.*, 006, **17**, 421.
25. (a) W. E. Motzer, *Environ. Forensics*, 2001, **2**, 301; (b) M. A. Greer, G. Goodman, R. C. Pleus and S. E. Greer, *Environ. Health Perspect.*, 002, **110**, 927.
26. N. C. Schroeder and K. R. Ashley, *J. Radioanal.Nucl. Chem.*, 2005, **263**, 567.
27. K. Bowman-James, A. Bianchi and E. Garcia-Espana, *Anion Coordination Chemistry*, Wiley-Blackwell, 2011.
28. R. Martinez-Máñez and F. Sancenón, *Chem. Rev.*, 2003, **103**, 4419.
29. (a) D. Quiñonero, C. Garau, C. Rotger, A. Frontera, P. Ballester, A. Costa and P. M. Deya, *Angew. Chem., Int. Ed.*, 2002, **41**, 3389; (b) M. Mascal, A. Armstrong and M. D. Bartberger, *J. Am. Chem. Soc.*, 2002, **124**, 6274; (c) I. Alkorta, I. Rozas and J. Elguero, *J. Am. Chem. Soc.*, 2002, **124**, 8593; (d) C. Garau, A. Frontera, D. Quinonero, P. Ballester, A. Costa and P. M. Deya, *ChemPhysChem*, 2003, **4**, 1344; (e) O. B. Berryman, V. S. Bryantsev, D. P. Stay, D. W. Johnson and B. P. Hay, *J. Am. Chem. Soc.*, 2007, **129**, 48; (f) M. Albrecht, C. Wessel, M. de Groot, K. Rissanen and A. Lüchow, *J. Am. Chem. Soc.*, 2008, **130**, 4600; (g) D. Y. Kim, N. J. Singh, J. W. Lee and K. S. Kim, *J. Chem. Theory Comput.* 2008, **4**, 1162; (h) A. Campo-Cacharrón, E. M. Cabaleiro-Lago and J. Rodríguez-Otero, *ChemPhysChem*, 2012, **13**, 570.
30. (a) A. Frontera, P. Gamez, M. Mascal, T. J. Mooibroek and J. Reedjik, *Angew. Chem., Int. Ed.*, 2011, **50**, 9564; (b) P. Ballester, *Acc. Chem. Res.*, 2013, **46**, 874; (c) H. T. Chifotides and K. R. Dunbar, *Acc. Chem. Res.*, 2013, **46**, 894.
31. J. C. Ma and D. A. Dougherty, *Chem. Rev.*, 1997, **97**, 1303.
32. R. E. Dawson, A. Henning, D. P. Weimann, D. Emery, V. Ravikumar, J. Montenegro, T. Takeuchi, S. Gabutti, M. Mayor, J. Mareda, C. A. Schalley and S. Matile, *Nat. Chem.*, 2010, **2**, 533.
33. Y. S. Rosokha, S. V. Lindeman, S. V. Rosokha and J. K. Kochi, *Angew. Chem., Int. Ed.*, 2004, **43**, 4650.
34. S. Guha and S. Saha, *J. Am. Chem. Soc.*, 2010, **132**, 17674.
35. S. Guha, F. G. Goodson, S. Roy, L. J. Corson and S. Saha, *J. Am. Chem. Soc.*, 2011, **133**, 15256.

36. S. Guha, F. G. Goodson, L. J. Corson and S. Saha, *J. Am. Chem. Soc.*, 2012, **134**, 13679.
37. M. R. Ajayakumar, P. Mukhopadhyay, S. Yadav and S. Ghosh, *Org. Lett.*, 2010, **12**, 2646.
38. F. G. Goodson, D. K. Panda, S. Ray, A. Mitra, S. Guha and S. Saha, *Org. and Biomol. Chem.*, 2013, **11**, 4797.
39. H. T. Chifotides, B. L. Schottel and K. R. Dunbar, *Angew. Chem., Int. Ed.*, 2010, **49**, 7202.
40. G. Aragay, A. Frontera, V. Lloveras, J. Vidal-Gancedo and P. Ballester, *J. Am. Chem. Soc.*, 2013, **135**, 2620.
41. C.-Z. Li, C.-C. Chueh, F. Ding, H.-L. Yip, P.-W. Liang, X. Li and A. K.-Y. Jen, *Adv. Mater.*, 2013, **25**, 4425.
42. (a) V. Gorteau, G. Bollot, J. Mareda and S. Matile, *Org. Biomol. Chem.*, 2007, **5**, 3000; (b) J. Mareda and S. Matile, *Chem. – Eur. J.*, 2009, **15**, 28.
43. A. Perez-Velasco, V. Gorteau and S. Matile, *Angew. Chem., Int. Ed.*, 2008, **47**, 921.
44. (a) B. Hille and W. Schwarz, *J. Gen. Physol.*, 1978, **72**, 409; (b) G. Eisenman and R. Horn, *J. Membr. Biol.*, 1983, **75**, 197; (c) D. A. Doyle, J. M. Cabral, R. A. Pfuetzner, A. Kuo, J. M. Gulbis, S. L. Cohen, B. T. Chait and R. MacKinnon, *Science*, 1998, **280**, 69.
45. (a) B. J. Rosenstein and P. L. Zeitlin, *Lancet*, 1998, **351**, 277; (b) D. Yamamoto, Y. Kiyozuka, Y. Uemura, C. Yamamoto, H. Takemoto, H. Hirata, K. Tanaka, K. Hioki and A. Tsubura, *J. Cancer Res. Clin. Oncol.*, 2000, **126**, 191.
46. Y. Zhang and P. S. Cremer, *Curr. Opin. Chem. Biol.*, 2006, **10**, 658.
47. E. V. Anslyn and D. A. Dougherty, *Modern Physical Organic Chemistry*, University Science Book, Sausalito, CA, 2006.
48. (a) D. Sun, S. V. Rosokha and J. K. Kochi, *J. Phys. Chem. B.*, 2007, **111**, 6655; (b) S. V. Rosokha and J. K. Kochi, *Acc. Chem. Res.*, 2008, **41**, 641.
49. R. S. Mulliken, *J. Am. Chem. Soc.*, 1952, **74**, 811.
50. S. Guha, F. S. Goodson, J. R. Clark and S. Saha, *Cryst. Eng. Commun.*, 2012, **14**, 1213.
51. A. Mitra, C. T. Hubley, D. K. Panda, R. J. Clark and S. Saha, *Chem. Commun.*, 2013, **49**, 6629.
52. S. T. Schneebeli, M. Frasconi, Z. Liu, Y. Wu, D. M. Gardner, N. L. Strutt, C. Cheng, R. Carmieli, M. R. Wasielewski and J. F. Stoddart, *Angew. Chem., Int. Ed.*, 2013, **52**, 13100.
53. M. R. Ajayakumar and P. Mukhopadhyay, *Chem. Commun.*, 2009, 3702.
54. S. V. Bhosale, S. V Bhosale, M. B. Kalyankar and S. J. Langford, *Org. Lett.*, 2009, **11**, 5418.
55. M. R. Ajayakumar, G. Hundal and P. Mukhopadhyay, *Chem. Commun.*, 2013, **49**, 7684.
56. M. R. Ajayakumar, D. Asthana and P. Mukhopadhyay, *Org. Lett.*, 2012, **14**, 4822.

57. (a) B. Abraham and L. A. Kelly, *J. Phy. Chem. B.*, 2003, **107**, 12534; (b) B. Abraham, S. McMasters, M. A. Mullan and L. A. Kelly, *J. Am. Chem. Soc.*, 2004, **126**, 4293.
58. (a) A. G. Griesbeck, A. Henz, J. Hirt, V. Ptatschek, T. Engel, D. Löffler and F. W. Schneider, *Tetrahedron*, 1994, **50**, 701; (b) A. G. Griesbeck, A. Henz, W. Kramer, J. Lex, F. Nerowski, M. Oelgemöller, K. Peters and E. M. Peters, *Helv. Chim. Acta*, 1997, **80**, 912; (c) A. G. Griesbeck, J. Neudörfl and A. de Kiff, *Beilstein J. Org. Chem.*, 2011, 7, 518.

CHAPTER 8

Naphthalene Diimide-based Photovoltaics

SUBASHANI MANIAM, HEATHER F. HIGGINBOTHAM, TOBY D. M. BELL AND STEVEN J. LANGFORD*

School of Chemistry, Monash University, Clayton, Victoria 3800, Australia
*Email: steven.langford@monash.edu

8.1 Introduction

Increasingly, scientific advances in the materials sciences require the creation of novel systems that are constructed from suitable molecular components capable of performing functions that mimic those executed by macroscopic electronic devices. Photovoltaic (PV) devices, for example, offer the possibility of obtaining cheap and easy to use energy from light sources and are currently available through dye-sensitised systems, or the more common and commercial silicon-based PV systems. For photovoltaic devices to be practical, the active component(s) must exhibit important features that include:[1,2]

(a) long-term stability under working conditions,
(b) ease of fabrication that allows for cheap and readily available materials,
(c) addressable with fast response times,
(d) high sensitivity and well-behaved redox properties,
(e) photophysical properties that allow for the absorption of visible to near infra-red light,

Monographs in Supramolecular Chemistry No. 21
Naphthalenediimide and its Congeners: From Molecules to Materials
Edited by G. Dan Pantoş

Published by the Royal Society of Chemistry, www.rsc.org

(f) provision for interaction with macroscopic electrodes and semiconductors,
(g) well-behaved solid-state properties, and
(h) an ability to be utilised in plastics or other media.

Electrochemical and photochemical inputs and outputs are among the easiest to interface to macroscopic systems, making them amenable for the eventual creation of pragmatic molecular photovoltaic devices. Hence, components that have the ability to act as efficient electron acceptors, and that can be functionalised easily so as to interface with electrodes, show promise as molecular electronic components for photovoltaic devices.[3,4] Furthermore, such devices require the use of both n-channel (electron-transporting) and p-channel (hole-transporting) type semiconductors as complementary elements in order to achieve low-power and high-speed performance required in electron-separated domains for photovoltaic assemblies.[5–7] Whilst there are many stable p-type organic semiconductors known, stable n-type organic semiconductors have proven difficult to obtain, so an opportunity exists to explore new n-type semiconductors for the development of this industry.[8,9]

8.1.1 Naphthalene Diimides

Naphthalene diimides (NDIs) are neutral, planar molecules, the basic structure of the most common member is shown in Figure 8.1. Its structure consists of a naphthalene core, with imide substituents at the 1,4,5,8 positions. Naphthalene diimides are part of the 'rylene family' of extended aromatic systems, the basic structures of which are also shown in Figure 8.1. While early NDIs of this type exhibited unremarkable properties such as little or no fluorescence, larger rylenes such as perylene diimides (PDIs)[4,10]

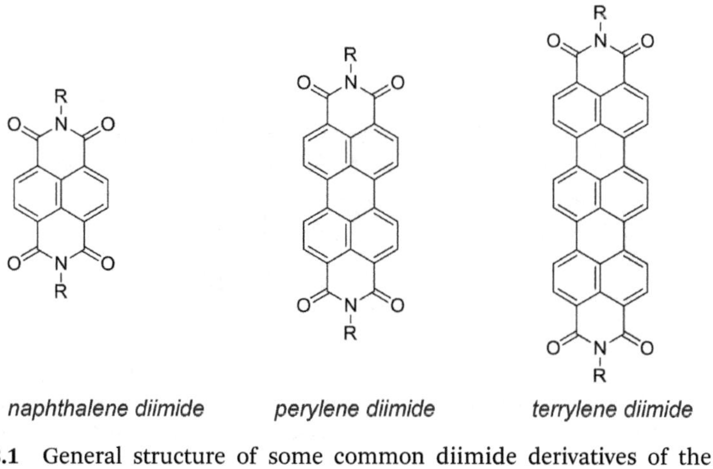

Figure 8.1 General structure of some common diimide derivatives of the rylene family.

and terrylene diimides (TDIs)[11,12] exhibit strong fluorescence, emitting in the red region of the spectrum and are currently fashionably used as fluorescent labels and in electronic devices.[13] As a result, NDIs remained relatively undiscovered as a functional building block in favour of the more brightly coloured and fluorescent larger relatives. Major disadvantages with respect to the larger rylenes, however, are their solubility, as they are prone to stacking leading to aggregates, and the difficulty of chemistry associated with preparation and functionalisation.

NDIs have been shown to be a versatile class of electron-deficient aromatic compounds with the potential to lead to useful photovoltaic systems as a result of their ability to form n-type semiconductor materials. Furthermore, the reduction potential of simple alkyl-substituted NDIs ($E^1_{red} = -1.20$ V *vs.* Fc/Fc^+)[14] are similar to that of ubiquinone, and as a result, NDIs have often been used as quinone alternatives in a range of applications such as photosynthetic models, solar energy and molecular electronics. Having a planar aromatic structure confers the ability of NDIs to stack in a cofacial arrangement (Figure 8.2) with itself or other aromatic molecules leading to potential conducting pathways and/or supramolecular ensembles with excimer properties.[15]

The radical anions of most NDIs are good chromophores, exhibiting intense and characteristic visible and near-infrared absorption bands (Figure 8.3).[16] The use of NDIs as a component in practical devices of the kind mentioned in this chapter requires that the properties of the anion radicals (and dianions) be well understood.

In situ forms of spectroelectrochemistry have been used to obtain electronic spectra and *ex situ* forms of measurements of electron paramagnetic resonance and fluorimetry that enable detailed characterisation of the anion radicals to be achieved. As a result, we can summarise the spectroelectrochemical properties of NDIs as shown in Figure 8.4.[16]

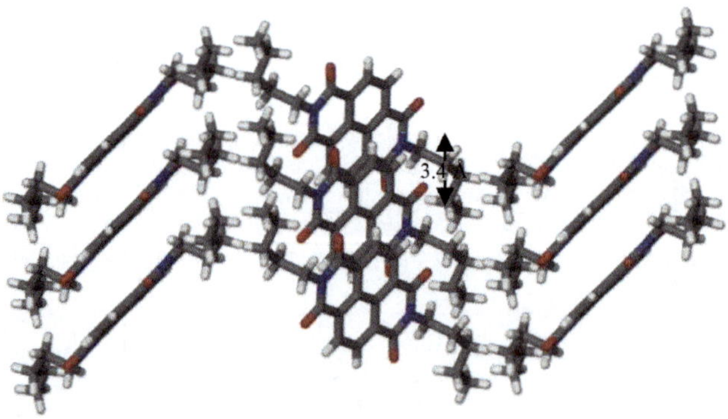

Figure 8.2 X-Ray crystal structure of a simple *N,N′*-dialkyl-substituted NDI illustrating strong ordering through the π–π stacking ability of the naphthalene core, and London dispersive forces of the alkyl groups. Reproduced from Andric *et al.* (2004), with permission from CSIRO Publishing.

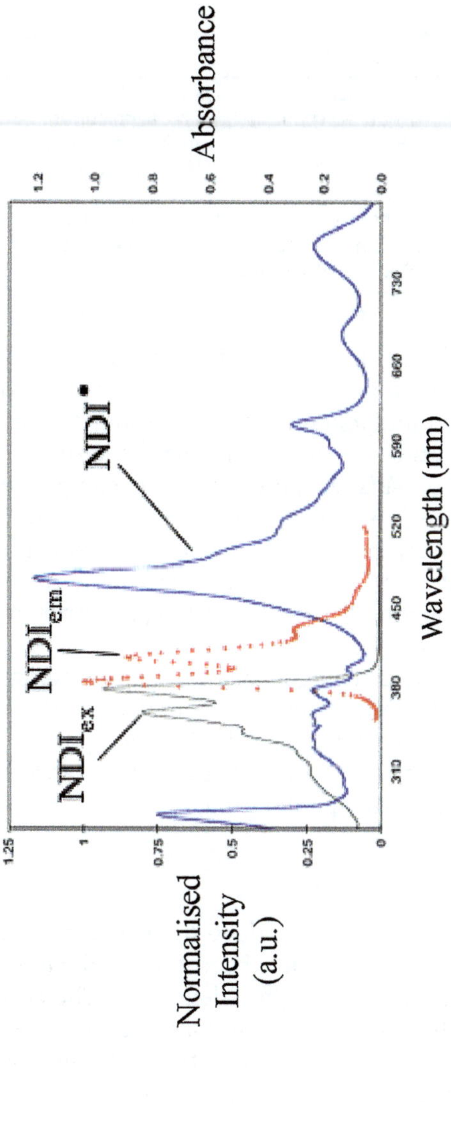

Figure 8.3 Absorption and emission characteristics of NDIs and their radical anions. Reproduced from *Chem. Soc. Rev.*, 2008, 37, 331–342 with permission from the Royal Society of Chemistry.

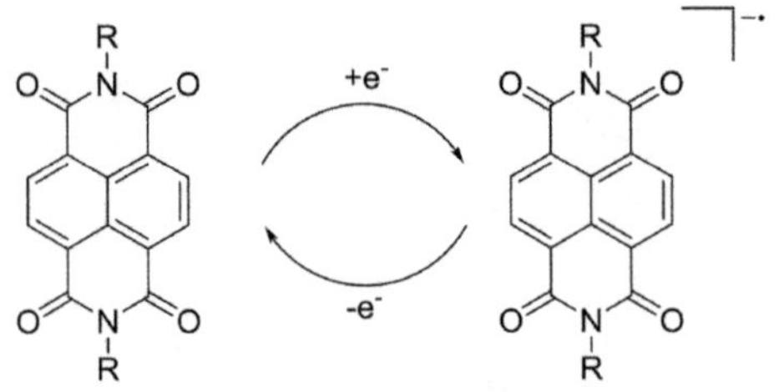

Technique	Output	
UV-Vis	Colourless	Purple
Electrochemistry	Reduction current	Oxidation current
EPR	Silent	Active and well structured
Fluorescence	Fluorescent	Weakly fluorescent

Figure 8.4 Summary of the spectroelectrochemical properties of NDIs. Reproduced from *Chem. Soc. Rev.*, 2008, **37**, 331–342 with permission from the Royal Society of Chemistry.

The use of NDIs as electron acceptors in photoinduced electron transfer processes will be addressed in greater detail in Section 8.3. However, at this point it is important to understand that the strong UV absorption characteristics of NDIs only utilise 3–5% of the solar spectrum.[16] In order to be more effective, a dye-sensitised approach using complementary chromophores can be employed in a similar way to that of nature, which employs strong visible light absorbers as both electroactive and photoactive components within its functioning biological machinery. An example of this approach is shown in Figure 8.5. Here, the zinc(II) porphyrin and its ability to axially coordinate pyridines have been used to construct an efficient photovoltaic device, whose charge separated state upon photoinduced electron transfer lasts for microseconds.[17] Changes in the connectivity between the metalloporphyrin donor and the NDI acceptor are able to modulate both electron transfer and charge recombination rates thereby allowing the fine tuning of the system's photophysics.

8.1.2 Core-substituted NDIs

A major discovery of NDIs, which heralded their recent popularity, involves the ability to tune their electronic absorption and emission properties through functionalisation of the naphthalene core. The addition of one or more substituents to the core of the naphthalene ring at the 2, 3, 6 and/or 7 position with electron-donating substituents results in brightly colourful, conducting, functional materials with vastly different photophysical properties to that of the parent NDI.[18] Synthetic advances have allowed the rapid generation of many new core-substituted NDI (cNDI) systems to be developed. For example, the electron conduction properties of cNDIs which incorporate strong electron-withdrawing groups are especially promising as organic thin-film transistors (OFET)[19] with the development of NDI polymers containing a soluble cNDI–bithienyl copolymer achieving excellent

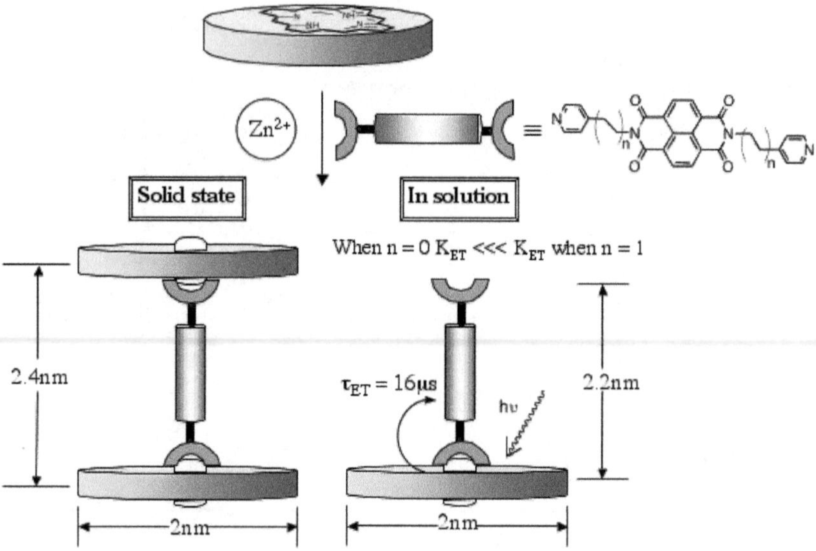

Figure 8.5 Assembly of a multichromophoric system utilising a porphyrin donor and NDI acceptor. The stoichiometry of the dyad assembly is phase dependent while the efficiency of the transduction, leading in one instant to a useful molecular battery, is component driven.

electron mobility when incorporated into printable electronic applications. The ability to tune the optoelectronic properties, stability, molecular packing, thin film morphology and charge carrying mobility of cNDIs by the inclusion of different substituents in the imide position or on the core is a powerful advantage.

By way of example, a system that achieves charge-separation lifetimes long enough for conversion into chemical work uses a rigid-rod scaffold containing a *p*-octiphenyl backbone with eight cNDIs attached to construct a π-stacked architecture.[20] In this case, the NDIs act as a component of an electron donor ensemble. The photoactive scaffolds are incorporated into egg yolk phosphatidylcholine large unilamellar vesicles encapsulating an acceptor quinone Q, and using a sacrificial electron donor EDTA and the pH detector 8-hydroxypyrene 1,3,6-trisulfonate, which was able to detect ET after irradiation of the cNDI rods at 635 nm (Figure 8.6).

8.2 Overview of Energy and Electron Transfer Processes

The key physical processes in photovoltaics are energy transfer and electron transfer. Following absorption of light (optical excitation), the energy often needs to move before it can be utilised. Electron transfer leads to the formation of charges (electrons and holes) that can then migrate to electrodes leading to current flow in a photovoltaic (PV) device. An excellent example of these processes at work can be found in nature in photosynthesis. The light harvesting complexes firstly absorb light and then transfer the

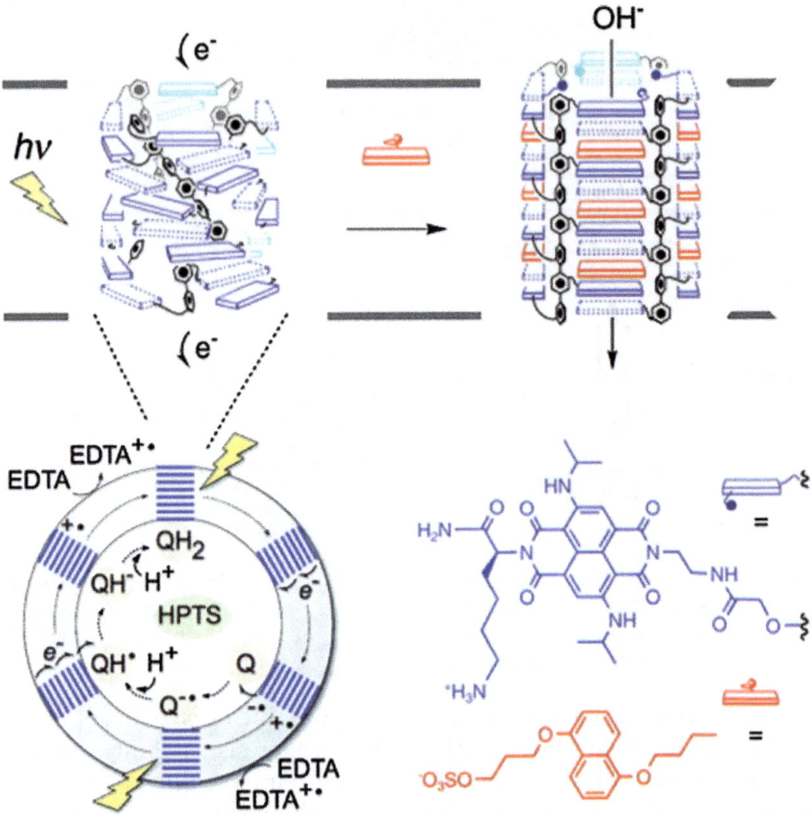

Figure 8.6 Rigid *p*-octiphenyl rods were used to create helical tetrameric π-stacks of blue, red-fluorescent naphthalene diimides that can span lipid bilayer membranes. In lipid vesicles containing quinone as electron acceptors and surrounded by EDTA as acceptors, transmembrane proton gradients arose through quinone reduction upon excitation with visible light.
From *Science*, 2006, **313**, 84–86. Reprinted with permission from AAAS.

excitation energy with nearly 100% efficiency to the photosynthetic reaction centres. It is here that a series of electron transfer steps lead to the separation of charges across the cellular membrane. The energy stored in this cross-membrane charge gradient is ultimately used to synthesise carbohydrates. The same principles apply in modern photovoltaic devices. Energy and electron transfer are used to convert light into charges, which are then free to migrate to electrodes to generate current.

8.2.1 Energy Transfer

Dipole–dipole mediated, or 'through space' energy transfer was first described by Förster.[21,22] The key result being that the efficiency, *E*, of through-space energy transfer decays with the sixth power of the distance between the energy donor and the acceptor. This is shown in eqn (8.1).

$$E_{\text{FRET}} = \frac{1}{1 + \left(\frac{R}{R_0}\right)^6} \tag{8.1}$$

In this equation, R is the distance between the donor and acceptor and R_0 is the so-called critical distance for Förster resonance energy transfer or FRET, and is the distance at which the probability of the excited donor relaxing by FRET to the acceptor is 50%. This is equivalent to saying that the rate constant for FRET is equal to the sum of all other rate constants deactivating the excited donor. Figure 8.7 shows the effect of FRET efficiency as a function of distance.

R_0 can be calculated from eqn (8.2a),

$$R_0{}^6 = \frac{9000\ln(10)\kappa^2\Phi_{\text{D}}J}{128\pi^5\eta^4N_{\text{A}}} \tag{8.2a}$$

and the rate constant for FRET from eqn (8.2b),

$$k_{\text{FRET}} = \frac{9000\ln(10)\kappa^2\Phi_{\text{D}}J}{128\pi^5\eta^4N_{\text{A}}\tau_{\text{D}}R^6} \tag{8.2b}$$

where J is the spectral overlap, Φ_{D} is the fluorescence quantum yield of the donor, κ is an orientation term, η is the refractive index of the medium, τ_{D} is the fluorescent lifetime of the donor and N_{A} is Avogadro's number. The

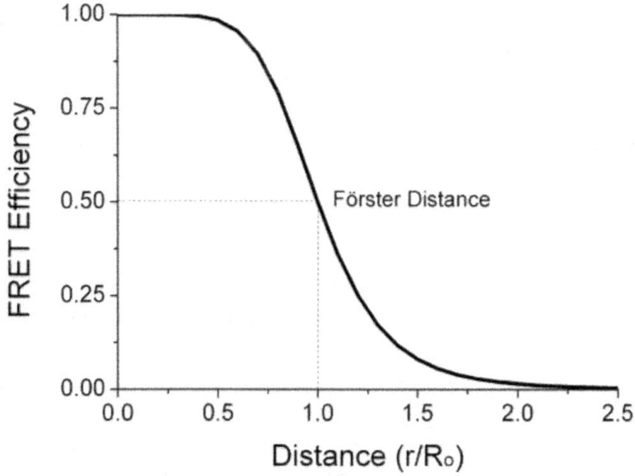

Figure 8.7 Plot of FRET efficiency as a function of distance.
Reprinted from *Discuss. Faraday Soc.*, 1959, **27**, 7–17 with permission from the Royal Society of Chemistry,

spectral overlap term refers to the extent of overlap between the donor's emission spectrum and acceptor's absorption spectrum, *i.e.* the range of wavelengths or energies where these processes can be resonant. The orientation term accounts for the directionality of the donor emission dipole and the acceptor absorption dipole and κ^2 can take values from 0 to 4.

Good small molecule FRET donor acceptor pairs typically have R_0 values in the 4–8 nm range. This has led to FRET being used to measure nanometre distances and has earned it the moniker of 'the spectroscopic ruler'. The only reported critical distance for an NDI-based FRET pair is 4.1 nm.[23] This is a promising value and the molecules involved are simple allyl cNDIs. Since the first brightly coloured cNDIs were synthesised,[24] the potential for these molecules to be used for FRET has been noted.[25]

Clearly there are a number of strategies to lengthen the critical distance and maximise FRET efficiency in NDIs over a given distance:

(a) increase the spectral overlap term,
(b) increase the quantum yield of the donor,
(c) arrange the donor and acceptor into a more favourable orientation.

As already mentioned, one of the most attractive properties of cNDIs is that their optical properties can be tuned by varying the nature and particularly the number of substituents on the NDI core. This means there is much potential for making NDI-based FRET pairs with increased spectral overlap by widening the range of resonant frequencies. Most cNDIs reported so far have Φ_D in the range of 0.2–0.6, which leaves room for improvement. More significantly, however, are the typical molar absorptivities of cNDIs. These are usually in the range of 10–20 000 $M^{-1}\,cm^{-1}$, which is low compared to many bright fluorophores. Increasing this value in the acceptor would also increase the spectral overlap. Given that the R_0 quoted above is for simple mono and di-allyl substituted cNDI pairs, it is apparent that longer critical distances should be achievable using cNDIs.

The orientation of the donor emission dipole and the acceptor absorption dipole affect k_{FRET} *via* the orientation term κ, which is given by eqn (8.3).

$$\kappa^2 = (\cos\theta_T - 3\cos\theta_D\cos\theta_A)^2 \tag{8.3}$$

Here, θ_T is the angle between the donor and acceptor transition dipoles and θ_D and θ_A are the angles between these dipoles and the vector connecting them. What is immediately apparent from eqn (8.3) is that a co-linear arrangement is best and yields a κ^2 value of 4. A parallel arrangement has $\kappa^2 = 1$ while a random arrangement equates to 0.67. $\kappa^2 = 0$ for perpendicular orientation of the dipoles. NDIs have much potential here too in the design of optimised systems, since the imide position is readily functionalised and joining NDIs to each other through the imides would place the dipoles in a co-linear arrangement such as in the example discussed below.[26,27] Furthermore, functionalisation at the imides has little effect on the overall optical properties, which are mostly controlled by core substitutions.

This 'decoupling' of spectral features and molecular architectural design is highly advantageous from a synthetic point of view.

8.2.2 Electron Transfer

The theory describing electron transfer was described by Hush, Marcus and others during the 1950s and 1960s.[28–30] The main factors affecting electron transfer were found to be the electronic coupling between the electron donor (D) and acceptor (A), the driving force for electron transfer and the reorganisation energy. This latter quantity is the work that has to be done by the molecules involved and the surrounding medium in response to the creation or movement of charges by the electron transfer reaction. By representing the energy surfaces of the reactant (D–A) and product (D^+–A^-) states as parabolas (Figure 8.8), the kinetic barrier to the electron transfer reaction, $\Delta G^{\ddagger}$ is determined from the driving force ΔG^0 and reorganisation energy, λ, as shown by eqn (8.4):

$$\Delta G^{\ddagger} = \frac{(\Delta G^0 + \lambda)^2}{4\lambda} \tag{8.4}$$

The rate of electron transfer k_{ET} in the non-adiabatic or 'high temperature' limit is given by eqn (8.5):

$$k_{ET} = \frac{2\langle H_{DA}\rangle^2}{h}\left(\frac{\pi^3}{4\lambda RT}\right)^{1/2} e^{\frac{-\Delta G^{\ddagger}}{RT}} \tag{8.5}$$

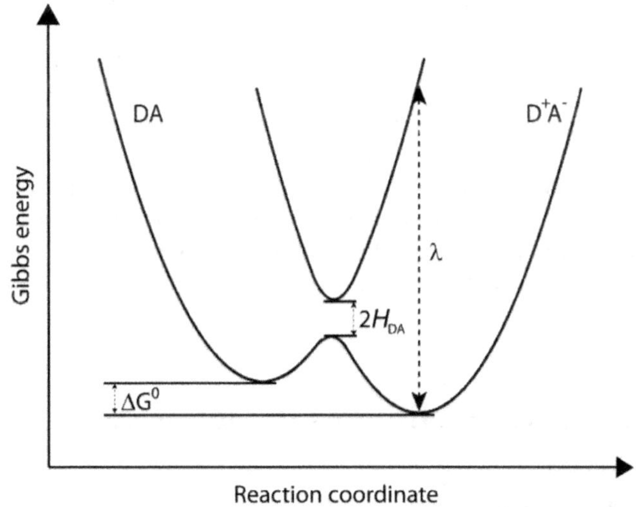

Figure 8.8 Diagram illustrating the key parameters in Marcus theory for ET. The parabolas represent the energy surfaces of the reactant (D–A) and product (D^+–A^-) states.
Reprinted from *Faraday Soc.*, 1961, **57**, 557–580 with permission from the Royal Society of Chemistry.

where H_{DA} is a Hamiltonian that describes the electronic coupling between the donor and acceptor. It is usually represented on an energy diagram as half of the magnitude of an 'avoided crossing' between the reactant and product parabolas as shown in Figure 8.8.

The interesting result of eqn (8.4) and (8.5) is that they predict a complex relationship between the rate constant and driving force for electron transfer. At first, k_{ET} increases with driving force and the kinetic barrier becomes smaller, eventually reducing to zero. From this point, further increases in driving force will lead to an increase in the barrier and a *reduction* in k_{ET}. These three regions of electron transfer are usually referred to as 'normal' ($\Delta G < \lambda$), 'barrierless' ($\Delta G = \lambda$) and 'inverted' ($\Delta G > \lambda$). Definitive experimental observation of the inverted region was a long time coming, but was eventually achieved by Closs and Miller in the 1980s.[31]

It is clear from eqn (8.5) that not only are the energy levels of the molecules involved very important, but so too are their redox values, since as a result of electron transfer, the donor is oxidised and the acceptor reduced. In the case of optically initiated electron transfer, the energy gap between the HOMO and LUMO levels dictates the amount of energy available and this needs to be greater than the energy required to form the charge transfer (CT) state from the ground state. The actual energy of the CT state is also highly dependent on the surrounding medium (especially its polarity) since charges are now in existence. It is here that the reorganisation energy is very important. It is usually divided into internal (molecular) and external (solvent) contributions and the latter often dominates in polar media meaning that the same donor and acceptor pair can show dramatically different behaviour in different situations.

It is clear that NDIs have much potential for application for electron transfer in D–A systems. As already noted, NDIs readily form an anion (see Figure 8.3) and can thus serve as electron acceptors. Being small molecules, they have broad solubility and can thus be used in polar environments where electron transfer is usually facilitated. Furthermore, the synthetic flexibility of NDIs can be used to tailor solubility, particularly by adding groups in the imide positions. This approach would also allow for optimisation of electron transfer processes. Many and varied molecular systems containing NDI moieties have already been synthesised and their properties explored. We now discuss some recent examples, focussing on NDIs in donor–acceptor systems and in photovoltaic applications.

8.3 Naphthalene Diimides in Donor–Acceptor Systems

The pursuit for new organic-based electron-transporting (n-type) and hole-transporting (p-type) materials has drawn attention to NDIs. More often than not, NDIs are tethered to an electron-rich molecule (p-type) to construct a donor–acceptor (D–A) system. In 2008, we wrote a comprehensive review of the literature of systems containing porphyrins and NDIs in photoinduced

CT studies.[32] Current research has moved towards optimisation of synthetic design toward favourable optoelectronic properties. These have included designing of D–A complexes with increased absorptivity (molar extinction coefficients) or absorptivity covering a wider range of the electromagnetic spectrum—towards black dyes.[18,23,24] Increases in emission lifetimes, quantum yields, stabilisation of charge separated states to amplify the efficiency of charge or electron mobility either in small molecules or polymeric systems are some of the key criteria for the development of NDI-based devices. For example, when substituted on the NDI core with derivatives of tetrathiafulvalene (TTF), **1** (Figure 8.9), the absorption spectrum of **1** covered close to the full electromagnetic spectrum (250–800 nm) from the visible through to the near-IR with clear CT bands from the TTF moiety to the NDI observed with λ_{max} 667 nm.[33] Another pathway for the formation of improved absorptivity and increased absorptivities covering more of the electromagnetic spectrum is the design of a metal-coordinated NDI system, **2** (Figure 8.9), creating a molecule with phosphorescence and a long-lived triplet state (τ_T) of 22.3 μs.[34]

Formation of a charge-separated state in covalently linked porphyrin-NDI dyads has been extensively investigated in the hope of creating light harvesting devices.[32,35,36] One such example is a zinc(II) tetraphenylporphyrin

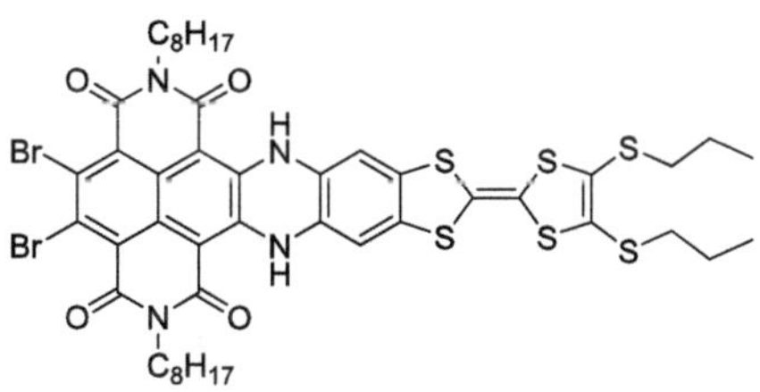

1

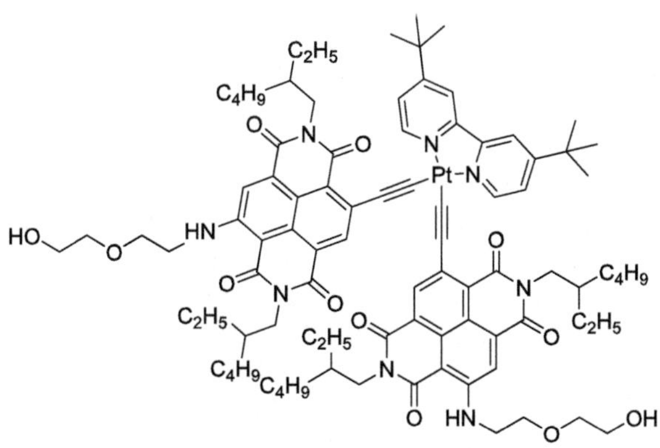

2

Figure 8.9 Structures of the TTF derivative **1** and Pt-coordinated NDI complex **2**. Reprinted from *Dalton Trans.*, 2011, **40**, 9085–9089 with permission from the Royal Society of Chemistry.

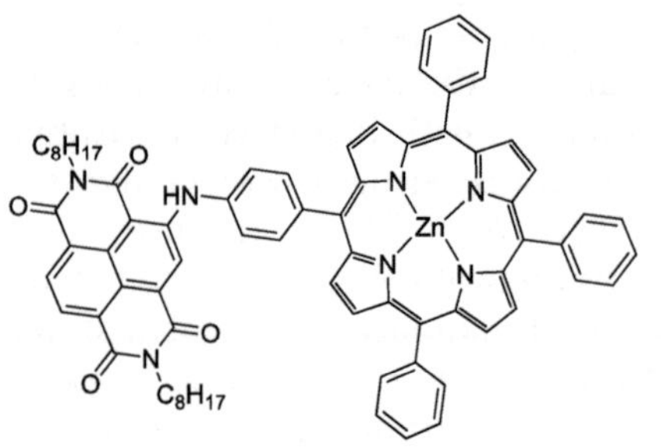

3

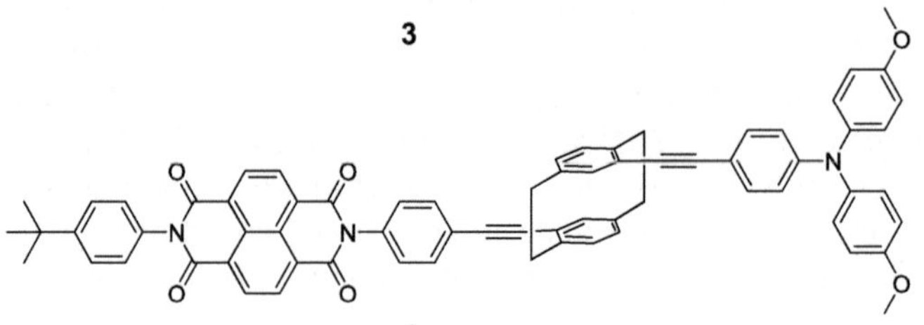

4

Figure 8.10 Structure of porphyrin–NDI dyad **3** and the triarylamine dyad **4**. Reprinted from *J. Photochem. Photobiol., A*, 2013, **251**, 167–174 with permission from Elsevier and *J. Phys. Chem. C*, 2012, **116**, 15265–15280. Copyright 2012 American Chemical Society.

amino substituted naphthalene diimide, **3** (Figure 8.10), which shows fast electron transfer from the S_2 state of the porphyrin identifying the possibility of light harvesting from a usually untapped part of the electromagnetic spectrum.[36]

Other examples of D–A systems include cyclophane bridged triarylamine-naphthalene diimides, *e.g.* **4**, with the ability to form both singlet and triplet charge-separated states. These states were found to be of similar energy levels and thus form an equilibrium upon photoexcitation.[37] The charge-separated state is easily characterised using UV-visible spectroscopy as these peaks do not overlap with those in the ground state (~350–400 nm) and their lifetime kinetics displayed a biexponential decay further identifying the two CT states: one fast, originating from the singlet charge-separation state and one long, originating from the triplet charge-separation state.

Naphthalene diimide-based D–A systems have also been incorporated in polymers and oligomers.[38–40] One such important example is the synthesis of 'molecular wires' consisting of Pt-acetylide repeat units end-capped with naphthalene diimides (**5**) (Figure 8.11(a)).[41,42] These NDI termini serve as exciton traps once the charge has shifted to the end of the chain using a 'hopping' mechanism (~27 ps), which then form a charge-separated state. Negative phonons were also investigated under a similar transport mechanism from the Pt-acetylides to the NDI termini (<10 ps), shifting the charge onto the NDI forming an anion radical (Figure 8.11(b)). In another example, a NDI polymeric structure using a BODIPY donor (**6**) gave a compound with

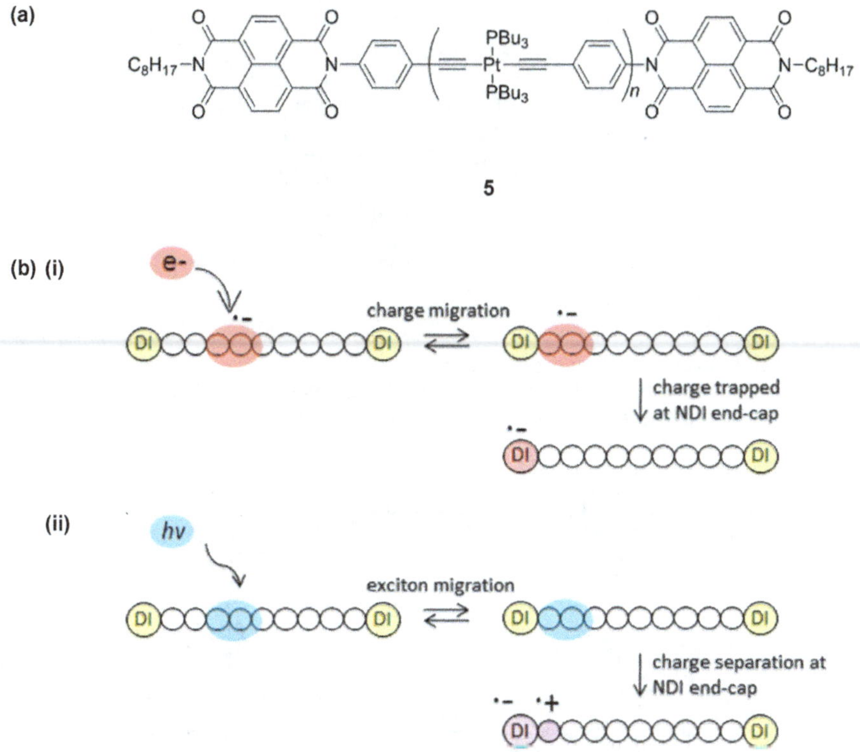

Figure 8.11 (a) Structure of NDI-based organometallic oligomers **5**, and (b) schematic of electron (i) and exciton (ii) transport in NDI-oligomers. DI represents the NDI end-group trap.
Reprinted with permission from *J. Am. Chem. Soc.*, 2011, **133**, 11289–11298. Copyright 2011 American Chemical Society.

broad absorption of the UV-visible spectrum with λ_{max} at 634 nm in chloroform, and a small optical band gap (Figure 8.12). Once cast in thin films, these polymers were shown to have electron mobility and the capacity to act as n-type semiconductors.[43]

Electron transfer in NDI D–A systems has also been used in a supramolecular context. This has led to the synthesis of electroactive rotaxanes,[44–46] pseudorotaxanes,[47] catenanes[26] and foldamers.[27,48] Often the interaction between the D–A units is the driving force for the formation of these unique architectures.[26] The presence of a D–A complex between electron-rich 1,5-dialkoxynaphthalene (DAN) and a NDI connected with long ethoxy chains produced a foldamer with an accordion-type structure in a DMSO–water mixture. As the water content was increased, the intensity of the CT band in the UV-visible spectrum also amplified denoting the NDI and DAN groups moving closer together.[48] A rotaxane consisting of a naphthalene crown threaded through a NDI-triazole-based 'dumbbell' displayed π–π stacking of the two units when the NDI was a neutral species. However, upon reduction of the NDI unit, the naphthalene shuttled off the NDI

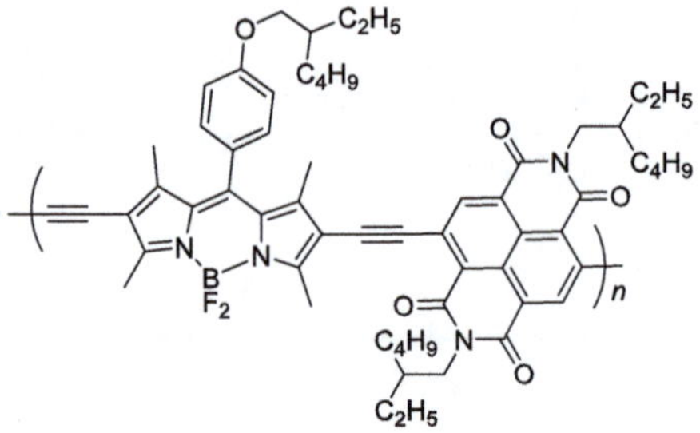

6

Figure 8.12 Structure of NDI polymer **6**.
Reprinted with permission from *Macromolecules*, 2011, **44**, 4767–4776. Copyright 2011 American Chemical Society.

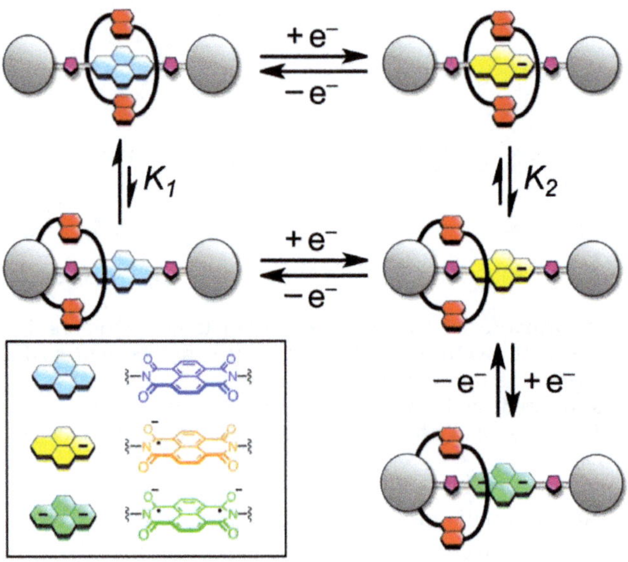

Figure 8.13 Proposed electrochemically induced shuttling motion of NDI-naphthalene-based rotaxane.
Reprinted with permission from *Org. Lett.*, 2012, **14**, 5188–5191. Copyright 2012 American Chemical Society.

moiety and moved along the length of the dumbbell as shown in Figure 8.13.[45]

As well as the mobility or transfer of electrons, NDIs have also been seen to participate in FRET processes.[22] Investigations into the energy transfer between a mono-amino NDI component to a di-amino NDI component in a bichromophoric molecule led to efficiencies of 96% and demonstrate clearly the complementarity available to differently-substituted NDIs for energy and electron transfer studies.[24] A few years later, the same group published a trio of dyads consisting of either one or two di-substituted NDI units covalently

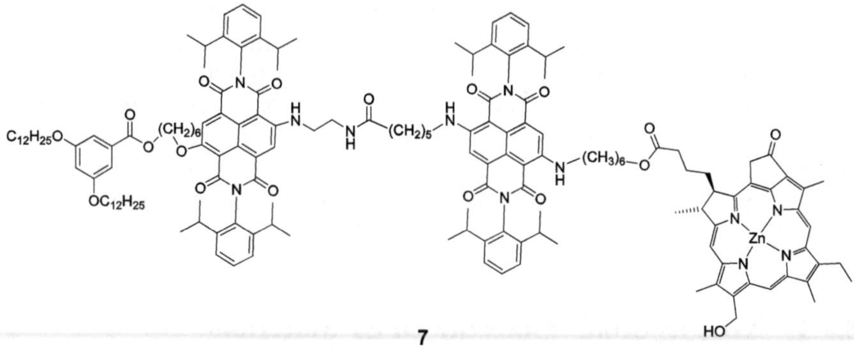

7

Figure 8.14 Triad cNDI construct (7) used to investigate photoinduced electron transport.
Reprinted with permission from *J. Am. Chem. Soc.*, 2008, **130**, 5929–5939. Copyright 2008 American Chemical Society.

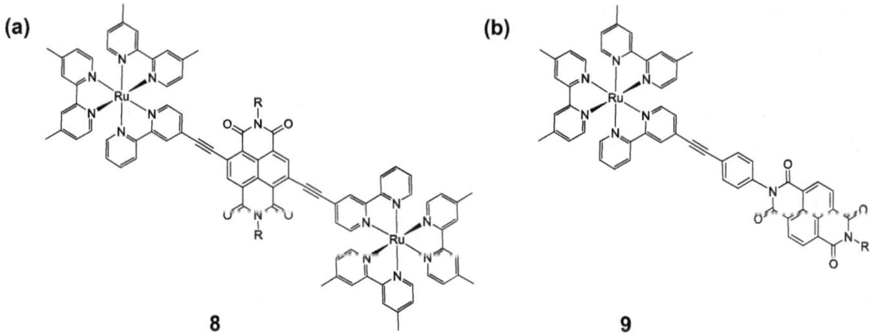

(a) 8 (b) 9

Figure 8.15 Connectivity effects (a) $Ru(bpy)_3$-cNDI-$R(bpy)_3$ **8** and (b) Ru(bpy)3-cNDI dyad **9**.

bound to a zinc porphyrin; as an example, compound 7 is shown in Figure 8.14. These dyes were found to have FRET efficiencies >99% when aggregated into rod-like structures, creating a potential light harvesting system.[25]

Care must also be taken in the connectivity of donor groups to naphthalene diimide acceptors. By way of example, the $Ru(bpy)_3$-cNDI-$R(bpy)_3$ dyads **8** and **9** (Figure 8.15) demonstrate the importance of electron coupling for ultrafast photoinduced ET. Simply by attaching the $Ru(bpy)_3$ donor to the core, the rate of electron transfer was increased by 1000 times compared to a similar NDI analogue with the same donor in the imide position.

8.4 Naphthalene Diimides in Organic Photovoltaic (OPV) Systems

Organic semiconducting materials are great options for low-cost electronics, facile processing and transparent large area printing for flexible display screens. In particular, semiconducting materials with high electron mobility and air stability are ideal candidates for OFETs. Thus, a lot of studies are done incorporating NDIs as n-channel components either as small molecules or tethered to polymers in developing OFETs. A brief understanding of NDI design and fabrication in organic transistors has recently been published.[49] NDIs are valuable building blocks developed for tuning both n-type and p-type properties, long-term ambient stability, and deposition onto flexible substrates with beneficial morphology.

8.4.1 NDIs as Small Molecule OFET Materials

The use of naphthalene diimides as n-type OFETs was first discovered in the mid-90s with reported electron mobilities in the order of 10^{-4} $cm^2 V^{-1} s^{-1}$.[50] Since then, many NDI-based aliphatic and aromatic substituents have found their way into OFETs. Compound **10** (Figure 8.16), for example, has electron mobilities of 6.2 $cm^2 V^{-1} s^{-1}$ and 0.41 $cm^2 V^{-1} s^{-1}$ when measured under inert and in ambient air conditions, respectively.[51] The air stability of NDIs has generally been improved by incorporating electron-withdrawing groups either at the *N*-imide position or on the naphthalene core. These substituents lower the LUMO level of the resulting compound and help prevent quenching by ambient oxidation.

The most common electron-withdrawing groups that have been incorporated into NDIs are perfluoroaryl substituents and simple alkyl chains (Figure 8.17).[52,53] Fluorine-substituted phenyl groups have been introduced at the *N*-imide positions resulting in air-stable n-channel OFETs with compound **16**, which has an electron mobility of 0.23 $cm^2 V^{-1} s^{-1}$ and is transparent and flexible.[54] Increasing the amount of $-CF_3$ substituents on phenyl groups effectively lowers the LUMO level, which also gives greater thermal stability. Interestingly, chlorine core-substituted NDIs in addition to fluorinated aliphatic substituents at the *N*-imide positions (Figure 8.18) have also been investigated.[55,56] While the fluorine substituents in **17** did not appreciably affect either the HOMO or LUMO energies, as there is a methylene space between the core and the fluorinated unit, the chlorine substituent on the core significantly reduced the LUMO while not affecting the HOMO energies. For example, for **20**, bearing four chlorine atoms, the

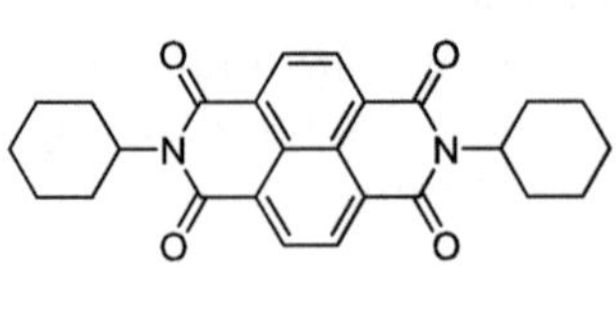

10

Figure 8.16 *N,N'*-Bis(cyclohexyl)naphthalene diimide **10** constructed for OFET applications.
Reprinted with permission from *Chem. Mater.*, 2008, **20**, 7486–7491. Copyright 2008 American Chemical Society.

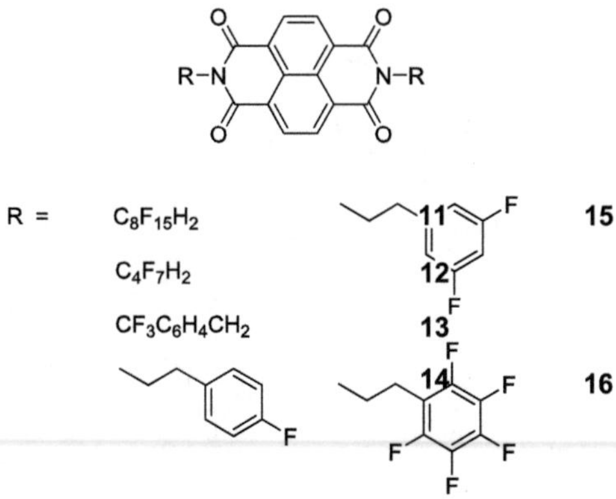

Figure 8.17 Chemical structures of fluorinated alkyl and aryl imide substituted NDIs.
Reprinted with permission from *Chem. Mater.*, 2009, **21**, 94–101. Copyright 2009 American Chemical Society.

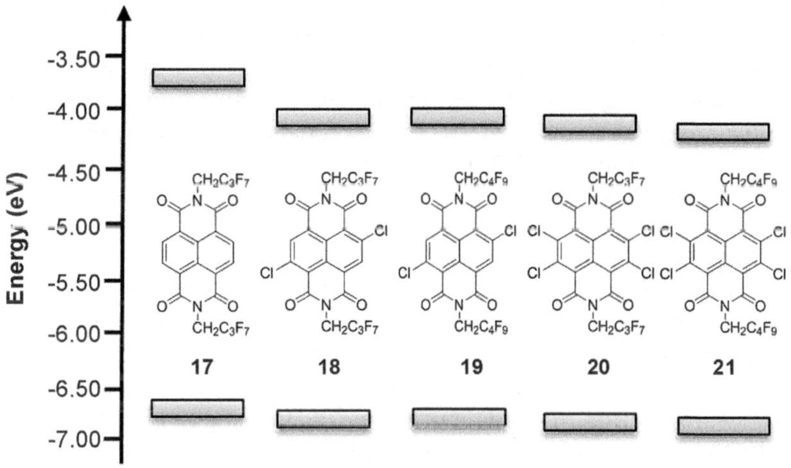

Figure 8.18 Chemical structures of non-core and chlorine substituted NDIs (**17**–**21**) with fluorinated aliphatic substituents at the *N*-imide positions and the corresponding HOMO and LUMO energies for the NDIs.
Reprinted with permission from *Adv. Funct. Mater.*, 2010, **20**, 2148–2156. Copyright © 2010 WILEY-VCH Verlag GmbH & Co. KGaA, Weinheim.

LUMO energy level was lowered from −3.72 eV (no chlorine atom) to −4.13 eV (Figure 8.18). In the presence of only two chlorine atoms on the core and methylene-perfluorobutyl groups at the *N*-imide positions, **19** yielded the best performing thin-film transistor displaying the highest mobilities reported: 1.26 $cm^2 V^{-1} s^{-1}$ and 1.43 $cm^2 V^{-1} s^{-1}$ in nitrogen and air, respectively, on octadecyltrimethoxysilane (OTS)-treated SiO_2. A similar halogenated NDI compound was studied at a single crystal level revealing a

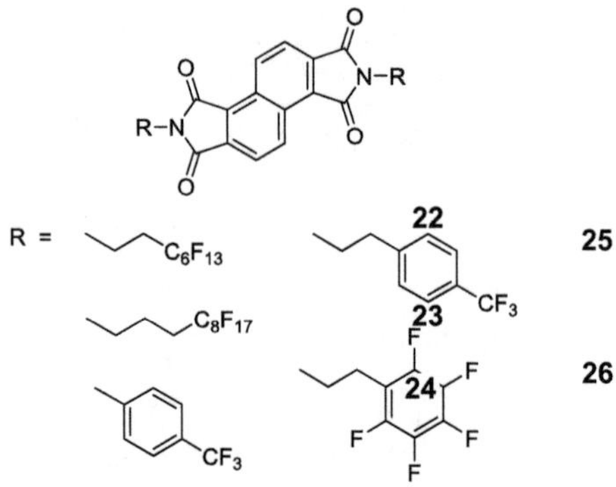

Figure 8.19 Angular-shaped NDIs (**22–26**) for n-type materials.
Reprinted from *Chem. Commun.*, 2012, **48**, 1254–1256 with permission from the Royal Society of Chemistry.

closely-packed π-system with a high packing density of 2.046 g cm^{-3}. Ribbon-shaped crystals with an n-channel mobility of 8.6 $cm^2\,V^{-1}\,s^{-1}$ in air were obtained and these fabricated devices degraded only 13% over 82 days. It was proposed that high electron mobility is created by the fast growth of these crystals along the *c*-axis, which is also the direction of the electron transport. Angular-shaped NDIs, which consist of fluorinated aliphatic and aromatic groups (**22–26**), have also been studied (Figure 8.19). These uniquely shaped NDIs formed small interconnected orderly packed grain upon deposition onto a substrate and exhibited semiconducting behaviour.[57]

NDIs with sulfur-rich moieties have also displayed air stability and n-type channel behaviour.[58,59] NDIs with fused sulfur heterocycles either as a five- or six-membered ring substituted with a cyano electron-withdrawing group have shown good solution processability in thin films producing electron mobilities from 10^{-6} to 0.26 $cm^2\,V^{-1}\,s^{-1}$.[58] Investigations of sulfur-rich NDIs in the absence of electron-withdrawing groups have been made and interestingly these compounds also showed great tolerance to ambient oxidation.[59] The cNDIs with 1,3-dithiole-2-thione-4,5-dithiolate (**27**) and 1,3-dithiole-2-one-4,5-dithiolate (**28**) (Figure 8.20) resulted in extended conjugation with superior intermolecular interactions through S–S contacts. Consequently, improved electron mobility was observed. For example, the electron mobility of compound **27** in a thin film increased from the order of 10^{-4} to 10^{-3} $cm^2\,V^{-1}\,s^{-1}$ after annealing at 80 °C for 30 seconds and further improved to 0.05 $cm^2\,V^{-1}\,s^{-1}$ upon annealing at 180 °C for 30 seconds.

Thus far, the discussion has focused on utilisation of NDIs for the purpose of n-type materials, however, NDIs have also been prepared with p-type or hole-transporting properties.[60] The molecular design involved extending π-conjugation along the NDI core by integrating a carbazole unit in **29**, which enhanced HOMO overlap. The addition of diisopropylphenyl groups at the I-imide positions prevent molecular packing along the *N*-imide

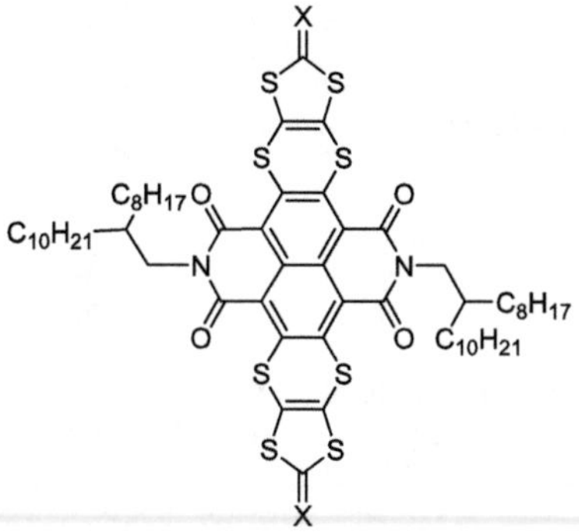

Figure 8.20 Structures of sulfur-rich NDIs **27** and **28**.
Reprinted from *J. Mater. Chem.*, 2011, **21**, 18042–18048 with permission from the Royal Society of Chemistry.

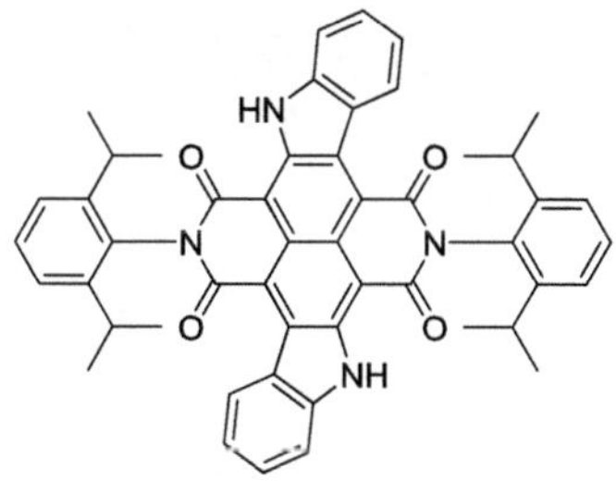

29

Figure 8.21 Structure of core-extended p-type NDI **29**.
Reprinted from *Chem. Commun.*, 2011, **47**, 11504–11506 with permission from the Royal Society of Chemistry.

axis (Figure 8.21). Thus, this material displayed p-type characteristics with a low HOMO energy level of −5.91 eV and a significant hole mobility of 0.56 $cm^2 V^{-1} s^{-1}$.

8.4.2 NDIs as OFET Materials in Polymers

Another popular means of introducing NDIs as acceptors in OFETs is in oligomeric or polymeric materials. Oligoarylene alternated NDI copolymers have been synthesised to investigate the electrochemical properties of these materials. The length of the oligoarylene was varied in each polymer, which significantly influenced the oxidation potential independent of the bis-NDI structure (Figure 8.22). Thus, the HOMO levels decreased from −5.45 eV for compound **32** to −5.50 eV and −5.84 eV for compounds **31** and **30**, respectively. These energy levels indicated that compounds **30** and **31** could behave both as n- and p-type carriers whereas compound **32** would

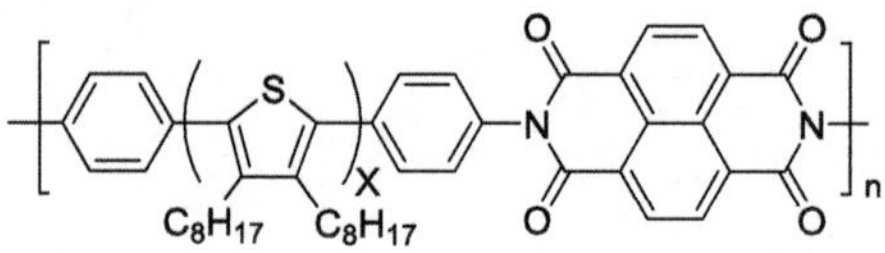

30
31
32

Figure 8.22 Structures of oligoarylene-NDIs **30–32**.
Reprinted from *Electrochim. Acta*, 2011, **56**, 10464–10472. Copyright 2011 with permission from Elsevier.

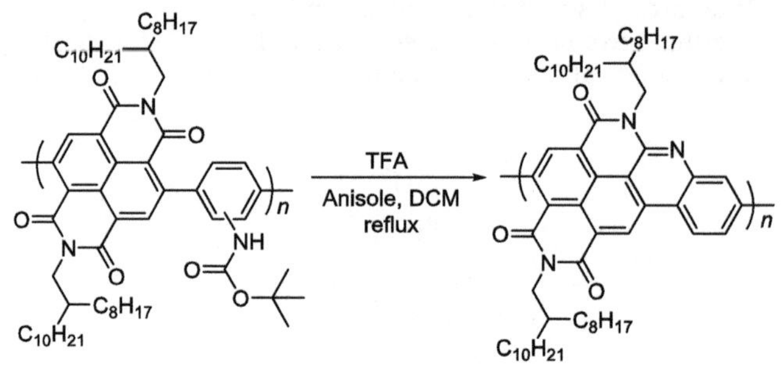

33 **34**

Scheme 8.1 Synthesis of ladder-based aryl-appended NDI polymer **34**.
Reprinted with permission from *Macromolecules*, 2011, **44**, 4721–4728. Copyright 2011 American Chemical Society.

distinctively show n-type character.[61] Luscombe and co-workers introduced n-type ladder-based polymers incorporating imine-bridged NDI heterocycles.[62,63] The ladder-based aryl-appended NDI polymer **34** was obtained through one step involving deprotection of di-*tert*-butyl dicarbonate (Boc) followed by intramolecular cyclisation (Scheme 8.1). This architecture led to strong π–π interactions along the polymer chain, however, all compounds were determined to be amorphous. They displayed average electron mobilities of 0.0026 $cm^2 V^{-1} s^{-1}$.

Adapting and further extending these sorts of architectures with thiophene and benzodithiophene groups (Figure 8.23) led to the development of new p-type materials.[64] These polymers were made into thin films and the absorption spectra for **35** and **36** were almost identical. Polymer **37** showed poor solubility and therefore aggregation was detected with a red shift in the absorption spectrum. The optical band gaps are estimated to be 1.97, 1.91 and 1.89 eV for **35**, **36** and **37**, respectively. Thin film OFETs of these materials exhibit hole mobility increasing only slightly after thermal annealing. Vapour annealing, however, increased the hole mobility of **36** (in dichloromethane) and **37** (*o*-dichlorobenzene) to 0.024 and 0.135 $cm^2 V^{-1} s^{-1}$, respectively, which is about 12 and 45 times higher than that with

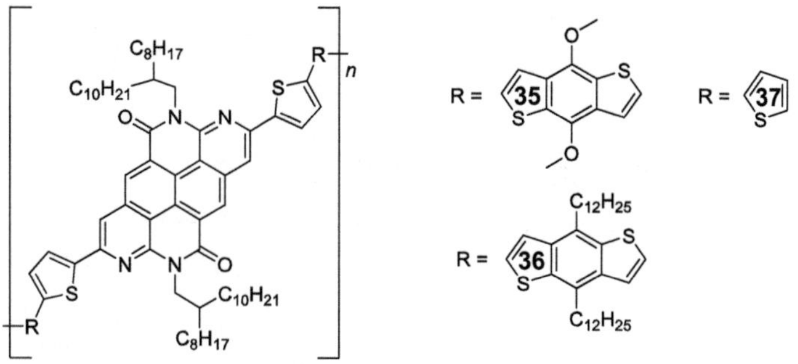

Figure 8.23 Core-extended NDIs for p-type materials.
Reprinted with permission from *Macromolecules*, 2013, **46**, 5504–5511.
Copyright 2013 American Chemical Society.

conventional thermal annealing. Enhancement of hole mobility was also seen in **36** when annealed from THF and toluene. Polymer **35** showed no difference between vapour and thermal annealing when poor solubility solvents were used. The hole mobility of **35** is low and not affected by thermal or solvent vapour annealing. A more fundamental study comparing the CT dynamics of a small molecule and a co-polymer of squaraine-naphthalene diimide was conducted by Völker *et al.*[65] In the small molecule, the local excitation of the squaraine moiety allows it to form the CT state followed by the charge-separated state within a calculated time of 2 ps. Charge recombination was determined to occur within 5.2 ps. For the polymer, the charge-separation state was not detected, which indicates that these states are likely confined within a D–A pair.

8.4.3 NDIs as Solar Cell Materials

NDIs have been integrated as n-type materials in tandem dye-sensitised solar cells (TDSSC),[66–68] all-polymer solar cells (OPV),[69–82] and photosystem supramolecular n/p-heterojunctions (SHJ).[83] In TDSSC, a NDI coupled to a perylene diimide (**38**, Figure 8.24) was used as the dyad with nanocrystalline NiO_2 as the electrode and cobalt-based polypyridyl as the electrolyte to give a high open circuit voltage, which yielded an overall efficiency of 0.14%, one of the highest recorded for this type of cell.[67] This efficiency was mainly attributed to the matched redox potential of the cobalt with the electrodes and the slow recombination rate of the dyad. Different variations of the cobalt-based electrolyte were also attempted in a p-type DSSC. More bulky ligands were introduced to the cobalt complex, and slower charge recombination reactions between the NiO_2 and the electrolyte were observed. Subsequently, photovoltaic properties utilising NDI **39** and the commonly used acceptor, fullerene (C_{60}) **40**, coupled with perylene diimide were conducted. Both the NDI and fullerene extended the charge-separation lifetime by five orders of magnitude compared to the parent perylene diimide dye. Interestingly, a NDI that is coupled directly to the perylene diimide without any spacer

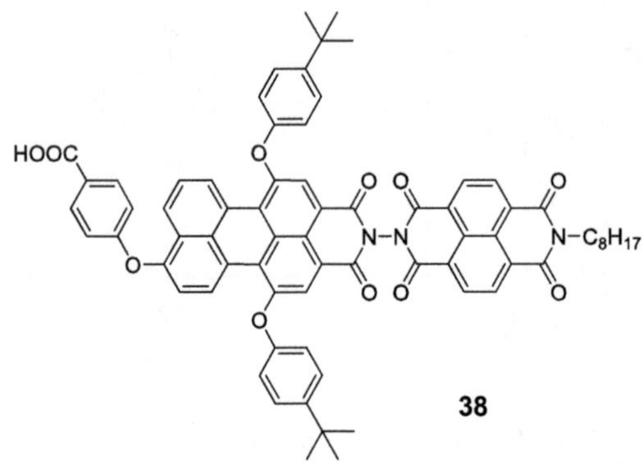

38

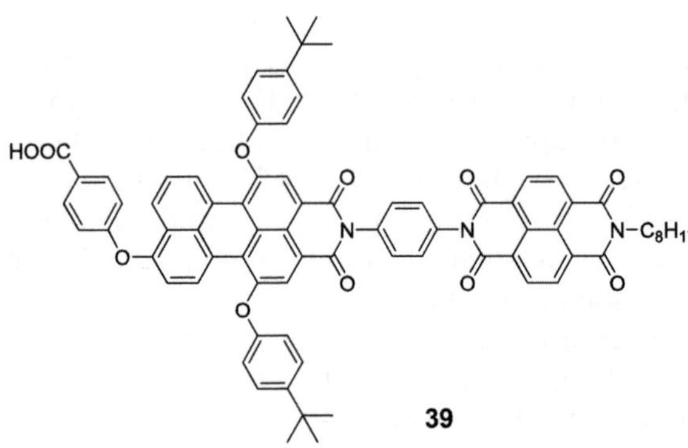

39

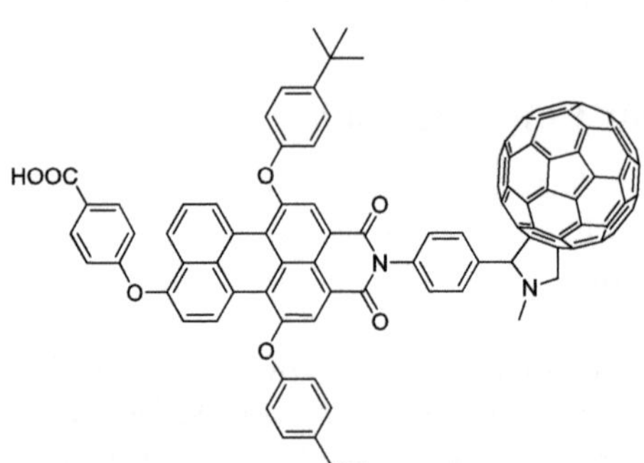

40

Figure 8.24 Perylene diimide dyad comprising NDIs **38** and **39**, and fullerene (C_{60}) **40**. Reprinted from *Energy Environ. Sci.*, 2011, **4**, 2075–2084 with permission from the Royal Society of Chemistry.

molecules displayed the best photovoltaic performance with an efficiency of 0.14% using a cobalt-based electrolyte.

In all-polymer solar cells, different device architectures such as bilayer and bulk heterojunction (BHJ) cells have utilised NDI as the acceptor component. In BHJ cells, NDI derivatives such as **42** are usually paired with the

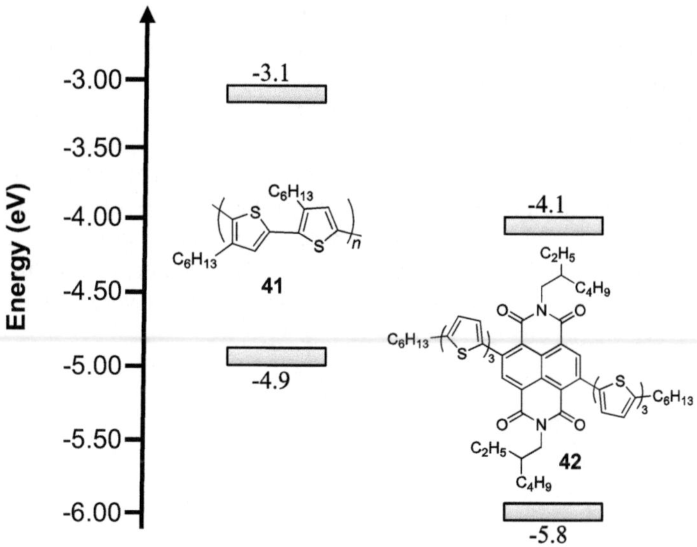

Figure 8.25 Molecular structure and the HOMO and LUMO energy levels of BHJ solar cells with donor P3HT **41** and acceptor NDI **42**.

donor component, poly(3-hexylthiophene), P3HT **41** (Figure 8.25).[70] The role of processing additives to improve device performance was established by blending **41** and **42** in the presence of only 0.2% of 1,8-diiodooctane as an additive. A marked 10-fold improvement in device performance was recorded from 0.14% to 1.5% as a result of the additive. The additive is believed to act as an electron- and hole-blocking layer, which prevents charge recombination. However, with an increased amount of the additive from 0.5–3.0%, the nano-morphology of the blend varied from a phase-separated state with aggregation of the NDI on a scale of >400 nm to visible nanorods of the NDI ~400 nm in size. A subsequent increase of the additive resulted in macrophase separation films with negligible photovoltaic performance.[69]

The photovoltaic properties of carbazole and benzothiophene polymers alternating with NDI units *via* copolymerization have been investigated.[73,74] Both polymers showed excellent thermal stabilities and broad absorptions in the visible region. The LUMO energy level of the benzothiophene and NDI copolymer was found to be −3.99 eV, thus it behaves as a typical n-type polymer, which can be used as an electron acceptor for all-polymer solar cells. Similarly, the carbazole-based copolymer also behaves as an n-type material in solar cells.

Other donors that are used with NDI to synthesise all-polymer solar cells are based on fluorene,[78] fused-thiophene[79,80] and 2,6-dialkylamino.[81] The co-polymer with fluorene gave a power conversion efficiency of 1.63%, which is one of the highest achieved with polymers of this kind. The design and synthesis of the fluorene-based polymer resulted in a high-lying LUMO energy with a high fill factor of 0.66. The effect of introducing 1,8-diiodooctane (DIO) as the additive was also investigated revealing that only 0.5% of DIO is required to obtain the maximum efficiency in this fluorene-based solar cell (Figure 8.26). The NDI-fused-thiophene polymers were synthesised *via* Stille coupling with excellent yields. The fused-thiophene groups within the NDI-copolymer backbone enabled tuning of the electronic structure and also improved charge mobilities within the active layer of the transistors.[79]

The angular-shaped NDI polymer **43** (Figure 8.27) gave a power conversion efficiency of ~0.32%. Polymers **43–45** recorded high V_{oc} values of 0.82–0.94 V compared to the linear-shaped NDI polymer, which had a value of only <0.6 V when blended with **41**. DFT calculations were also performed

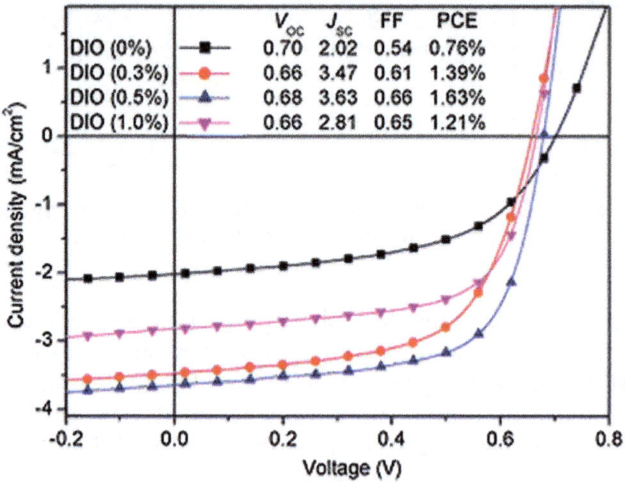

Figure 8.26 *J–V* curves of solar cells based on **41**: poly(fluorene-alternate-naphthalene diimide), with different ratios of DIO, as an additive. Reprinted from *Chem. Commun.*, 2012, **48**, 5283–5285 with permission from the Royal Society of Chemistry.

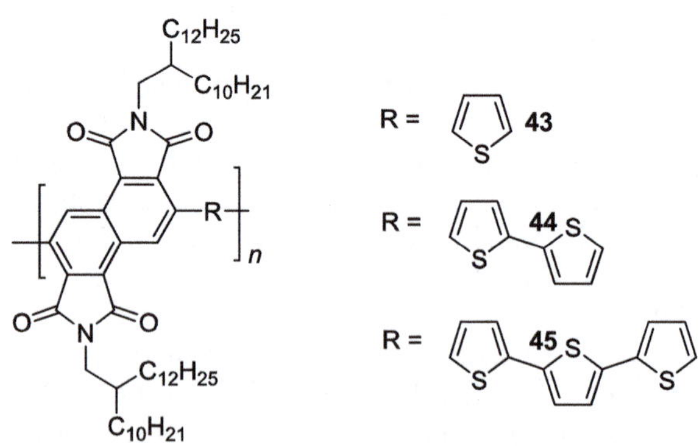

Figure 8.27 Angular-shaped NDI polymers **43–45** for all-polymer solar cells. Reprinted with permission from *J. Polym. Sci., Part A: Polym. Chem.*, 2013, **51**, 1999–2005. Copyright © 2013 Wiley Periodicals, Inc.

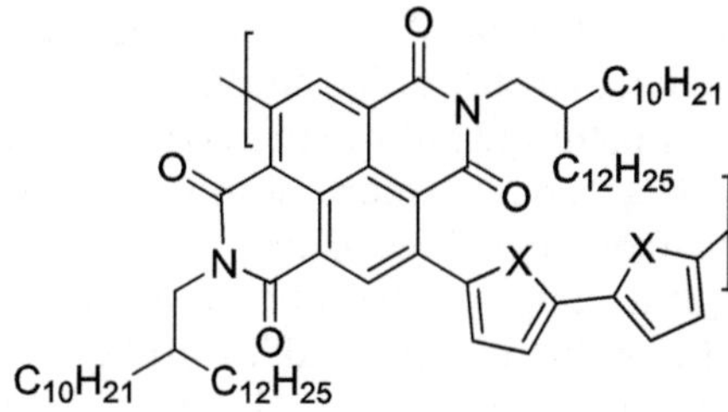

46 X = S

47 X = Se

Figure 8.28 Poly(NDI-alternate-biselenophene) **47** as an alternative to the thiophene derivative **46** for solar cells.

to evaluate the difference in frontier molecular orbital energy levels, which indicated that the LUMO energy of the angular-shaped NDI (−3.31 eV) is much higher than the linear-shaped monomer (−3.67 eV).[75] Poly(NDI-alternate-biselenophene) (Figure 8.28) displayed high electron mobility up to 0.07 $cm^2 V^{-1} s^{-1}$, which makes it an excellent candidate to be incorporated in solar cells in comparison to its thiophene derivative. When blended with **41**, a photovoltaic power of 0.9% was recorded with 19% of the photocurrent originating from near-infrared light harvesting by the NDI polymer.[77] Alternatively, with fewer selenophene units attached to the NDI and a blend with a thiazolothiazole-based polymer as the donor, this gave an efficiency of 3.3%, which is the highest known efficiency to date for all-polymer solar cells. This significant improvement in performance was attributed to the lamellar crystalline morphology of the NDI–selenophene polymer, which allows balanced electron and hole transport in the polymer/polymer blend.[76]

More creative but elaborate engineering was required to construct a photosystem SHJ.[83] It adapts a zipper-type supramolecular assembly with π-stacking of various *N*-imide and core-substituted NDIs along one axis connected by rigid-rod scaffolds. The long-range assembly of SHJ is important to ensure high charge mobility and to prevent charge recombination. Since the SHJ was made up of a layer-by-layer architecture, the photovoltaic performance of this device was highly influenced by aspects such as its critical thickness and the hydrophobic substituent at the NDI core.

8.4.4 NDIs as Ambipolar Materials

In all the cases above, NDIs have been introduced as n-type or p-type sources for OFETs, photovoltaic cells and in solar cells; however, they have been discovered and also tailored to behave as ambipolar materials, which can transport both holes and electrons.[82,84] By varying the length of the thiophene side chain attached to a NDI unit, the HOMO energies were able to be modulated without a concomitant shift in the LUMO energy levels. With only one thiophene unit, electron transport dominates, however, with more thiophene units, this facilitates hole transport, thus modulation of the amount of thiophene units helps to fine-tune the ambipolar behaviour. Ortiz *et al.* have looked at the problems with making ambipolar OFETs and stress that NDIs are great candidates to understand the architecture and electronic structure required for ambipolar materials.[85] Some of the requirements were a planar structure to increase intramolecular π–π stacking, HOMO energies close to the gold Fermi level and decreasing reorganisation energies for both holes and electrons to accommodate charge transport processes. As an example, they have showed the results of a series of first- and second-generation NDIs (some PDIs in the first-generation series) with fused thiophene rings at the imide position. Terthiophene **48** in Figure 8.29 displayed the best ambipolar behaviour with maximum mobilities of 2.01×10^{-2} and 6.79×10^{-3} $cm^2\,V^{-1}\,s^{-1}$ for electrons and holes, respectively, when fabricated *via* drop casting followed by annealing at 90 °C.

A top-gate OFET was fabricated to study the ambipolar performance of NDI **49** (Figure 8.30).[86,87] This compound consists of two NDIs bridged on the core by a dithienopyrrole. Thin films of **49** were deposited by spin coating followed by annealing at 100 °C for 15 minutes. OFETs fabricated with **49** exhibit electron and hole mobilities of 1.1 and 0.012 $cm^2\,V^{-1}\,s^{-1}$, respectively, thus they operate both as p- and n-channel devices. Batch to batch testing of the same electronic devices were tested to obtain consistency of device preparation. Within different batches, the electron mobility values varied from 0.8 (±0.2) to 1.2 (±0.3) $cm^2\,V^{-1}\,s^{-1}$ and the average hole mobility values varied from 0.007 (± 0.003) to 0.009 (± 0.004) $cm^2\,V^{-1}\,s^{-1}$. Variants of structure **49** were incorporated into polymer-based semiconductors, which

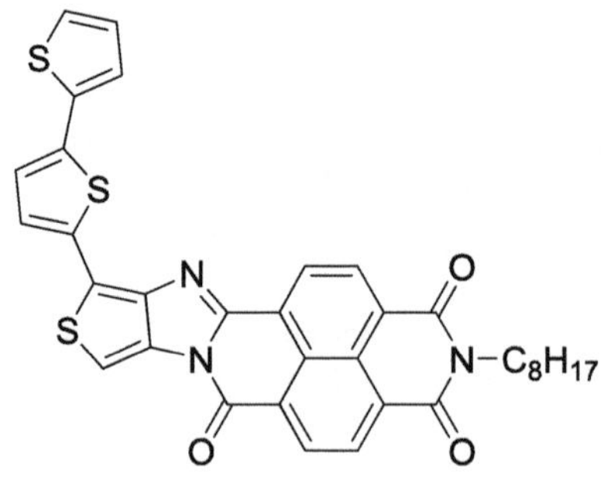

48

Figure 8.29 Tailored molecular structure of thiophene-fused NDI **48** for ambipolar behaviour.
Reprinted with permission from *Chem. – Eur. J.*, 2012, **18**, 532–543.

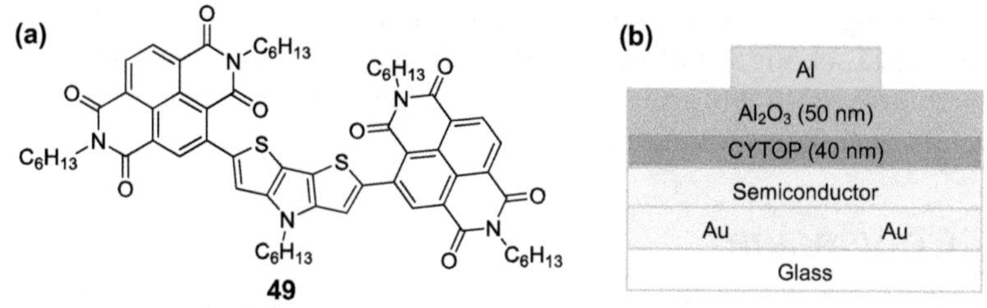

Figure 8.30 (a) Molecular structure of **49** and (b) device structure of a top-gate OFET. Reprinted with permission from *Chem. Mater.*, 2011, **23**, 3408–3410. Copyright 2011 American Chemical Society.

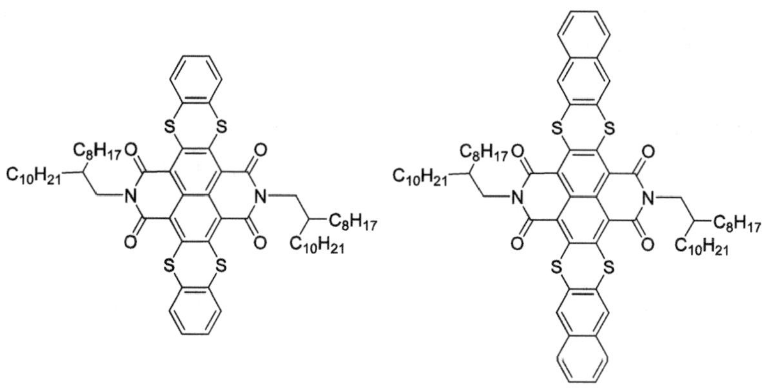

50 **51**

Figure 8.31 cNDIs **50** and **51** as ambipolar materials. Reprinted from *Chem. Sci.*, 2012, **3**, 2530–2541 with permission from the Royal Society of Chemistry.

exhibited moderate crystallinity in thin films and fibres.[88] It was found that the LUMO energy levels of all nine copolymers were almost identical, lying between −3.88 and −3.79 eV. Three of the polymers with strong electron-donating thiophene groups achieved electron mobilities of 0.006–0.02 $cm^2 V^{-1} s^{-1}$ and hole mobilities of greater than 10^{-3} $cm^2 V^{-1} s^{-1}$ with high HOMO levels, more than −5.4 eV with ambipolar charge transport behaviour.

Other ambipolar materials seen in the study used core-extended NDIs.[59,89,90] Two of them (**50** and **51**) (Figure 8.31) exhibited a very narrow band gap of less than 1.9 eV and the HOMOs were close to the work function of gold (−5.1 eV).[89] Thin films of compounds **50** and **51** displayed ambipolar behaviour under a nitrogen atmosphere with a balance of hole and electron mobilities, which further increased with annealing. The highest hole and electron mobilities for **50** were, after annealing at 120 °C, 2.6×10^{-3} $cm^2 V^{-1} s^{-1}$ and 2.9×10^{-3} $cm^2 V^{-1} s^{-1}$, respectively, whereas **51** showed hole and electron mobilities of 0.047 $cm^2 V^{-1} s^{-1}$ and 0.016 $cm^2 V^{-1} s^{-1}$, respectively, after annealing at 140 °C. High charge mobilities were attributed to the morphological change within the thin films when the temperature was increased with both **50** and **51** forming a platelet-like structure, which is beneficial for charge transporting.[59,90]

8.5 Conclusion

Naphthalene diimides show immense promise as components for photovoltaic devices, including aspects of photosynthetic mimicry and organic field-effect transistors. Their favourable electrochemical, photophysical and semiconducting properties bode well for significant advances to be made. These advances must also be supported by new chemistries that expand the core and a systematic approach to understanding the necessary features to move away from an iterative approach.

References

1. S. Ahmad, E. Guillen, L. Kavan, M. Graetzel and M. K. Nazeeruddin, *Energy Environ. Sci.*, 2013, **6**, 3439–3466.
2. M. Graetzel, R. A. J. Janssen, D. B. Mitzi and E. H. Sargent, *Nature*, 2012, **488**, 304–312.
3. C. L. Chochos, N. Tagmatarchis and V. G. Gregoriou, *RSC Adv.*, 2013, **3**, 7160–7181.
4. E. Kozma and M. Catellani, *Dyes Pigm.*, 2013, **98**, 160–179.
5. C. Duan, F. Huang and Y. Cao, *J. Mater. Chem.*, 2012, **22**, 10416–10434.
6. A. F. Eftaiha, J.-P. Sun, I. G. Hill and G. C. Welch, *J. Mater. Chem. A*, 2014, **2**, 1201–1213.
7. Y. Li, *Acc. Chem. Res.*, 2012, **45**, 723–733.
8. P. Sonar, J. P. F. Lim and K. L. Chan, *Energy Environ. Sci.*, 2011, **4**, 1558–1574.
9. R. Stalder, J. Mei, K. R. Graham, L. A. Estrada and J. R. Reynolds, *Chem. Mater.*, 2014, **26**, 664–678.
10. C. Huang, S. Barlow and S. R. Marder, *J. Org. Chem.*, 2011, **76**, 2386–2407.
11. T. Weil, E. Reuther, C. Beer and K. Muellen, *Chem. – Eur. J.*, 2004, **10**, 1398–1414.
12. Q. Bai, B. Gao, Q. Ai, Y. Wu and X. Ba, *Org. Lett.*, 2011, **13**, 6484–6487.
13. X. Zhan, A. Facchetti, S. Barlow, T. J. Marks, M. A. Ratner, M. R. Wasielewski and S. R. Marder, *Adv. Mater.*, 2011, **23**, 268–284.
14. M. Bauscher and W. Maentele, *J. Phys. Chem.*, 1992, **96**, 11101–11108.
15. G. Andric, J. F. Boas, A. M. Bond, G. D. Fallon, K. P. Ghiggino, C. F. Hogan, J. A. Hutchison, M. A. P. Lee, S. J. Langford, J. R. Pilbrow, G. J. Troup and C. P. Woodward, *Aust. J. Chem.*, 2004, **57**, 1011–1019.
16. S. V. Bhosale, C. H. Jani and S. J. Langford, *Chem. Soc. Rev.*, 2008, **37**, 331–342.
17. K. P. Ghiggino, J. A. Hutchison, S. J. Langford, M. J. Latter, M. A. P. Lee and M. Takezaki, *Aust. J. Chem.*, 2006, **59**, 179–185.

18. N. Sakai, J. Mareda, E. Vauthey and S. Matile, *Chem. Commun.*, 2010, **46**, 4225–4237.
19. F. Wuerthner and M. Stolte, *Chem. Commun.*, 2011, **47**, 5109–5115.
20. S. Bhosale, A. L. Sisson, P. Talukdar, A. Fuerstenberg, N. Banerji, E. Vauthey, G. Bollot, J. Mareda, C. Roeger, F. Wuerthner, N. Sakai and S. Matile, *Science*, 2006, **313**, 84–86.
21. T. Fõrster, *Ann. Phys.*, 1948, **2**, 55–75.
22. T. Fõrster, *Discuss. Faraday Soc.*, 1959, **27**, 7–17.
23. T. D. M. Bell, S. Yap, C. H. Jani, S. V. Bhosale, J. Hofkens, S. F. C. De, S. J. Langford and K. P. Ghiggino, *Chem. – Asian J.*, 2009, **4**, 1542–1550.
24. F. Würthner, S. Ahmed, C. Thalacker and T. Debaerdemaeker, *Chem. – Eur. J.*, 2002, **8**, 4742–4750.
25. C. Roger, Y. Miloslavina, D. Brunner, A. R. Holzwarth and F. Würthner, *J. Am. Chem. Soc.*, 2008, **130**, 5929–5939.
26. D. Cao, M. Amelia, L. M. Klivansky, G. Koshkakaryan, S. I. Khan, M. Semeraro, S. Silvi, M. Venturi, A. Credi and Y. Liu, *J. Am. Chem. Soc.*, 2010, **132**, 1110–1122.
27. E. Takahashi, H. Takaya and T. Naota, *Chem. – Eur. J.*, 2010, **16**, 4793–4802.
28. R. A. Marcus, *J. Chem. Phys.*, 1956, **24**, 966–978.
29. R. A. Marcus and N. Sutin, *Biochim. Biophys. Acta, Rev. Bioenerg.*, 1985, **811**, 265–322.
30. N. S. Hush, *Trans. Faraday Soc.*, 1961, **57**, 557–580.
31. J. R. Miller, L. T. Calcaterra and G. L. Closs, *J. Am. Chem. Soc.*, 1984, **106**, 3047–3049.
32. S. J. Langford, M. J. Latter and C. P. Woodward, *Photochem. Photobiol.*, 2006, **82**, 1530–1540.
33. M. Jaggi, B. Schmid, S.-X. Liu, S. V. Bhosale, S. Rivadehi, S. J. Langford and S. Decurtins, *Tetrahedron*, 2011, **67**, 7231–7235.
34. Y. Liu, W. Wu, J. Zhao, X. Zhang and H. Guo, *Dalton Trans.*, 2011, **40**, 9085–9089.
35. N. Banerji, S. V. Bhosale, I. Petkova, S. J. Langford and E. Vauthey, *Phys. Chem. Chem. Phys.*, 2011, **13**, 1019–1029.
36. B. Robotham, K. A. Lastman, S. J. Langford and K. P. Ghiggino, *J. Photochem. Photobiol., A*, 2013, **251**, 167–174.
37. C. Kaiser, A. Schmiedel, M. Holzapfel and C. Lambert, *J. Phys. Chem. C*, 2012, **116**, 15265–15280.
38. V. Senkovskyy, R. Tkachov, H. Komber, M. Sommer, M. Heuken, B. Voit, W. T. S. Huck, V. Kataev, A. Petr and A. Kiriy, *J. Am. Chem. Soc.*, 2011, **133**, 19966–19970.
39. P. Piyakulawat, A. Keawprajak, J. Wlosnewski, M. Forster and U. Asawapirom, *Synth. Met.*, 2011, **161**, 1238–1244.
40. C. J. Kudla, D. Dolfen, K. J. Schottler, J.-M. Koenen, D. Breusov, S. Allard and U. Scherf, *Macromolecules*, 2010, **43**, 7864–7867.
41. J. M. Keller and K. S. Schanze, *Organometallics*, 2009, **28**, 4210–4216.

42. J. M. Keller, K. D. Glusac, E. O. Danilov, S. McIlroy, P. Sreearuothai, A. R. Cook, H. Jiang, J. R. Miller and K. S. Schanze, *J. Am. Chem. Soc.*, 2011, **133**, 11289–11298.
43. B. C. Popere, P. A. M. Della and S. Thayumanavan, *Macromolecules*, 2011, **44**, 4767–4776.
44. S. K. Dey, A. Coskun, A. C. Fahrenbach, G. Barin, A. N. Basuray, A. Trabolsi, Y. Y. Botros and J. F. Stoddart, *Chem. Sci.*, 2011, **2**, 1046–1053.
45. d. R. H.-P. Jacquot, J. Iehl, C. J. Bruns, P. L. McGrier, M. Frasconi, A. A. Sarjeant and J. F. Stoddart, *Org. Lett.*, 2012, **14**, 5188–5191.
46. A. M. Cagulada and D. G. Hamilton, *J. Am. Chem. Soc.*, 2009, **131**, 902–903.
47. A. K. Mandal, M. Suresh and A. Das, *Org. Biomol. Chem.*, 2011, **9**, 4811–4817.
48. S. De and S. Ramakrishnan, *Chem. – Asian J.*, 2011, **6**, 149–156.
49. F. Wuerthner and M. Stolte, *Chem. Commun.*, 2011, **47**, 5109–5115.
50. J. G. Laquindanum, H. E. Katz, A. Dodabalapur and A. J. Lovinger, *J. Am. Chem. Soc.*, 1996, **118**, 11331–11332.
51. D. Shukla, S. F. Nelson, D. C. Freeman, M. Rajeswaran, W. G. Ahearn, D. M. Meyer and J. T. Carey, *Chem. Mater.*, 2008, **20**, 7486–7491.
52. H. E. Katz, J. Johnson, A. J. Lovinger and W. Li, *J. Am. Chem. Soc.*, 2000, **122**, 7787–7792.
53. H. E. Katz, A. J. Lovinger, J. Johnson, C. Kloc, T. Siegrist, W. Li, Y. Y. Lin and A. Dodabalapur, *Nature*, 2000, **404**, 478–481.
54. B. J. Jung, J. Sun, T. Lee, A. Sarjeant and H. E. Katz, *Chem. Mater.*, 2009, **21**, 94–101.
55. J. H. Oh, S.-L. Suraru, W.-Y. Lee, M. Koenemann, H. W. Hoeffken, C. Roeger, R. Schmidt, Y. Chung, W.-C. Chen, F. Wuerthner and Z. Bao, *Adv. Funct. Mater.*, 2010, **20**, 2148–2156.
56. M. Stolte, S. L. Suraru, F. Wuerthner, J. H. Oh, Z. Bao, J. Brill, M. Koenemann, J. Qu, U. Zschieschang and H. Klauck, *Proc. SPIE*, 2010, **7778**, 777804.
57. S.-C. Chen, Q. Zhang, Q. Zheng, C. Tang and C.-Z. Lu, *Chem. Commun.*, 2012, **48**, 1254–1256.
58. Y. Hu, X. Gao, C.-A. Di, X. Yang, F. Zhang, Y. Liu, H. Li and D. Zhu, *Chem. Mater.*, 2011, **23**, 1204–1215.
59. L. Tan, Y. Guo, G. Zhang, Y. Yang, D. Zhang, G. Yu, W. Xu and Y. Liu, *J. Mater. Chem.*, 2011, **21**, 18042–18048.
60. S.-L. Suraru, U. Zschieschang, H. Klauk and F. Wuerthner, *Chem. Commun.*, 2011, **47**, 11504–11506.
61. P. Gawrys, A. Zoltowska, M. Zagorska and A. Pron, *Electrochim. Acta*, 2011, **56**, 10464–10472.
62. M. M. Durban, P. D. Kazarinoff and C. K. Luscombe, *Macromolecules*, 2010, **43**, 6348–6352.
63. M. M. Durban, P. D. Kazarinoff, Y. Segawa and C. K. Luscombe, *Macromolecules*, 2011, **44**, 4721–4728.
64. Y. Li, G. Zhang, Z. Liu, X. Chen, J. Wang, C. A. Di and D. Zhang, *Macromolecules*, 2013, **46**, 5504–5511.

65. S. F. Võlker, A. Schmiedel, M. Holzapfel, C. Bohm and C. Lambert, *Phys. Chem. Chem. Phys.*, 2013, **15**, 19831–19844.
66. L. Favereau, J. Warnan, Y. Pellegrin, E. Blart, M. Boujtita, D. Jacquemin and F. Odobel, *Chem. Commun.*, 2013, **49**, 8018–8020.
67. P. L. Le, A. L. Smeigh, E. Gibson, Y. Pellegrin, E. Blart, G. Boschloo, A. Hagfeldt, L. Hammarstrom and F. Odobel, *Energy Environ. Sci.*, 2011, **4**, 2075–2084.
68. J. Warnan, J. Gardner, P. L. Le, J. Petersson, Y. Pellegrin, E. Blart, L. Hammarstrom and F. Odobel, *J. Phys. Chem. C*, 2014, **118**, 103–113.
69. G. Ren, E. Ahmed and S. A. Jenekhe, *Adv. Energy Mater.*, 2011, **1**, 946–953.
70. E. Ahmed, G. Ren, F. S. Kim, E. C. Hollenbeck and S. A. Jenekhe, *Chem. Mater.*, 2011, **23**, 4563–4577.
71. G. Ren, E. Ahmed and S. A. Jenekhe, *J. Mater. Chem.*, 2012, **22**, 24373–24379.
72. D. K. Susarova, P. A. Troshin, D. Hoeglinger, R. Koeppe, S. D. Babenko, R. N. Lyubovskaya, V. F. Razumov and S. N. Serdar, *Sol. Energy Mater. Sol. Cells*, 2010, **94**, 803–811.
73. J. Chen, M.-M. Shi, X.-L. Hu, M. Wang and H.-Z. Chen, *Polymer*, 2010, **51**, 2897–2902.
74. X. Hu, L. Zuo, H. Pan, F. Hao, J. Pan, L. Fu, M. Shi and H. Chen, *Sol. Energy Mater. Sol. Cells*, 2012, **103**, 157–163.
75. S.-C. Chen, Q. Zheng, Q. Zhang, D. Cai, J. Wang, Z. Yin and C. Tang, *J. Polym. Sci., Part A: Polym. Chem.*, 2013, **51**, 1999–2005.
76. T. Earmme, Y.-J. Hwang, N. M. Murari, S. Subramaniyan and S. A. Jenekhe, *J. Am. Chem. Soc.*, 2013, **135**, 14960–14963.
77. Y.-J. Hwang, G. Ren, N. M. Murari and S. A. Jenekhe, *Macromolecules*, 2012, **45**, 9056–9062.
78. M. Zhao, K. Hashimoto and K. Tajima, *Jpn. J. Appl. Phys.*, 2013, **52**, 05DA15.
79. M. Yuan, M. M. Durban, P. D. Kazarinoff, D. F. Zeigler, A. H. Rice, Y. Segawa and C. K. Luscombe, *J. Polym. Sci., Part A: Polym. Chem.*, 2013, **51**, 4061–4069.
80. Q. Xu, J. Wang, S. Chen, W. Li and H. Wang, *eXPRESS Polym. Lett.*, 2013, **7**(842–851), 810.
81. R. Fernando, Z. Mao and G. Sauve, *Org. Electron.*, 2013, **14**, 1683–1692.
82. H. Krueger, S. Janietz, D. Sainova, D. Dobreva, N. Koch and A. Vollmer, *Adv. Funct. Mater.*, 2007, **17**, 3715–3723.
83. S.-I. Sakurai, J. Areephong, L. Bertone, N.-T. Lin, N. Sakai and S. Matile, *Energy Environ. Sci.*, 2011, **4**, 2409–2416.
84. Y. Zhao, Y. Guo and Y. Liu, *Adv. Mater.*, 2013, **25**, 5372–5391.
85. R. P. Ortiz, H. Herrera, C. Seoane, J. L. Segura, A. Facchetti and T. J. Marks, *Chem. – Eur. J.*, 2012, **18**, 532–543.
86. L. E. Polander, S. P. Tiwari, L. Pandey, B. M. Seifried, Q. Zhang, S. Barlow, C. Risko, J.-L. Bredas, B. Kippelen and S. R. Marder, *Chem. Mater.*, 2011, **23**, 3408–3410.
87. S. P. Tiwari, J. Kim, K. A. Knauer, D. K. Hwang, L. E. Polander, S. Barlow, S. R. Marder and B. Kippelen, *Org. Electron.*, 2012, **13**, 1166–1170.

88. X. Guo, F. S. Kim, M. J. Seger, S. A. Jenekhe and M. D. Watson, *Chem. Mater.*, 2012, **24**, 1434–1442.
89. H. Luo, Z. Cai, L. Tan, Y. Guo, G. Yang, Z. Liu, G. Zhang, D. Zhang, W. Xu and Y. Liu, *J. Mater. Chem. C*, 2013, **1**, 2688–2695.
90. L. Tan, Y. Guo, Y. Yang, G. Zhang, D. Zhang, G. Yu, W. Xu and Y. Liu, *Chem. Sci.*, 2012, **3**, 2530–2541.

CHAPTER 9

Perylenetetracarboxylic Diimide and its Covalently-linked Molecular Arrays

YINGYUAN ZHAO[a] AND XIYOU LI*[b]

[a] School of Petroleum and Chemical Engineering, Dalian University of Technology, China; [b] College of Science, China University of Petroleum (East China), Qingdao, Shandong 266580, China
*Email: xiyouli@upc.edu.cn

9.1 Introduction

Perylenetetracarboxylic diimides, also called perylene diimides and abbreviated as PDIs, are the fruit flies of photochemists. The parent PDIs exhibit a flat π-system composed of two naphthalene units. Each of the naphthalene units is attached to an imide group and connected to the other naphthalene unit by two C sp^2–sp^2 single bonds. Their unique properties, such as wide absorption and emission bands in the visible region, high fluorescence quantum yield, good optoelectronic properties, exceptional thermal and photochemical stabilities, and easy modification of molecular structures, have attracted researchers from different branches of chemistry and material science.

Since the pioneering work of Kardos, who first reported *N,N*-dimethyl PDI in 1913,[1] PDIs have been the subject of tremendous investigation. After 1950,

Monographs in Supramolecular Chemistry No. 21
Naphthalenediimide and its Congeners: From Molecules to Materials
Edited by G. Dan Pantoş

Published by the Royal Society of Chemistry, www.rsc.org

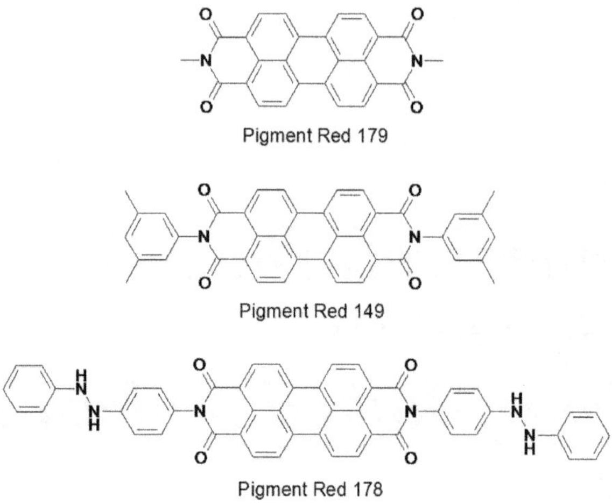

Figure 9.1 Chemical structures of three PDIs widely used as industrial pigments. Reprinted with permission from ref. 5. Copyright 2011 American Chemical Society.

several members of the PDI family, such as Pigment Red 179, Red 178 and Red 149 (Figure 9.1), found their way into industrial-scale production.[2] As a group of high performance pigments, PDIs possess excellent migrational, chemical, thermal, and photo stabilities as well as high tinctorial strength[2,3] and exhibit variable colors from red to violet, and even shades of black.[4] Today, PDIs are still recognized as an important class of high-performance pigments, particularly in the automobile industry.[2,5]

However, the intrinsic insolubility of PDIs in conventional organic solvents restricted the exploration of their photophysical properties and applications in other fields. Thanks to Geissler's work in 1959, the potential of PDIs as fluorescent dyes with high fluorescence quantum yields and exceptional photo-stability was discovered,[6] which opened the door to fundamental research on the applications of PDIs as a functional component in artificial photosynthesis,[7–12] as well as many other newly developed applications.

Nowadays, PDIs have been investigated as active materials in various areas. They have been proven to be valuable fluorophores in various electronic and optical applications, such as organic field-effect transistors (OFETs),[13–19] organic photovoltaic cells (OPVs),[20–25] optical power limiters,[26] fluorescent solar collectors,[27] dye lasers,[28,29] and fluorescent sensors.[30–33] Furthermore, they have also been utilized in living cell staining[34,35] and photo-acoustic imaging of deep tumors.[36] All these applications depend on the excellent electronic or photophysical properties of PDIs. For example, PDIs feature a relatively low reduction potential, which enables their use as electron acceptors for organic electronics. Their fairly electron-deficient character also endows PDIs with high photochemical stability towards photo-oxidation, the major destructive mechanism for dyestuffs. PDIs are also one of the most promising electron-transport materials available now, with a low electron affinity close to that of C_{60} derivatives.[37,38] The absorption and fluorescence spectra of PDIs are mostly in the visible region with large molar extinction coefficients and exceptional high fluorescence quantum yields, which endow PDIs with good solar light harvesting properties. X-Ray diffraction experiments of several single crystals revealed that the parent PDIs exhibit flat π-systems. Hence, the PDIs tend to form π-stacks in solid pigment particles or in concentrated solutions.[39–44]

The chemical structure of PDI is shown in Figure 9.2. PDIs with different chemical and physical properties can be readily obtained by introducing substituents selectively at the imide nitrogen positions,[4,45–47] the 1, 6, 7, 12 positions of the hydrocarbon core (bay-positions),[48–50] or the 2, 5, 8, 11 positions (*ortho*-positions).[51–54] Inspired by the light harvesting system of natural photosynthesis, various supramolecular systems based on PDIs by utilizing weak noncovalent intermolecular interactions, such as hydrogen bonding,[55] metal coordination,[56] electrostatic,[57] hydrophobic[58] and π–π interactions,[13] have been constructed and their photophysical properties investigated. These noncovalent forces proved to be useful to control the structure and photophysical properties of the supramolecular system. However, investigations towards understanding the relationship between structure and spectroscopic, as well as electrochemical, properties have been retarded because of the flexible supramolecular structures formed based on weak intermolecular interactions. Therefore, the development of covalently connected PDI molecular arrays with rigid structures is necessary for revealing the relationship between the structure and photophysical properties of the PDI assembly. Until now, PDI molecular arrays with parallel,[58–61] orthogonal,[11,62] linear[12] or even star-shaped configurations[63] have been successfully prepared by using proper spacers and their photophysical properties have been investigated. In this chapter, we will introduce the development of the synthesis and molecular structure modification of PDIs first, and then will summarize the recent development of the construction of covalently linked PDI molecular arrays and their photophysical properties.

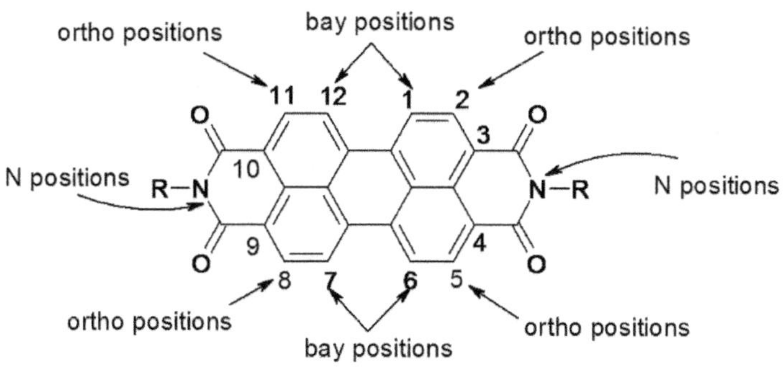

Figure 9.2 Molecular structure of PDI with position numbers.

9.2 Modification of the Molecular Structure of PDI

Chemical modification of the molecular structure of PDIs is very important to obtain desirable photophysical and electronic properties as well as solubility for PDIs. The starting material for synthesizing PDIs is normally perylenetetracarboxylic dianhydride (PTCDA) in both laboratory research and the industrial production process. PTCDA can be easily obtained with analytical purity even on a technical scale.[5,64] As shown in Scheme 9.1, acenaphthene is first oxidized to provide 1,8-naphthalene dicarboxylic acid anhydride, which is further treated with ammonia to give naphthalene-1,8-dicarboxylic imide. Subsequently, PDI is obtained by dimerizing these naphthalene imide molecules in a molten alkali. The PTCDA is finally synthesized by saponification of PTCDI with hot concentrated sulfuric acid. Although insoluble and with high melting points, symmetrical PDI pigments, such as Pigment Red 179, Pigment Red 178 and Pigment Red 149, can be easily obtained in high yields by the condensation of PTCDA with various aliphatic and aromatic primary amines. Because most of the applications require PDIs with reasonable solubility in conventional organic solvents in order to achieve a high purity,[3,5] various synthetic methods have been developed for the purpose of introducing solubilizing groups at the imide nitrogen positions, bay-positions or *ortho*-positions.

9.2.1 Introducing Substituents at N-Positions

The introduction of substituents at the N-positions has been extensively studied. The synthesis takes advantages of the fact that the two nitrogen

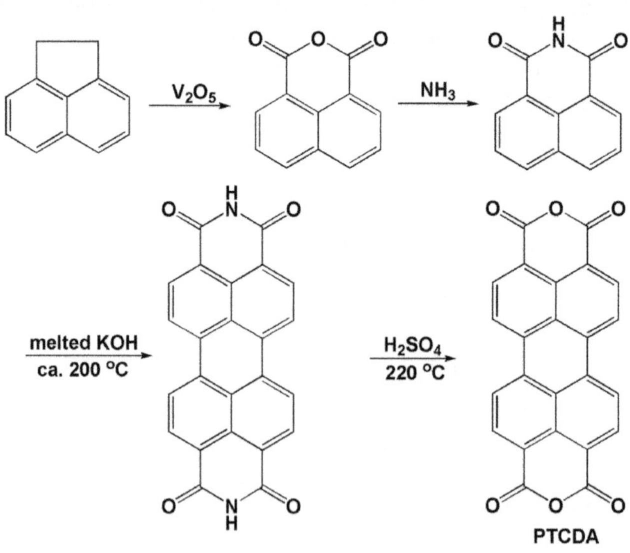

Scheme 9.1 Synthesis of PTCDA.[64]

positions at the imides of PDI are the π-orbital wave function nodes,[65] leading to enormous options for modifying the structures of the two side chains without a significant change to the photophysical properties. Generally, we can introduce the same groups (symmetrical substitution) or different groups (asymmetrical substitution) at the two N-positions. Symmetrical substitution is relatively easy to realize by the reaction of PTCDA with corresponding primary amines under inert gas in a proper solvent (Scheme 9.2). In the 1990s, Langhals and coworkers introduced the so-called swallow-tail-substituents to the imide nitrogen positions and successfully improved the solubility in conventional organic solvents.[64] The condensation of PCTDA with reactive primary aliphatic amines can be carried out under mild reaction conditions without a catalyst at room temperature. But for less reactive bulky aliphatic or aromatic amines, it is necessary to use molten imidazole or quinoline as a solvent. Sometimes, anhydrous zinc acetate must be used as a catalyst. Other metal salts, such as iron, lead, or copper salts, can also be used as catalysts, but the efficiency is always less than that of zinc salts. The purification of these symmetrical PDIs is relatively easy and the isolated yields are normally higher than 90%.[64]

The synthesis of asymmetrical PDIs with different substituents at the imide positions is not as easy as that of symmetrical PDIs. Attempts to synthesize asymmetrical PDIs by adding two different amines (either simultaneously or sequentially) to react with PTCDA are usually unsuccessful because the reactivities of the amines with PTCDA are different. Typically, only traces of the desired products are observed, with the dominant species being the symmetrical PDIs.[66] Currently, the general method to obtain asymmetrical PDIs is by condensation of a primary amine with perylene monoimide monoanhydride. There are currently three general methods to synthesize perylene monoimide monoanhydride. As shown in Scheme 9.3, the first synthetic method for perylene monoimide monoanhydride from imidization of PTCDA is actually the same as preparing the respective

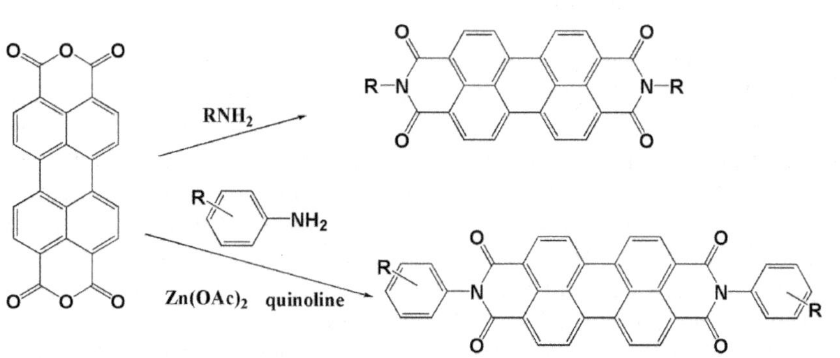

Scheme 9.2 Symmetrical substitution at the N-positions (top: dialkyl and bottom: diaryl).

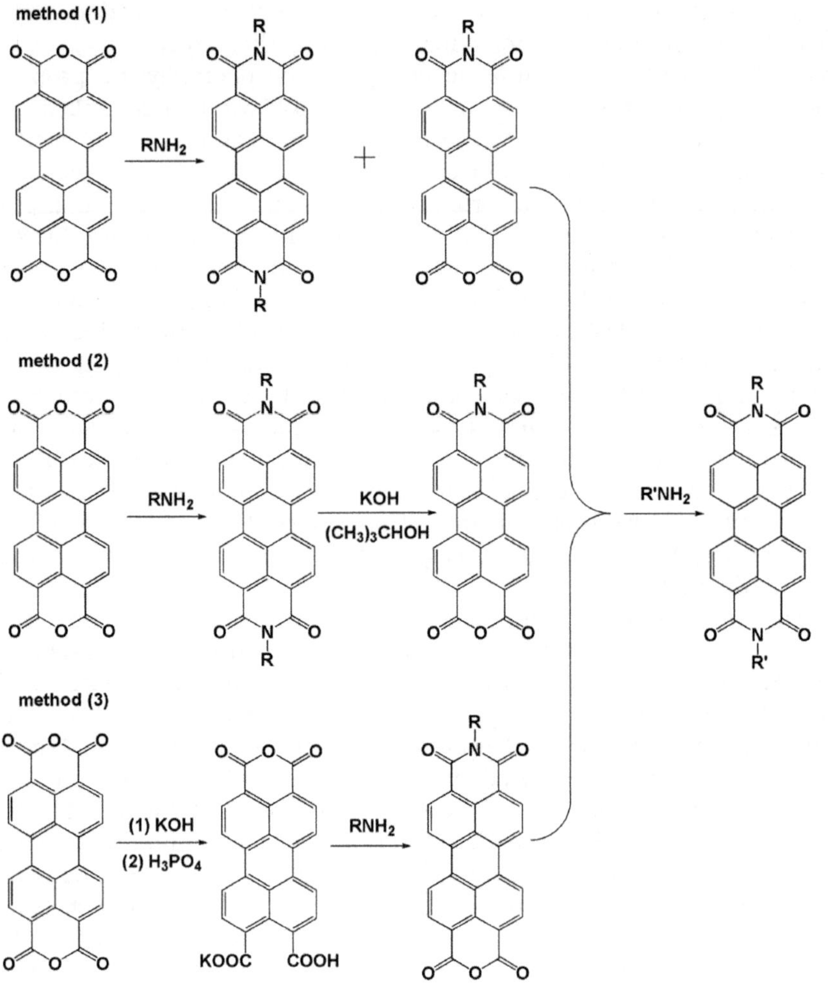

Scheme 9.3 The three methods for preparing PDIs with different substituents on the imide nitrogen positions.[5]

symmetrical PDIs. It is challenging to synthesize a perylene monoimide monoanhydride directly from PTCDA with the respective primary amine. Typically, diimides are the predominant products, but the yields of perylene monoimide monoanhydride can be increased by controlling the reactant ratio, reaction temperature and time.[67] Another method (Scheme 9.3, method 2) relies on partial hydrolysis of symmetrical PDIs, a reaction that gives yields of about 50%. This method is most widely utilized because of the higher yield and easier purification. The third practical method was reported by Tam-Chang and coworkers.[68] The PTCDA was first converted to a mixed anhydride–dicarboxylate salt by a partial hydrolysis reaction, followed by imidization reactions to obtain perylene monoanhydride monoimide in *ca.* 60–90% yield.

9.2.2 Introducing Substituents at the Bay Positions

The introduction of substituents at the 1, 6, 7, and 12 positions (bay-positions) of PDIs can be used to increase solubility and modify the electronic and optical properties. These bay-functionalized PDIs are usually synthesized from the corresponding halogenated PDIs. Tetrachloro and dibromo PDI derivatives are key intermediates for the synthesis of a wide range of other bay-substituted derivatives. 1,6,7,12-Tetrachloro PDIs are obtained by imidization of a tetrachloro PTCDA derivative, which is synthesized by tetrachlorination of PTCDA with sulfuric acid and chlorine. Seybold and coworkers at BASF first incorporated four phenoxy groups by nucleophilic displacement of chlorine substituents by heating the corresponding tetrachloro species with phenol and potassium carbonate in *N*-methylpyrrolidone.[69] 1,6,7,12-Tetrafluoro PDIs can be obtained by the reaction of 1,6,7,12-tetrachloro PDI with potassium fluoride in sulfolane at elevated temperatures by using (*N*,*N*-dimethylimidazolidino)tetramethylguanidinium chloride or 18-crown-6 as a catalyst (Scheme 9.4).[70] However, it has been proved that the introduction of other nucleophiles is difficult.[71]

Brominated PDI intermediates are used more frequently to prepare the bi-substituted PDIs. There are generally two pathways to obtain brominated PDIs from PTCDA. The first one was developed by Böhm and coworkers, who brominated the PTCDA first and then reacted it with the corresponding amines.[72] The second strategy was to introduce substituents at the imide positions first and then the obtained PDIs were brominated at room temperature in solvents such as dichloromethane. The reaction conditions of the second method are relatively mild, thus, it is popular in laboratory

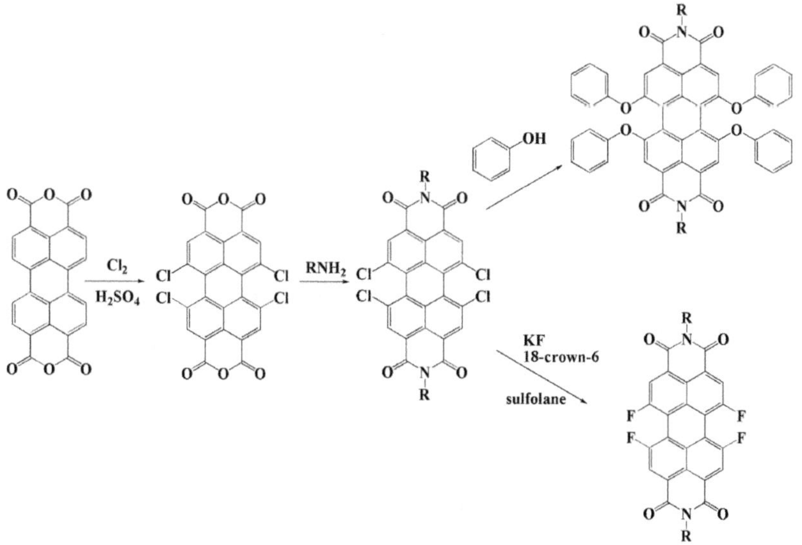

Scheme 9.4 Preparation of PDIs by chlorination.[5]

synthesis. Dibrominated PDIs are usually the predominant products accompanied by mono-, tri- and even tetrabrominated PDIs. The product mixture can be easily separated by column chromatography, but the two dibromo PDI regioisomers (1,7- and 1,6-dibromo PDIs) are difficult to separate and can only be distinguished from one another by high resolution (>400 MHz) ^{1}H NMR spectroscopy. 1,7-Dibrominated PDIs can be relatively easily separated by repetitive recrystallization.[73] Though the products obtained by bromination are always a mixture, we can increase the ratio of desired products by changing the conditions. For example, in the former method, the yields of tri- and tetrabrominated PDIs can be increased by increasing the reagent amount including bromine and sulfuric acid, elevating the reaction temperature and prolonging reaction time.[74] In the latter method, decreasing the bromination temperature to *ca.* 24 °C can promote the formation of monobrominated PDIs, while heating to *ca.* 50 °C can increase the yield of dibrominated PDIs to 90% (Scheme 9.5).[75]

The most popular chemical reaction used to prepare bay-substituted PDIs from brominated PDIs is nucleophilic substitution. Fluorine, carbon, cyano, oxygen, nitrogen and sulfur nucleophiles can couple to bromo-PDIs providing novel PDI derivatives with a variety of optical and electronic properties. Besides the symmetrical substitution at the 1,7-positions, as shown in Scheme 9.6, asymmetrical substituted and monosubstituted compounds at the 1,7-positions can also be synthesized (Scheme 9.7).[50,76–78]

Metal-catalyzed cross-coupling reactions can also be used to obtain various functionalized PDIs. Among these cross-coupling reactions, Suzuki

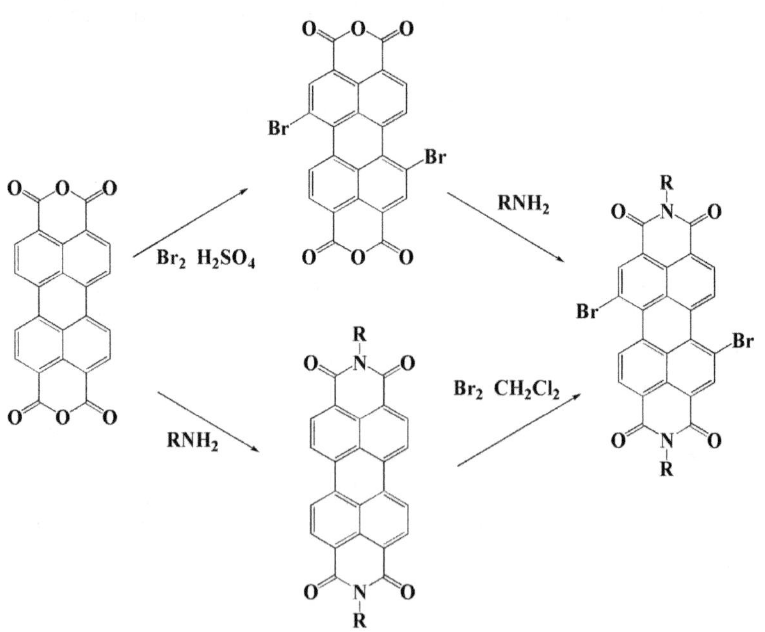

Scheme 9.5 Two methods for preparing 1,7-dibrominated PDI.

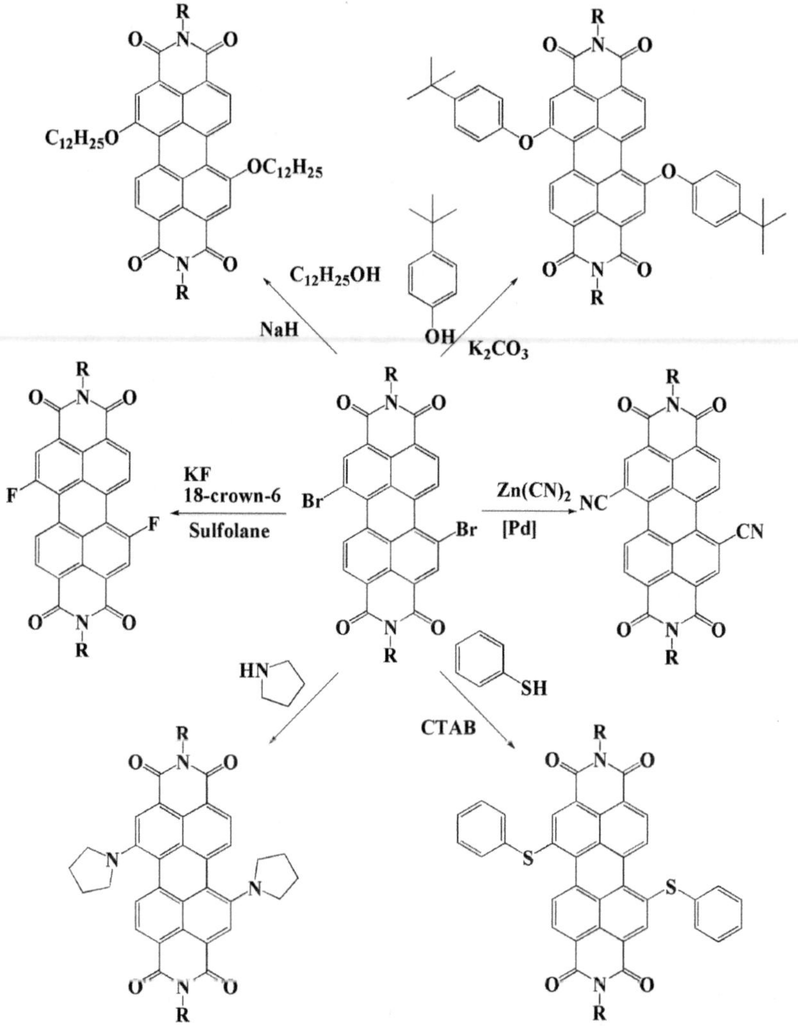

Scheme 9.6 Preparation of symmetrical substituted PDIs with nucleophilic substitutions.[3]

coupling is one of the most employed reactions, which involves the palladium-mediated coupling of organic electrophiles, such as aryl halides, with organoboron compounds in the presence of a base. One example is the Suzuki reaction of tetrabromo PDI and phenylboronic acid in the presence of KF, Ag_2O and $Pd(PPh_3)_4$ in THF to afford 1,6,7,12-tetraphenyl PDIs (Scheme 9.8).[79]

The Stille reaction provides another valuable method for the synthesis of PDI derivatives,[49] allowing the design of molecules with potential research interests in solar cells and field-effect transistors. For example, the Stille reaction of 1,7-dibromo PDI with 2,6-bis(tri-*n*-butylstannyl)dithieno[3,2-*b*:2′,3′-*d*] thiophene in toluene with $Pd(PPh_3)_4$ as a catalyst afforded

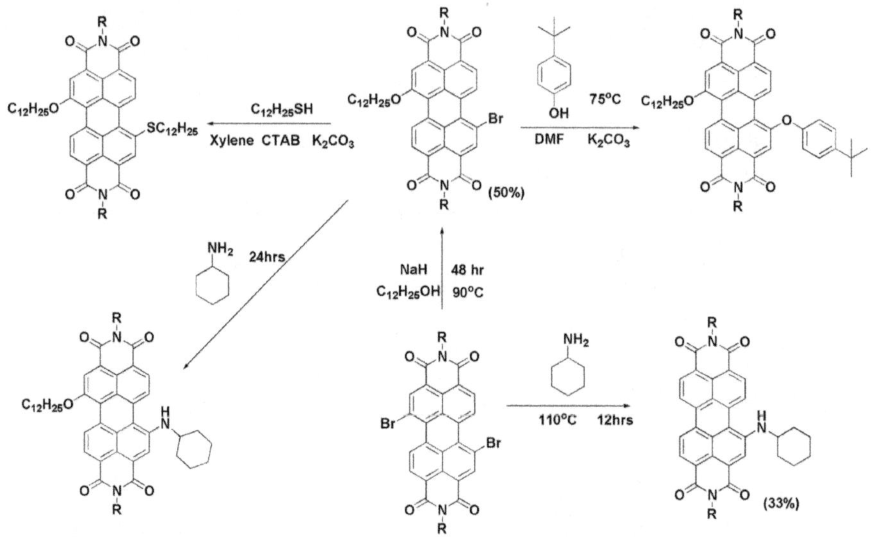

Scheme 9.7 Preparation of asymmetrical substituted PDIs with nucleophilic substitutions.[50]

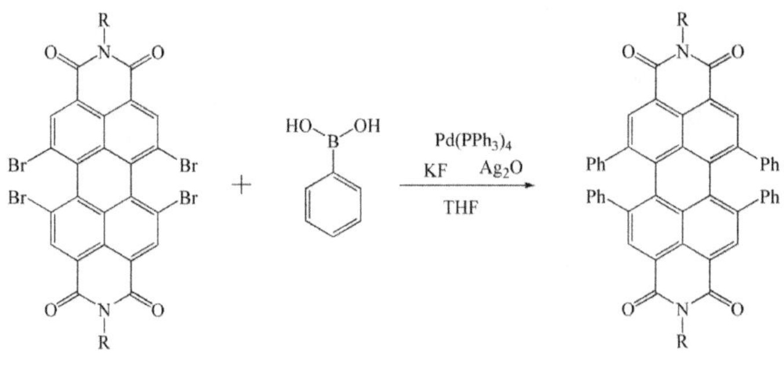

Scheme 9.8 1,6,7,12-Tetraphenylperylene diimide derivative prepared by Suzuki coupling.
Reprinted with permission from ref. 79. Copyright 2006 American Chemical Society.

poly{[*N*,*N*′-bis(2-decyl-tetradecyl)-3,4,9,10-perylene diimide-1,7-diyl]-*alt*-(dithieno[3,2-*b*:2′,3′-*d*]thiophene-2,6-diyl)} with a yield of 84% (Scheme 9.9).[15]

Bay-substitution of PDIs can also be achieved *via* Sonogashira reaction. Reaction of 1,7-dibromo PDI with the corresponding alkynyl compounds in the presence of $Pd(PPh_3)_4$ and CuI–Et_3N in THF provided bisalkynyl-substituted PDIs in *ca.* 90% yield. These bisalkynyl-substituted PDIs can further be used to expand PDIs to coronene-2,3,8,9-tetracarboxylic acid diimides in the presence of DBU (1,8-diazabicyclo[5.4.0]undec-7-ene) (Scheme 9.10).[80–83]

Substitutions at the bay positions and expansion of the PDI core do not necessarily have to start from the halogenated derivatives. Recently, Yue and

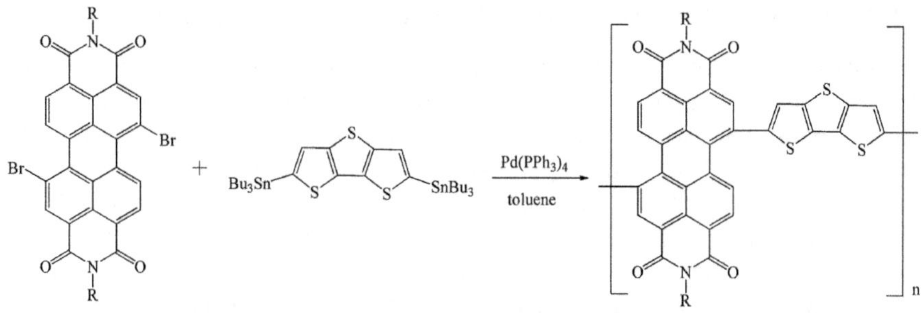

Scheme 9.9 Thiophene-substituted PDI polymer prepared by Stille coupling.[15]

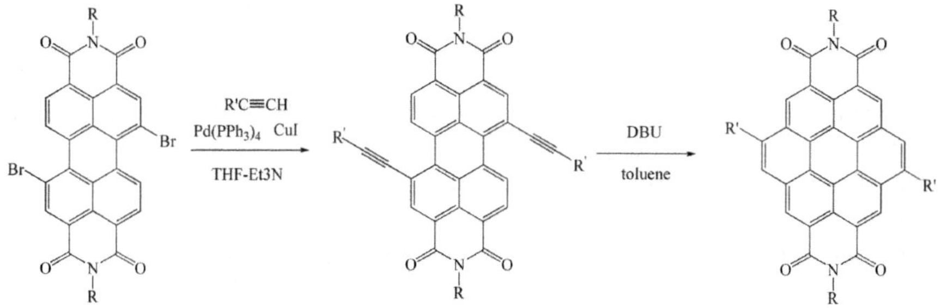

Scheme 9.10 Functionalization of PDI at the bay positions by Sonogashira reaction.

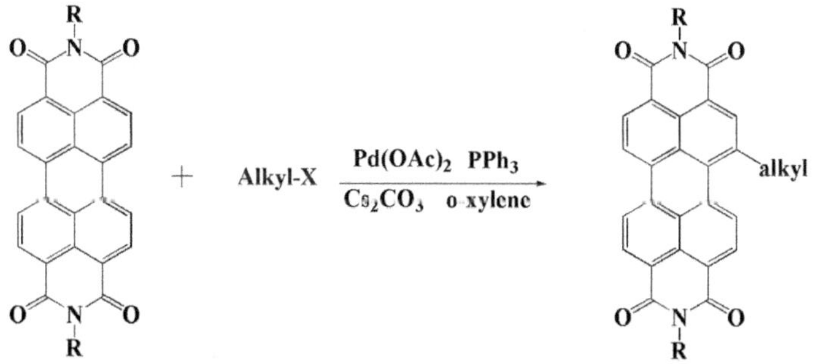

Scheme 9.11 Pd-catalyzed alkylation of PDIs with alkyl halides.
Reprinted with permission from ref. 84. Copyright 2011 American Chemical Society.

coworkers reported a method for the direct functionalization of the C–H bonds of PDI to construct the C–C bond under the system of $Pd(OAc)_2$ as the catalyst, PPh_3 as the ligand, and Cs_2CO_3 as the base in *o*-xylene.[84] This synthetic route can be applied to a wide range of substrates, especially alkanes with fluorocarbon chains and hydrophilic oligo(ethylene glycol) chains (Scheme 9.11).

9.2.3 Introducing Substituents at the *ortho*-Positions

The introduction of substituents at the *ortho*-positions (2,5,8,11-positions) of PDIs was unavailable until Nakazono and coworkers employed the Murai–Chatani–Kakiuchi protocol: ruthenium-catalyzed C–H bond activation and an addition strategy, affording the formation of C–C bonds at the *ortho* positions.[51,85,86] One example is the successful installation of *n*-hexyl substituents at the 2,5,8,11-positions of PDIs by heating a mixture of bis(*N*-ethylpropyl)PDI and 1-hexene in mesitylene in the presence of [$RuH_2(CO)(PPh_3)_3$] (Scheme 9.12).[85] Another example, the reaction of bis(*N*-ethylpropyl)-PDI with phenyl boronic acid neopentyl glycol ester with a mesitylene and pinacolone mixture as a reaction medium and [$RuH_2(CO)(PPh_3)_3$] as a catalyst afforded 2,5,8,11-tetraphenyl PDI in 83% yield (Scheme 9.12).[51]

Borylation reactions provide a valuable method for the synthesis of 2,5,8,11-tetraboronate PDIs, which can act as starting materials for introducing groups at the *ortho*-positions of PDI (Scheme 9.13). More importantly, this method provides an alternative method for the introduction of heteroatoms and other functionalities at *ortho*-positions in addition to the formation of C–C bonds. In 2011, iridium-catalyzed[87] and ruthenium-catalyzed[88] direct boronization of PDIs at the 2,5,8,11-positions were developed by Teraoka and Battagliarin, respectively. Mono-, di- and triboronate PDIs can also be prepared by controlling the reaction conditions. Introduction of heteroaryl groups at the 2,5,8,11-positions of PDIs is difficult by Ru-catalyzed direct arylation with arylboronic esters, but, Suzuki–Miyaura cross-coupling

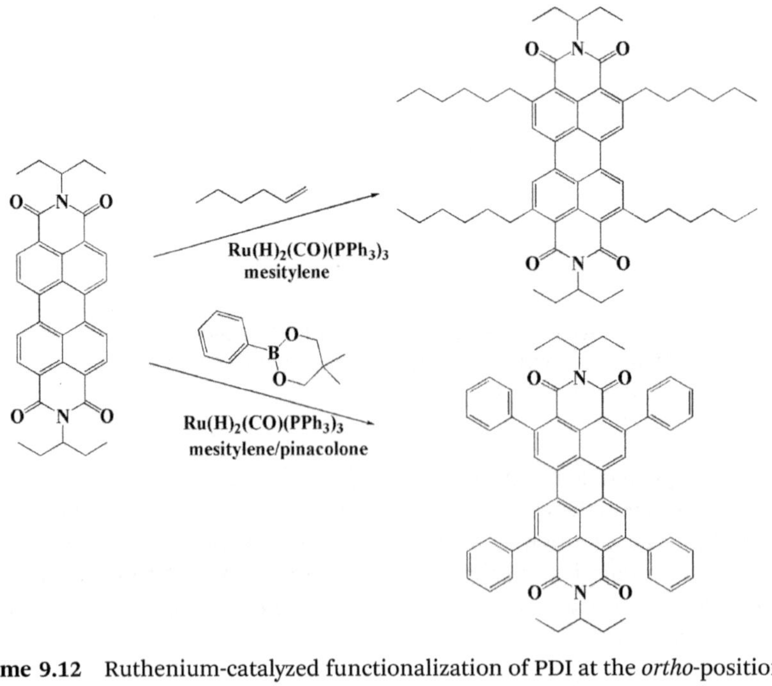

Scheme 9.12 Ruthenium-catalyzed functionalization of PDI at the *ortho*-positions.[85]

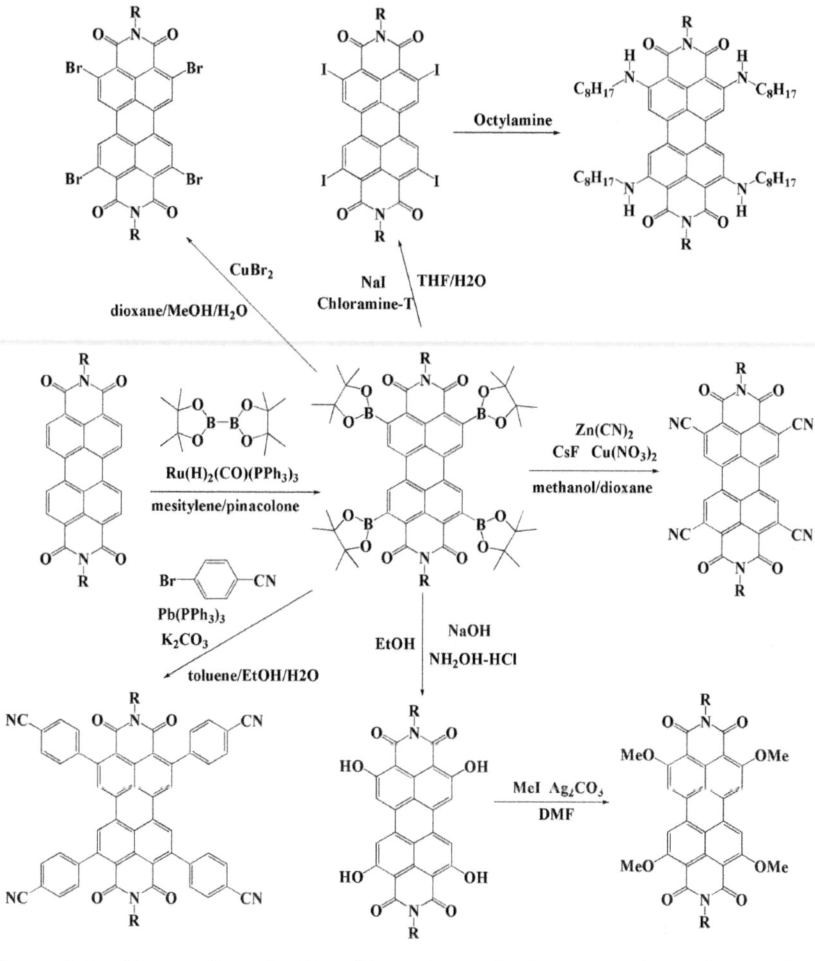

Scheme 9.13 Preparation of PDIs with various substituents at the *ortho*-positions.

of the obtained tetraboronate PDIs can be carried out easily, affording heteroarylated PDIs.[87] Oxidation of arylboronic acid pinacol esters affords OH groups.[89] Thus, tetrahydroxy PDI can be prepared from the oxidation of tetraboronate PDI. The obtained tetrahydroxy PDI can further be converted into tetramethoxy PDI by methylation of the OH groups.[87] Another possibility is copper-catalyzed reactions involving the synthesis of tetrahalogenated and tetracyanated PDIs.[52]

More recently, the synthesis of bay- and *ortho*-substituted PDIs has been developed.[90] In 1992, the first perchloro-substituted PDIs (Cl_8-PTCDI) were obtained by chlorination of perylene-3,4:9,10-tetracarboxylic diimide in chlorosulfonic acid at 80 °C.[91] It was not until 2010 that Würthner and coworkers demonstrated that Cl_8-PTCDI is a high-performance n-channel organic semiconductor.[92] PDIs substituted at the bay- and *ortho*-positions can start from bay-substituted tetrachloro-PDIs, as shown in Scheme 9.14. The obtained Br_4Cl_4-PDIs can be used as precursors for getting fluoride-,

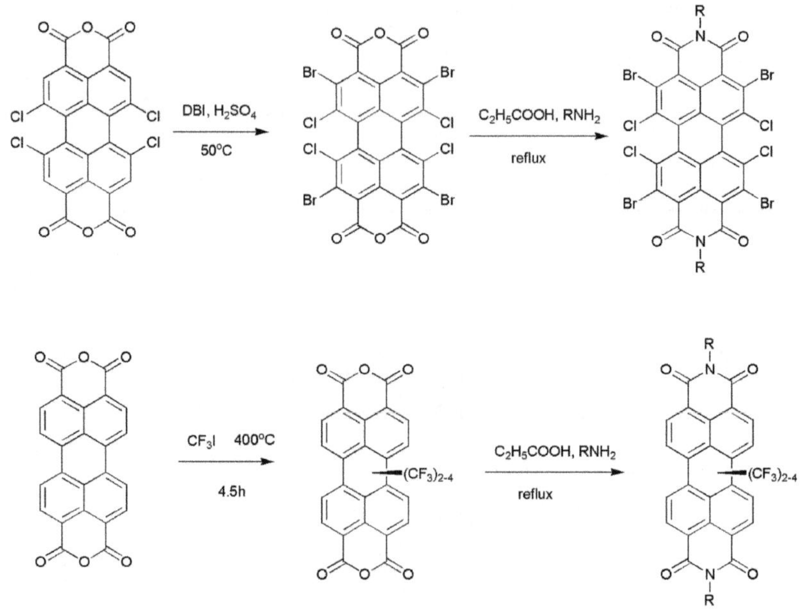

Scheme 9.14 The two methods for preparing bay- and *ortho*-substituted PDIs.

cyanide-, amine- and thiophenol-substituted PDIs.[93,94] In 2015, Strauss and coworkers reported a simultaneous bay- and *ortho*- substitution method for PTCDA (Scheme 9.14).[54]

9.3 Basic Properties of PDIs

9.3.1 Structure

The properties of PDIs are closely related to their unique structures. It has been confirmed that the parent PDIs exhibit flat π-systems.[44] The introduction of substituents at the N-positions and *ortho*-positions has no obvious influence on the highly planar structure of the perylene core.[87] This is in sharp contrast to the PDIs modified at the bay-positions, in which the perylene backbone is considerably twisted (Figure 9.3). The propeller-like twisting of the two naphthalene half units is considered to be related to the sterical strain at the bay-positions. Therefore, it is not surprising that along with an increase in substituent number, the twisting angle between the two naphthalene rings will be larger. This has been proven by the crystal structures of PDIs. The torsion angle of 1,7-diphenoxy PDI is about 13.7°,[95] whereas a tetraphenoxy PDI has a torsion angle of 25°.[96] It has also been confirmed that the linking atoms between the substituents and the perylene ring play a dominating role in determining the twisting angles between the two naphthalene rings. For example, the twisting angle of 1,7-di(dodecyloxy) PDI is 15°, but this angle increases to 23° for 1,7-di(thiododecyl) PDI.[97]

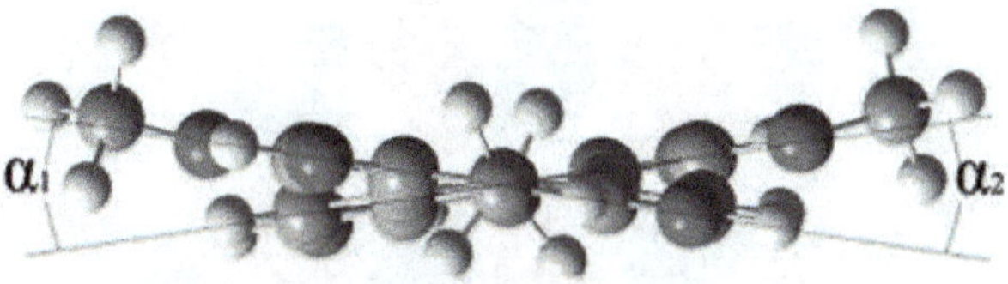

Figure 9.3 Side view of the minimized structure of 1,7-disubstituted PDIs. Reprinted with permission from ref. 97. Copyright © 2007 WILEY-VCH Verlag GmbH & Co. KGaA, Weinheim.

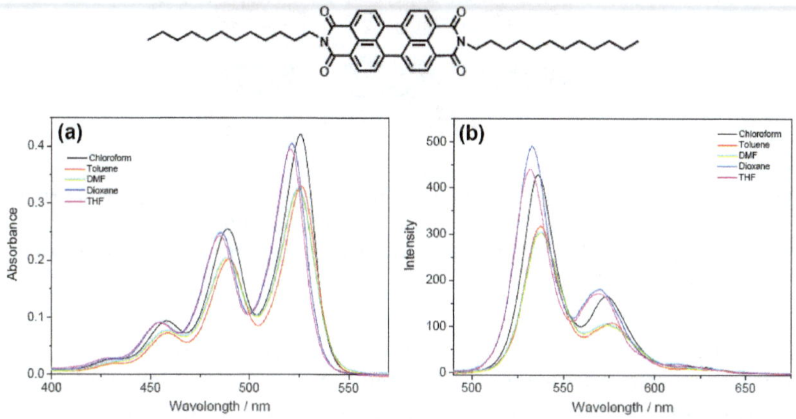

Figure 9.4 (A) UV/Vis absorption and (B) emission spectra of a typical PDI (5.0 μM) in different organic solvents.
Reprinted with permission from ref. 99. Copyright 2006 American Chemical Society.

On the other hand, the number of atoms connected directly to the linking atom also affects the twisting angles. The twisting angle of 1,7-dimethylamino PDI (21.1°) is much larger than that of 1,7-dimethoxy PDI (15°). This is because two carbon atoms connect directly to the nitrogen atom in the dimethylamino group, whereas only one carbon atom connects with the oxygen atom in the methoxyl group. The former brings large repulsions between the hydrogen atoms at the 6 and/or 12 positions of the perylene ring with the substituents at the 1,7 positions.[98]

9.3.2 Spectroscopic Properties

Generally, PDIs exhibit a vibronically resolved absorption with large extinction coefficients (from 400 to 600 nm) and strong yellow-green fluorescence in the visible region. Figure 9.4 shows the absorption and emission spectra of a typical PDI without any substituents on the perylene core in different organic solvents.[99] Generally, the imide substituent has a negligible influence on the absorption and emission properties of PDIs. The absorption and

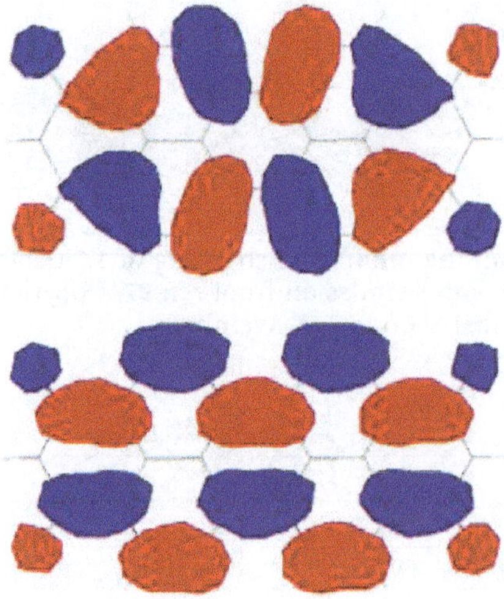

Figure 9.5 HOMO (top) and LUMO (bottom) of perylene bisimides. Note that both frontier orbitals exhibit nodes at the imide nitrogens.
Reproduced from ref. 3 with permission from the Royal Society of Chemistry.

emission peaks vary by less than 5 nm with different N-substituents.[5] This is because the nitrogen atoms are located on the nodal planes of both the HOMO and LUMO (Figure 9.5), which reduces the coupling between the PDI units and the imide substituents to a minimum. The symmetries of these orbitals also indicate that the optical transition is polarized along the long molecular axis (a line connects two imide nitrogen atoms, N–N). The fluorescence quantum yields of N-substituted PDIs are close to 100%. The fluorescence lifetime for PDI monomers in solution is about 4 ns.[59] However, when phenyl substituents are at the N-positions of PDI, the fluorescence quantum yield decreases to 70%, which can be attributed to the vibronic motions ("loose bolt effect")[100] because the substituents are not fixed in an orthogonal conformation. If electron-rich substituents such as alkoxyphenyl are connected at the N-positions, the fluorescence quantum yield can drop to 5%.[101–103] This drop can be ascribed to a photoinduced electron transfer from the electron-rich substituent to the electron-deficient PDI unit.

In contrast, core substitution shows a much more significant effect on both the absorption and emission properties of PDIs due to the stronger electronic coupling between the PDI unit and the substituents. Substituent effects at the bay-positions are substantial due to the large contribution to the HOMO by these positions. For example, the absorption and emission spectra of 1,7-diphenoxy PDI are bathochromically shifted by *ca.* 20 and 40 nm, respectively, with respect to that of the unsubstituted PDI dyes. More pronounced spectral changes take place if the substitution of the aromatic cores with stronger π-donors, such as pyrrolidino groups, afford

1,7-dipyrrolidino PDIs with a green color due to the bathochromic shift of over 150 nm relative to the unsubstituted analogues. Electron-donating substituents lead to not only a bathochromic shift on the absorption spectra, but also considerable peak broadening and an unresolved vibronic fine structure simultaneously, which can be attributed to both the participation of the side groups in the conjugation of the PDI core and the twisting in molecular conformation of the PDI ring.[50] The extremely small fluorescence quantum yields of these kinds of compounds are caused by intramolecular charge transfer.[98] By contrast, for electron-withdrawing groups, such as fluoro,[104] chloro,[105] bromo,[73] and cyano substituents,[106,107] limited spectral changes take place with only a few nanometers' shift and there is no significant charge-transfer character. The fluorescence quantum yields remain high for these compounds.

Compared with bay-positions, substitution at the *ortho*-positions has been believed to be less effective in modifying the electronic nature of PDIs. Installation of alkyl chains at the 2,5,8,11-positions of PDIs does not significantly change the absorption and emission properties, but somewhat lowers the molar extinction coefficients of the lowest energy absorption band. The fluorescence quantum yields of these compounds are almost the same as the parent PDIs.[85] With the introduction of electron-withdrawing groups, such as cyano and halogen groups, hypsochromic shifts in the absorption and emission are observed. This is because these electron-withdrawing groups have increased the HOMO–LUMO gap by stabilizing the HOMO. Substantial decreases in the fluorescence quantum yields are also observed for these compounds.[52] Incorporation of extended conjugated groups in the *ortho*-positions also has a significant impact on the spectroscopic properties of PDIs. Slight bathochromic shifts in the absorption and emission spectra are always observed. Moreover, the fluorescence quantum yields are much lower than the parent PDIs. Taking tetrakis(*p*-(*N*,*N*-dimethylamino)phenyl) PDI as an example, the λ_{max} has red-shifted about 4 nm and the absorption spectrum exhibits a broadened band around 600–700 nm. Furthermore, fluorescence is perfectly quenched in this compound.[51]

9.3.3 Redox Properties

PDIs are fairly electron-deficient molecules, which are easy to reduce and rather difficult to oxidize. For most PDI derivatives, two reversible reduction waves corresponding to the formation of the radical anions and dianions and one reversible oxidation wave can be observed in appropriate solvents. Substituents at the N-positions have little effect on the redox potentials because of the nodes of both HOMO and LUMO on the nitrogen atoms. By contrast, substituents at the bay-positions significantly affect the redox potentials. When the substituents are electron-withdrawing groups, such as cyano and bromo, the PDIs will be easily reduced. This is because of the inductive effect of electron-withdrawing groups on the HOMO and LOMO orbitals. When electron-donating substituents are introduced, the reduction potential is increased, and the reversible oxidation wave appears at a lower potential. For the strong electron-donating substituents, such as pyrrolidinyl, the reduction potential of PDI increased by 0.3 V, and sometimes two reversible oxidation waves can be found.[108,109] Substituents at the *ortho*-positions also have pronounced effects on the redox properties of PDIs. Similar to that observed for bay-position-substituted compounds, substitution by chlorine, bromine, and especially by the cyano group at the *ortho*-positions leads to strong oxidant dyes.[52] As expected, positive shifts in the oxidation and reduction potentials are also caused by other electron-withdrawing aryl groups.[87] Electron-donating aryl groups lead to negative shifts of reduction potentials.[51] PDI derivatives are widely used to establish long-lived charge-separated states in photoinduced electron transfer, enabling one to readily follow the dynamics of electron transfer processes. The redox potentials of PDI can be tuned efficiently by appropriate substituents resulting in derivatives that are either good electron acceptors or donors.[3,5]

9.3.4 Solubility and Aggregation

PDIs, initially used as pigments, cannot dissolve in conventional organic solvents because of their intrinsic insolubility. Concentrated sulfuric acid can dissolve nearly all PDI pigments *via* protonation while the pigments can be recovered by diluting the acid solution with water.[65] Most current research on PDIs is related to organic electronics, photo-induced processes, and supramolecular organization, which require PDIs with reasonable solubility in conventional organic solvents. Substitution at the N-positions, bay-positions and *ortho*-positions can improve the solubility of PDI derivatives. After modification, halogenated solvents, such as dichloromethane and chloroform, are typically good solvents for these PDI derivatives.[5] Very high solubility can be realized by choosing proper substituents. Water-soluble PDIs can also be obtained by simple incorporation of hydrophilic groups.[110]

The packing of PDIs in the solid state has been extensively studied since the early 1980s in order to understand how to control the colors of PDI pigments. Crystal structures reveal that the flat PDI cores are arranged in π-stacks showing that the principal axes of the molecules are parallel to one another at a distance of *ca.* 3.4 Å, similar to the interlayer distance in graphite.[3,41,42] Substituents at the imide positions can significantly affect the stacking distance and the longitudinal and transverse offsets between the neighboring dyes. For the ethyl and some benzyl substituents, screw-type stacking of PDIs has been observed because of the rotational offsets.[44] The interaction of the π-systems in the crystal lattice leads to the color changes of PDI pigments in the solid state. A simple empirical relationship for the dependence of the crystal color on the longitudinal and transverse offsets of the dyes, which relates the bathochromic shift and the band broadening to the extent of the π–π contact area between the stacked chromophores, was proposed by Hädicke and co-workers.[39–42] For example, Pigment Red 178 exhibits a brilliant red shade because the large transverse offsets bring the π–π

interactions to a minimum, leading to a pigment with little crystallochromic shifts compared with their solution spectra. As for Pigment Red 179, smaller transverse shifts and more perturbation take place, creating a maroon-type pigment. Slipping in the longitudinal direction is less effective in reducing the coupling between neighboring π-systems. Therefore, pigments without or with only small transverse offsets are often black due to the strongest electronic interaction between neighboring molecules, such as Pigment Black 32 (Figure 9.6). Because of intramolecular steric interactions, substitution at the bay-positions usually leads to a distortion of the π-systems from planarity and disruption of the π-stacking of PDIs. Though introduction of substituents in the *ortho*-positions does not cause any distortion of the flat perylene core, intermolecular electronic interactions in the crystal can be disturbed by the introduced substituents. For example, a powder sample of 2,5,8,11-tetrahexyl PDI emits at 635 nm with a quantum yield of up to 0.59.[95]

Supramolecular organization of PDIs to form complex functional architectures is essential for the application of PDIs. Due to the strong π–π interactions between the molecules, most of the PDI derivatives have the tendency to form one-dimensional (1D) materials. The substituents at the imide nitrogen position can provide PDIs with various functional groups but without significant alteration of the electronic properties and geometry of the PDI skeleton. This provides many options for optimizing the molecular structure to achieve shape-defined self-assembly with desired optoelectronic properties for various device applications. One rule for side-chain design is to utilize steric hindrance. Linearly configured side chains are usually employed for 1D self-assembly, while the large, bulky, branched side chains usually possess strong steric hindrance, restricting extended 1D stacking.

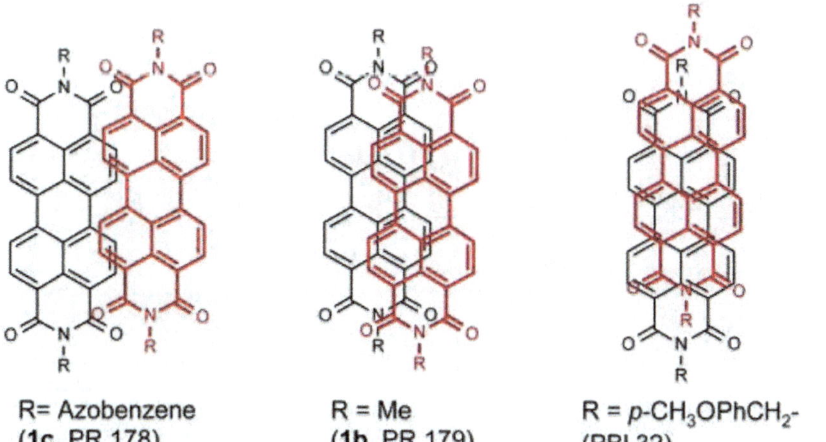

Figure 9.6 Transverse and longitudinal displacements of the stacked π-systems in the crystals of red (PR 178), maroon (PR 179) and black (PBI 32) PDI pigments. Reproduced from ref. 3 with permission from the Royal Society of Chemistry.

For example, *N*,*N*′-di(dodecyl)-perylene-3,4,9,10-tetracarboxylic diimide self-assembles into one-dimensional (1D) nanobelts while *N*,*N*′-di(nonyldecyl)-perylene-3,4,9,10-tetracarboxylic diimide aggregates into zero-dimensional (0D) nanoparticles (Figure 9.7).[99,111] Another rule for side-chain design is to introduce other interactions besides π–π interactions. For example, chiral configured side chains are usually employed for making PDI molecules chiral in nature, which can guide the PDI molecules to aggregate into helical fibers. The direction of the helical fibers depends on the chirality of the substitutes (Figure 9.8).[46] Other interactions, such as metal–ligand

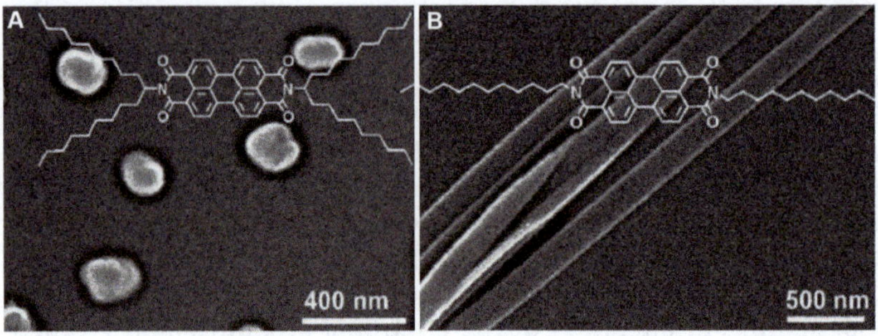

Figure 9.7 SEM images show the side-chain effect on the self-assembly of PDI. Reprinted with permission from ref. 111. Copyright 2008 American Chemical Society.

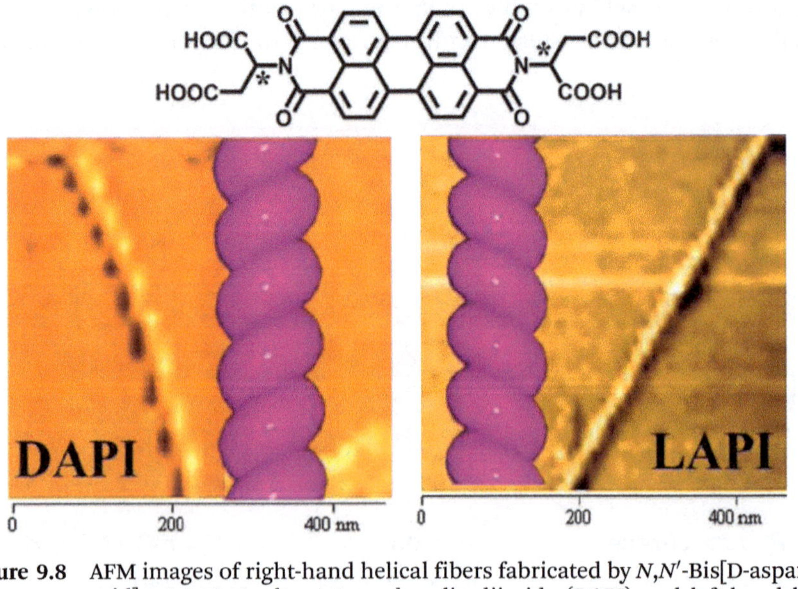

Figure 9.8 AFM images of right-hand helical fibers fabricated by *N*,*N*′-Bis[D-aspartic acid]-3,4,9,10-perylenetetracarboxylic diimide (DAPI) and left-hand helical fibers fabricated by *N*,*N*′-Bis[L-aspartic acid]-3,4,9,10-perylenetetracarboxylic diimide.
Reprinted from ref. 46 with permission from the Royal Society of Chemistry.

coordination, hydrophobic and hydrophilic interactions, hydrogen bonds, and ionic interactions can also be employed by proper side-chain design.[3]

Bay-substitution of PDI rings brings large steric hindrance not only within the molecules, but also between the neighboring molecules in aggregates, which causes a twisting of the PDI molecular plan and provides an opportunity to control the structure of PDI aggregates with different bay substituents. Supramolecular architectures of PDIs have been summarized excellently in the review articles of Würthner,[3,112] Wasielewski[113] and Zang.[114] Therefore, the supramolecular systems of PDI will not be discussed in this chapter again. We will review only the development of PDI materials constructed by covalent bonds.

9.4 Construction of PDI Molecular Arrays by Covalent Bonds

In natural photosynthesis, organisms optimize solar energy conversion through organized assemblies of photofunctional chromophores and catalysts within proteins that provide specifically tailored environments for chemical reactions.[115] As with their natural counterparts, artificial photosynthetic systems for practical solar fuel production must collect light energy, separate charge, and transport charge to catalytic sites where multielectron redox processes will occur. One of the best understood photosynthetic systems is the light-harvesting apparatus of purple non-sulfur bacteria, which typically consists of two pigment–protein complexes: light-harvesting complexes I and II (LHC I and LHC II).[115] Generally, multiple steps of intra- and inter-complex excitation energy transfer from LHC II to LHC I and finally to the reaction center (RC) result in the charge-separated state (CSS) after multiple efficient electron transfer steps. The precise arrangement of chromophores and reaction centers leads to inter- or intra-complex energy transfer and charge separation at the RC with a quantum efficiency as high as 95% despite involving complex structures composed of many different chlorophyll molecules.[116–118]

Scientists with different disciplines of chemistry try to duplicate the photoinduced processes in the light-harvesting systems of natural photosynthesis in simple chemical systems. Spontaneous self-organization of organic chromophores has been a very active field in artificial photosynthesis since the important role of chlorophyll aggregates in photosynthesis in purple bacteria was revealed.[112,119,120] In order to achieve efficient solar energy collection with the molecular arrays of organic dyes, a great amount of knowledge about the relationship between the structures and photophysical properties of molecular aggregates must be accumulated. However, structure–property relationships cannot be established in supramolecular systems because the structures of the flexible supramolecular systems cannot be fully resolved. To solve this problem, the design and synthesis of model systems with the chromophores bonded into a rigid structure are necessary. Covalent synthesis of functional building blocks with well-defined molecular geometries and donor–acceptor distances allows a strict explanation of structure–property relationships and precise control over energy and electron transfer.

9.4.1 General Concepts

Before we start to discuss the interactions between the PDI subunits in a molecular array, some general concepts need to be described first.

Exciton theory. This theory was first described by Kuhn[121] and was further clearly established by Förster based on the contributions of many scientists.[122] The molecular exciton model has been described as an interpretative tool for the study of the spectra and photochemistry of composite molecules. The results of the interactions of chromophores may be a splitting (repulsion) of the color-determining occupied and unoccupied orbitals and a synchronization (coupling) of the movement of electrons. Two new electronic transitions will be produced from the original two identical electronic transitions due to the splitting. As shown in Figure 9.9, one of them is a hypsochromic shift (β) and the other is a bathochromic shift (α). The strengths of the coupling will decide the intensity of the splitting. However, only one of these electronic transitions is allowed, which is decided by the orientation of the transition moments of the chromophores. Generally, parallel transition dipoles (Figure 9.10(a)) lead to the β transition while the α transition with a lower energy belongs to the in-line transition dipoles (Figure 9.10(b)). Oblique transition dipoles (Figure 9.10(c)) and coplanar inclined transition dipoles (Figure 9.10(d)) can lead to the

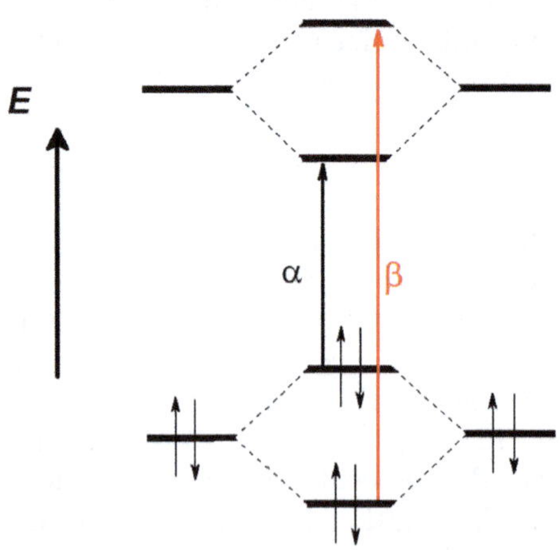

Figure 9.9 The interaction of chromophores causes the splitting of originally, energetically identical color-determining orbitals to two new pairs with the electronic transitions α and β.
Reprinted with permission from ref. 126. Copyright © 2005 Verlag *Helv. Chim. Acta* AG, Zürich, Switzerland.

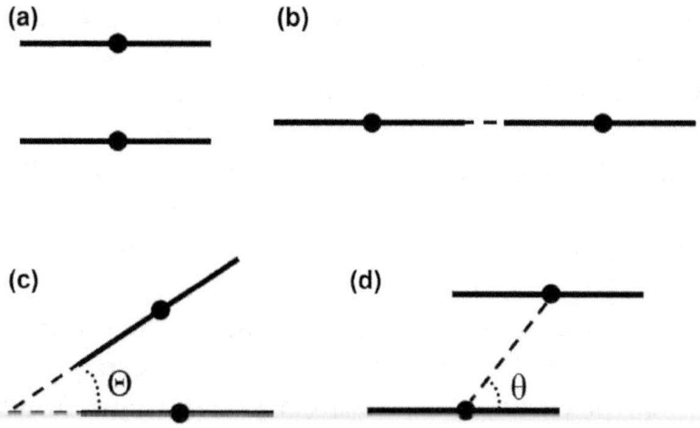

Figure 9.10 Orientation of the transition moments of bichromophoric dyes in space: (a) parallel transition dipoles; (b) in-line transition dipoles; (c) oblique transition dipoles with enclosed angle Θ; (d) coplanar slipped transition dipoles with angle θ.
Reprinted with permission from ref. 126. Copyright © 2005 Verlag *Helv. Chim. Acta* AG, Zürich, Switzerland.

alteration of the transition. As shown in Figure 9.10(c), if the enclosed angle (Θ) is in the region between 0 and 90°, the β transition is preferred. When the angle is exactly 90°, the moments are orthogonal, therefore, there is no exciton effect. At angles larger than 90° until 180° (180° = in-line transition dipoles), the α transition becomes more favorable. A continuous change of the exciton effects can also be attained by a shift of the chromophores with respect to each other. As shown in Figure 9.10(d), this case covers continuously the variation of angle θ between polarization axis and the line of the molecule's centre. Thus, 0° corresponds to Figure 9.10(a) and 90° corresponds to Figure 9.10(b). The β transition will be more favorable if the value of θ is between 0° and 54.7°. At exactly 54.7°, the exciton splitting is zero. Finally, if the angle θ is between 54.7° and 90°, the α transition is preferred.

H-aggregates and J-aggregates. Dye aggregates can be classified as "H-" and "J-type" on the basis of their spectroscopy changes with respect to their monomer. Following exciton theory, the molecules are arranged in a "side by side" way or "face to face" way for planar molecules ($0° \leq \theta < 54.7°$) in "H-type" aggregates, and a "head to tail" way in "J-type" aggregates ($54.7° < \theta < 90°$).[123] The "J-type" aggregates usually have a narrow red-shifted absorption (α transition) and emission band with a small Stokes shift and relative high fluorescence quantum yields.[124] "H-type" aggregates present a blue-sifted absorption band (β transition) and exhibit, in most cases, low or no fluorescence.[125] H-aggregates have roughly cofacial orientations, which enhance electronic communication between adjacent chromophores and cause efficient energy and/or charge transport.

Because of the variety in available PDI molecule modification methods, the covalently linked aggregates of PDI can be constructed by introducing linkages at either the N-positions or PDI core-positions. It is worth noting that there is only one electronic transition in the visible region for PDI, which is polarized in the direction of the long axis (N–N connection line).[126] This has actually significantly simplified the explanation of the spectroscopic properties of PDI aggregates. The N-atoms are ideal linking positions for the construction of large PDI arrays because of the frontier molecular orbital nodes at these positions, which render the chromophore insensitive to substituent effects. In addition to exciton coupling, conjugation effects may be introduced when substituents are linked at the bay- and *ortho*-positions of the PDI ring, which will make changes to the spectroscopic properties more complicated. According to the relative configuration of the PDI subunits, the molecular array of the PDI can be categorized as: cofacial, orthogonal, linear or cyclic, each of which provides a definite relative configuration for the interacting PDI subunits and then a relationship between the structure and property can be established with ease. These covalently linked PDI complexes can serve as model systems with which to study charge transport,[127,128] excimer formation,[59,129] and energy migration.[130–132]

9.4.2 Cofacial Arrays of PDI

Cofacially π-stacked arrangements of chromophores are found in many biologically important systems such as DNA and the primary electron donor within photosynthetic reaction center proteins.[58] The construction of PDI arrays with cofacial structures and measuring their photophysical properties can provide us with direct knowledge about the structure–property relationship, because the structure parameters, such as the center-to-center distance between the neighboring PDI units, the longitudinal and transverse displacement, the dihedral between the PDI planes (Θ), the rotational displacement, and other structure parameters can be easily measured. This information will help us to understand the principles of the high efficient photoinduced processes in natural photosynthesis and develop novel artificial photosynthetic systems.

The simplest cofacial array is a covalently linked PDI dimer. The dimer can be linked by one spacer to construct a "U-type" structure or by two spacers to give a circle structure. Comparatively, the circle structure is more rigid than the U-type structure. According to the relative position of chromophores, the cofacial arrays can be roughly divided into "H-aggregates" and "J-aggregates".

9.4.2.1 H-aggregates

Xanthenes are usually used as spacers for constructing H-aggregates of PDIs, as shown in Figure 9.11. The transition moments of PDI chromophores are constrained to conformations close to a parallel orientation by the rigid structure of the xanthene scaffold. The presence of the carbon and oxygen atoms between the two phenyl moieties of the xanthene linker and the rotational freedom around the xanthene–imide bonds allow for conformational changes, therefore, the structures of these compounds can be regulated by introducing different substituents.[133,134] According to PM3 quantum chemical calculations, the xanthene spacer allows for a distance between the cofacially arranged PDI units of *ca.* 4.5 Å.[52] This kind of PDI H-aggregate was first reported by Wasielewski and coworkers in 2002,[135] as shown in Figure 9.11 (compound **1**). Compared with the corresponding monomer, the absorption spectra of **1** show a hypsochromic intense band at 511 nm. This hypsochromic shift with respect to its monomer is consistent with the geometry of the "H-type" PDI dimer. The parallel transition dipole

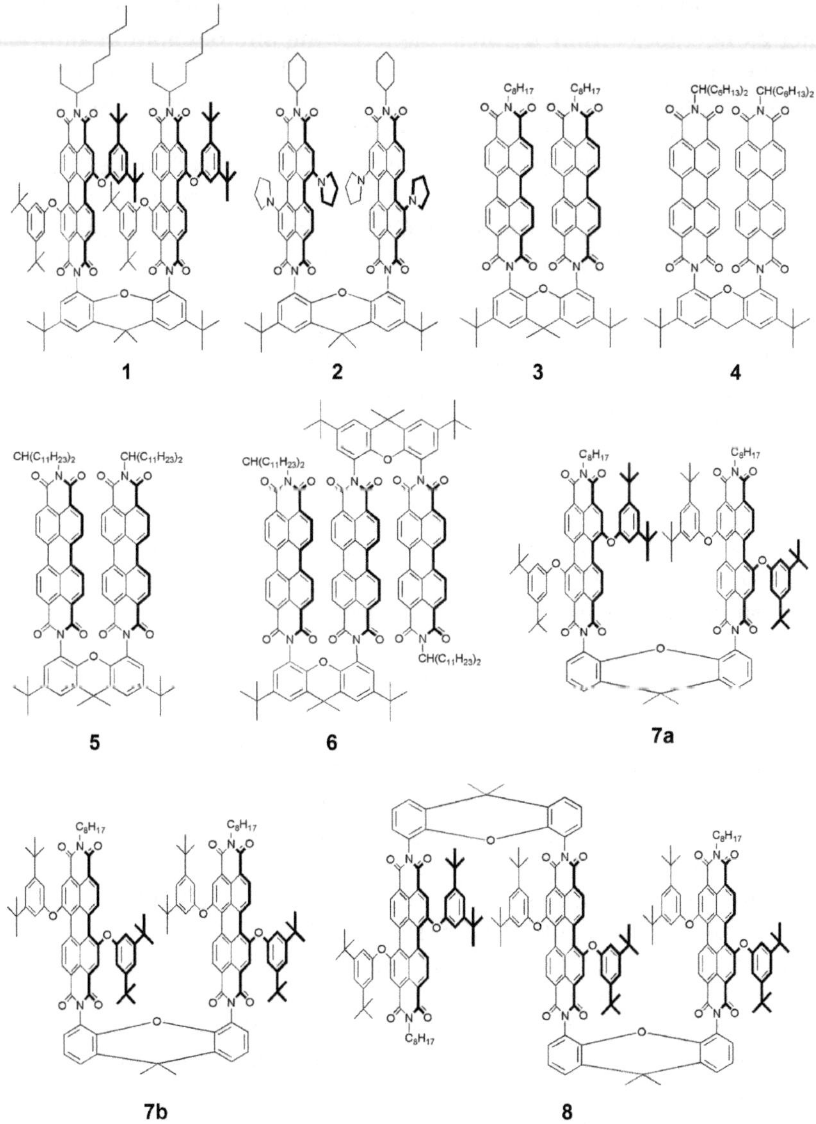

Figure 9.11 Molecular structures of H-aggregates linked by xanthene spacers.

orientation, which is enforced by the xanthene spacer, is stable and not solvent dependent. A broad excimer emission at long wavelength range is observed for this compound. In the same year, the same group reported another PDI H-aggregate.[58] As shown in Figure 9.11 (compound **2**), the PDI subunits in this compound have two pyrrolidinyl groups at the bay positions. A similar absorption change to that of **1** has been observed for this compound, but the fluorescence of **2** is completely quenched. It is reported that the photoexcitation of cofacial orientated arene dimers usually leads to excimer formation.[136] But the fluorescence quenching in **2** is attributed to an intradimer photoinduced electron transfer. The symmetric cofacial PDI dimer **2** undergoes symmetry breaking following photoexcitation and yields complete charge separation between the two halves of the dimer. This predication has been proved by ultrafast transient absorption spectra and the spectroelectrochemistry of **2** in both MTHF and toluene.

In order to systematically investigate the photoinduced process and the structure–property relationship in "H-type" PDI aggregates, more PDI aggregates linked by a xanthene scaffold were designed and synthesized. When planar PDI derivatives without any substituents at the bay positions are attached to the xanthene spacer, these compounds most probably have the optimal geometry and strong exciton coupling, such as compounds **3**, **4**, **5** and **6**.[59,129,137] However, because of the steric interaction, when branched substituents are attached to the N-positions, the optimal geometry for exciton coupling will be interfered with and the transition dipoles are no longer parallel. For example, because of the larger substituent 12-tricosanyl groups, the absorption of **5** has different vibronic peak spacings and intensity ratios from that of **3**. In toluene, only an excimer-like state emission band is observed for **3**, but the emission of **5** contains both a red-shifted featureless component as well as two resolved shoulders on the blue edge of the main band. This is because the transition dipoles are not parallel in **5**, the transition from the lower exciton state to the ground state becomes partially allowed.[138] The lifetime of the excimer-like state in **3** ($\tau = 9.2 \pm 0.1$ ns) is shorter than that of **5** ($\tau = 28.6 \pm 0.1$ ns) due largely to the increased non-radiative decay in **3** because the sterically less-demanding *n*-octyl groups permit stronger electronic interactions between the two PDI molecules. Compared with **5**, trimer **6** supplies a larger π-stacked system and the *t*-butyl groups on the xanthene provide enhanced steric interactions with the bulky 12-tricosanyl groups. The interaction among PDI subunits is further reduced in trimer **6**. Therefore, in the UV/Vis absorption spectrum, trimer **6** exhibits a slightly larger energy splitting between the first and second vibronic bands than **5**. It is reported that a larger π-stacked PDI oligomer induces lower fluorescence quantum yields.[139] However, the fluorescence behavior of trimer **6** is similar to that of dimer **5**. These suggest that the formation and decay of the excimer-like state within **6** most likely involves a strong interaction between only two of its PDI subunits. Excimer emissions are always observed in the fluorescence of these cofacially structured PDI arrays. To gain further insight into the relationship between conformational behaviors and electronic structures of π-stacked PDIs, the excimer formation dynamics of these intramolecular π-stacked PDIs were examined by single-molecule fluorescence spectroscopy.[137] The single-molecule fluorescence intensity and lifetime trajectories of **5** and **6** in solution and in a PMMA polymer matrix were monitored. The energy level of the excimer-like state and the corresponding fluorescence lifetime are sensitive to molecular configuration and the degree of overlap between two PDIs. If the molecules possess a more stable excimer conformation, the fluorescence lifetime of the coupled excimer-like state is longer. The fluorescence lifetimes of dimer **5** and trimer **6** in the solid state are 10 ns and 20 ns longer than those in solution, respectively, due to a more stable excimer conformation in the polymer matrix.[137] Scheme 9.15 shows the possible change of molecular conformations of **5** and **6**. After the rapid formation of an excimer-like state upon photoexcitation, the lifetime distributions of dimer **5** and trimer **6** are analogous, consisting of three sections: strongly coupled excimer-like, weakly coupled excimer-like, and monomer emission lifetimes. Differently, the strongly coupled excimer-like state lifetime distribution of trimer **6** subdivides into two distributions. The one with a mean lifetime of 38.1 ns is assigned to three well-stacked PDIs. For dimer **5**, the three sections with mean lifetimes of 22.9, 9.6, and 5.4 ns are attributed to excimer-like conformers with good orbital overlap, a relatively poor orbital overlap and the monomer.

The structure of dye **4** is similar to **5**, therefore, it is reasonable to assume that the optical properties of **4** are similar to those of **5**. Veldman has studied the photophysical properties of **4** under different conditions (solvent, temperature) and the triplet state formation process was predicted. Previous studies on triplet excited-state population in photoexcited electron donor–acceptor dyads have shown that enhanced intersystem crossing (ISC) can be induced *via* population of the radical (ion) pair. Two mechanisms of ISC *via* the radical pair have been proposed: radical pair intersystem crossing (RP-ISC)[140] and spin–orbit intersystem crossing (SO-ISC).[141] Under direct photoexcitation, the fluorescence quantum yields of PDI monomers in solution are high ($\Phi_F \geq 95\%$) and triplet quantum yields are low ($\Phi_F < 0.1\%$ and

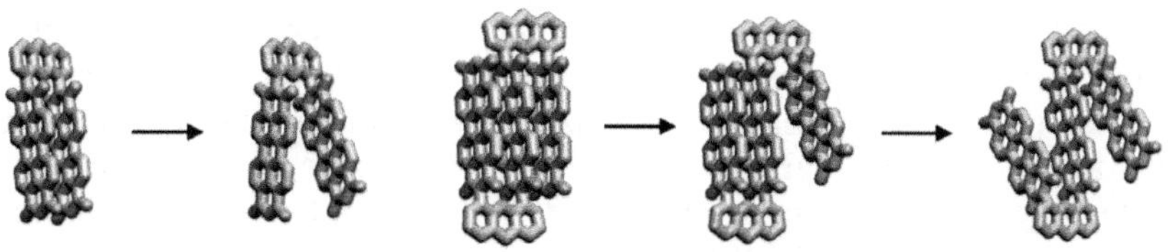

Scheme 9.15 Hypothetical change of molecular conformation. Left: from a strongly coupled excimer-like to a weakly coupled excimer-like structure in **5**. Right: from a strongly coupled excimer-like state to a dimer-like excimer state and a weakly coupled excimer-like state.
Reprinted with permission from ref. 137. Copyright 2010 American Chemical Society.

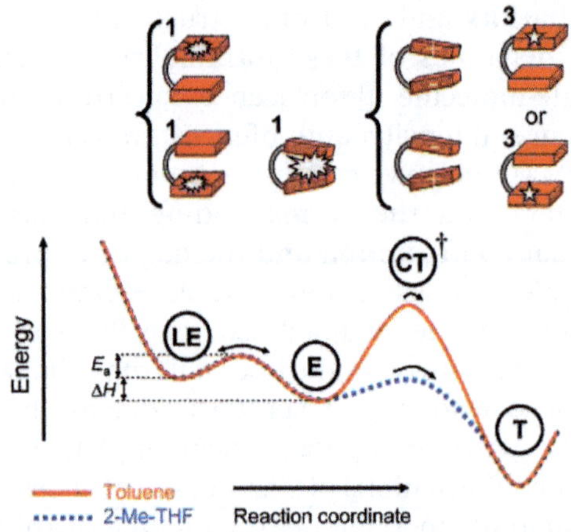

Figure 9.12 Schematic diagram showing the lowest singlet exciton state (LE), the excimer-like state (E), the highly polar transition or intermediate state (CT), the localized triplet excited state (T), and the energy barriers between these states in TOL (solid line) and MTHF (dashed line). Reprinted with permission from ref. 129. Copyright 2008 American Chemical Society.

~0.5%) because SO-ISC is slow.[29,142] The quantum yield for triplet formation in cofacially stacked PDI dimers can be orders of magnitude higher than that of monomeric PDI chromophores, because intersystem crossing is enhanced by the presence of this CT transition state or intermediate state ($PDI^{\bullet +}$–$PDI^{\bullet -}$) and occurs much faster and more efficient in solvents of higher polarity. Figure 9.12 describes the formation and decay processes of the various singlet and triplet excited states of **4** following photoexcitation in toluene (relatively nonpolar) and MTHF (relatively polar). The starting point is the vibrationally relaxed lowest singlet exciton state (LE) formed by photoexcitation that structurally resembles the ground state. Conformational relaxation of the LE state leads to the formation of an intramolecular excimer (E) state, likely having a reduced distance between the two PDI units, enabling stronger delocalization of π-electrons. The stacking of PDI chromophores creates a CT transition or intermediate state ($PDI^{\bullet +}$–$PDI^{\bullet -}$) at energies just above the emissive singlet states. Intersystem crossing to the triplet (T) state is enhanced by a nearby CT transition state. The increase of k_{ISC} in more polar environments is a consequence of the lowering of the CT transition state, which causes the increased conversion of LE/E into T.[124]

Cofacially arranged dimers and trimers of PDI with bay substitution have also been prepared. In **7a** and **7b** (two isomers of **1**), 3,5-di-*t*-butylphenoxy groups are connected at the 1,7-positions of PDI. Normally, the two bay-substituted PDI dimers show typical absorption bands of H-aggregates with a blue-shifted maximal absorption band. The optimal parallel cofacial geometry and strong interactions between the chromophores in these PDI dimers or trimers benefit the emission from the excimer-like state.

The staggered conformation of the phenoxy groups in **7a** permits the π systems of the two chromophores to interact more strongly, resulting in an excimer-like state with a lower energy emission relative to that of **7b**. The steric interactions of the eclipsing phenoxy groups in **7b** distort its excited-state structure from the parallel cofacial geometry and then lead to somewhat increased transition probability for emission from the lower exciton state in **7b**. The lifetime of excimer emission of PDI varied dramatically with the structure change. The decreased emission lifetime reflects a tight association and a more favorable interaction. The lifetimes of **7a** ($\tau = 21 \pm 1$ ns) and **7b** ($\tau = 19 \pm 1$ ns) are larger than that of **3** ($\tau = 9.2 \pm 0.1$ ns) but smaller than that of **5** ($\tau = 28.6 \pm 0.1$ ns), suggesting the strengths of the electronic interactions between the two PDI molecules in their excimer-like states are intermediate to those of **3** and **5**. As for compound **8**, the trimer structure enforces the chromophore rigidity. The excimer-like emission from **8** has a fluorescence quantum yield that is about 3–4 times greater than those of **7a** and **7b**, while the fluorescence lifetime of **8** is only about two times longer than those of **7a** and **7b**, suggesting that its geometry differs from that of **7a** and **7b**. The observed excimer-like states of trimers **6** and **8** are the result of dimeric interactions within these trimers. The addition of a third chromophore does not ensure that the entire trimer ensemble has structural features that optimize long-range electronic interactions.[143]

Besides xanthene, other spacers, such as triptycene, triazine and phenylene ethynylene, have also been reported (Figure 9.13). Similar to the xanthene scaffold, the triptycene bridge is also rigid, but the distance between the cofacially arranged PDI units is smaller. Dimer **9** has a π-stacking distance of 4.2 Å at the bridge imide end and 3.5 Å at the alkyl imide end. Rapid excimer formation is observed in $\tau = \sim 2$ ps for dimer **9**.[144] The PDI chromophores linked by the triazine and phenylene ethynylene scaffold are flexible and an equilibrium between the face-to-face and non-cofacial conformation in solution was observed.[14,145,146] Taking dimer **10** as an example, the blue-shifted main absorption band indicates a strong face-to-face π–π interaction in solution. The fluorescence spectrum of **10** in toluene shows a weak broad band around 685 nm with a long fluorescence lifetime (20.5 ns), which can be ascribed to the excimer-like states.[147] However, a strong monomeric emission at *ca.* 580 nm was also observed, which suggested the presence of an equilibrium between the face-to-face and non-cofacial conformation of **10** in solution. This appears to contradict the results of the absorption spectrum, which shows exclusively the absorption of a face-to-face stacked dimer but merely has to do with the detection limits of the two spectroscopic methods. Monomeric PDI has a fluorescence quantum yield near 100% and is much higher than that of the excimer. A few percent of monomeric PDI will easily be detected by fluorescence, but not by absorption because the extinction coefficients of the various states of PDIs are on the same order of magnitude.[123]

The construction of PDI H-aggregates was also attempted by direct overlapping of the chromophores in cyclophanes, as shown in Figure 9.14.

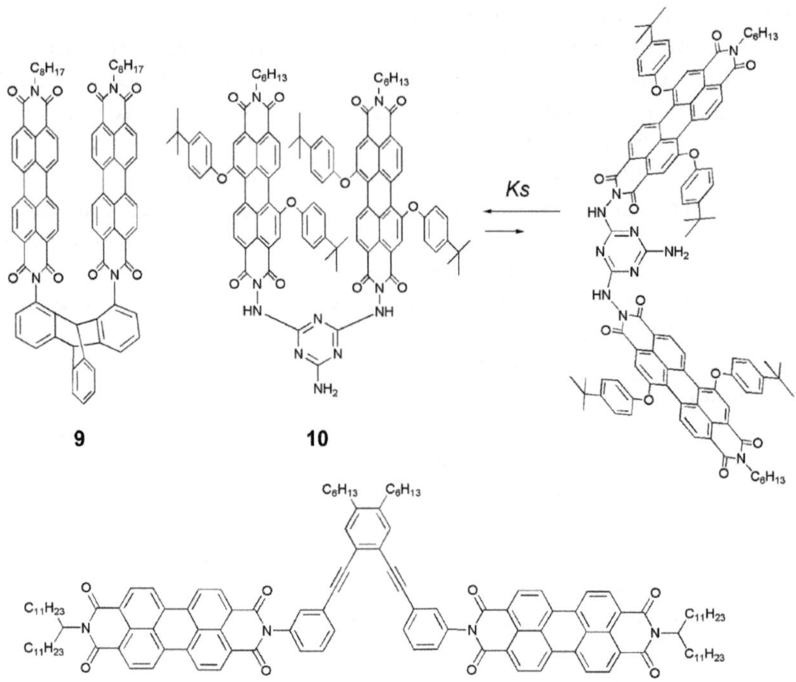

Figure 9.13 Molecular structures of **9**, **10** and **11** with triptycene, triazine and phenylene ethynylene as spacers, respectively, and the equilibrium between non-cofacial and face-to-face conformations for **10** in solution.

Cyclophane **12** was prepared from PTDCA and alkyl diamine under dilution conditions.[148] As predicated, there is a pronounced hypsochromic shift of the absorption of **12** compared with the corresponding monomer, which is typical for H-aggregates. However, a surprisingly bathochromically shifted fluorescence with a rather high fluorescence quantum yield ($\Phi = 40\%$) is observed for **12**. The fluorescence was ascribed to the J-aggregates of **12** rather than the H-aggregates by the authors. This can be interpreted as a consequence of the slipping from H to J stacking of the chromophores after the vertical optical excitation. The flexible structure of compound **12** brings uncertainty for the explanation of its photophysical properties. Therefore, cyclophanes with more rigid structures have been prepared.

In compounds **13**, **14**, **15a**, **15b** and **16**, the PDI subunits are fixed by short and rigid spacers, and restricted to a face-to-face stacked structure.[60,149] The slippage along the long axis of the PDI ring is completely prevented. In the UV/Vis spectra, the main absorption bands of all these compounds are blue-shifted in comparison with their monomeric counterparts. This is the evidence of the formation of H-aggregates. The calculated average interplanar distance for **13** is *ca.* 3.8 Å, which is close to that of PDI aggregates in solution and films. The distance for the "U-type" dimer is larger, such as

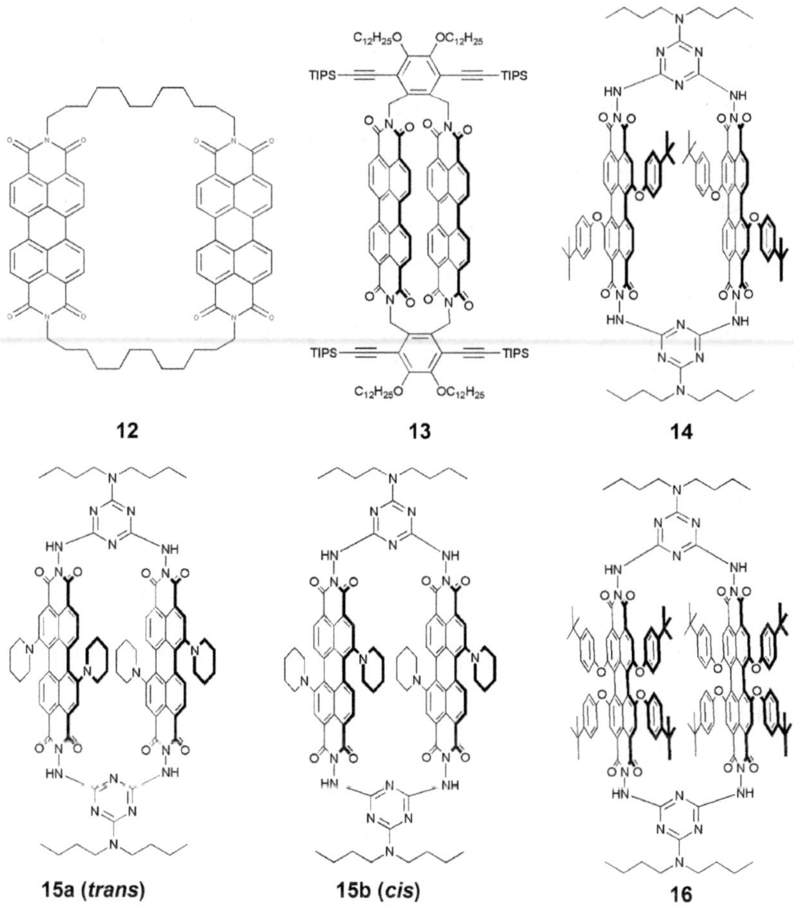

Figure 9.14 Structures of PDI cyclophanes.

4.4 Å for compound **5** and 4.5 Å for compound **3**. Along with the decreases of the PDI interplanar distance (**3** > **5** > **13**), the degree of electronic coupling increases, the structural mobility decreases, and the excimer formation times decrease.[149]

Both the electronic absorption and fluorescence spectra of the cyclophanes were found to change along with the number and nature of the side groups at the bay positions of the PDI ring. Tiny structure differences can induce significant changes in the optical properties. A smaller interplanar distance between the two PDI planes and an in-plane torsion angle between the long axis of the two PDI rings lead to a stronger π–π interaction between the two PDI rings. Therefore, the main absorption band for *trans* isomer **15a** blue-shifted by about 40 nm while that for *cis* isomer **15b** blue-shifted by about 55 nm with respect to the monomer. This result indicates that the π–π interaction between the two PDI units in the *cis* isomer **15b** is significantly larger than that of the *trans* isomer **15a**. This result is also supported by DFT calculations. No emission was detected for cyclophanes **15a** and **15b**, indicating the presence of charge separation between the two PDI units in the dimer.[58] The introduction of four substituents at the bay positions induces a large degree of deviation from the ideal face-to-face stacking geometry for the two PDI units in cyclophane **16**. Thus an obvious absorption corresponding to the transition to the lower-energy exciton state was found in the electronic absorption spectrum. The detected emission bands of cyclophanes **14** and **16** are attributed to the excimer-like states.[123,135] The excited states of the two cyclophanes have remarkable electron-transfer characteristics, as evidenced by the solvent-polarity-dependent fluorescence quantum yields and lifetimes. Interestingly, a pretty long fluorescence lifetime has been detected for cyclophane **14** in toluene solution (38.1 ns), which seems the longest lifetime of the excimer emission of PDIs. This is most probably due to the rigid molecular structure of cyclophane **14**, which hinders structural relaxation during the decay of the excited states.

9.4.2.2 J-aggregates

Over the past several decades, there has been a tremendous effort in understanding the structure–property relationship of J-type aggregates for the purpose of developing novel functional materials.[150] Significant progress has been made in this field; however, even the highly sophisticated NMR spectroscopy and electron microscopy techniques available nowadays cannot reliably resolve the structural details of the molecular packing in most of the molecular aggregates, and thus the much desired structure–property relationship, even though for the first J-aggregate, pseudoisocyanines aggregate, is still a matter of debate 75 years after its discovery, and will remain a topic of future research.[151]

The research results on the emission properties of supramolecular J-type aggregates of PDI are quite controversial in the literature. Most of the reported J-type aggregates in the literature show a broad emission band in the red region with low fluorescence quantum yields, which is similar to that of H-type aggregates.[152] Meanwhile, several reported J-type aggregates present narrow emission bands with high fluorescence quantum yields and small Stokes shifts, which are called "genuine J-aggregating PDIs" because they have similar photophysical properties to that of the J-type aggregates of pseudoisocyanines.[153] To reveal the relationship between the structure and photophysical properties of PDI aggregates, synthesizing covalently linked PDI J-aggregates with well defined structures is necessary. However, the covalently linked J-type aggregates of PDI are not available until the appearance of the covalently linked slip-stacked "bare" PDI dimers (J-aggregates) based on xanthene scaffolds (**17** and **18**) as shown in Figure 9.15.[61] Meanwhile, several other "J-type" bay-substituted PDI dimers (**19** and **20**) and trimers (**21** and **22**) with slipped face-to-face stacked molecular structures have also been synthesized.[154] The long axis of the two π–π stacked PDI molecules are slipped by 4.3, 7.9, 5.1 and 7.1 Å for **17**, **18**, **19**

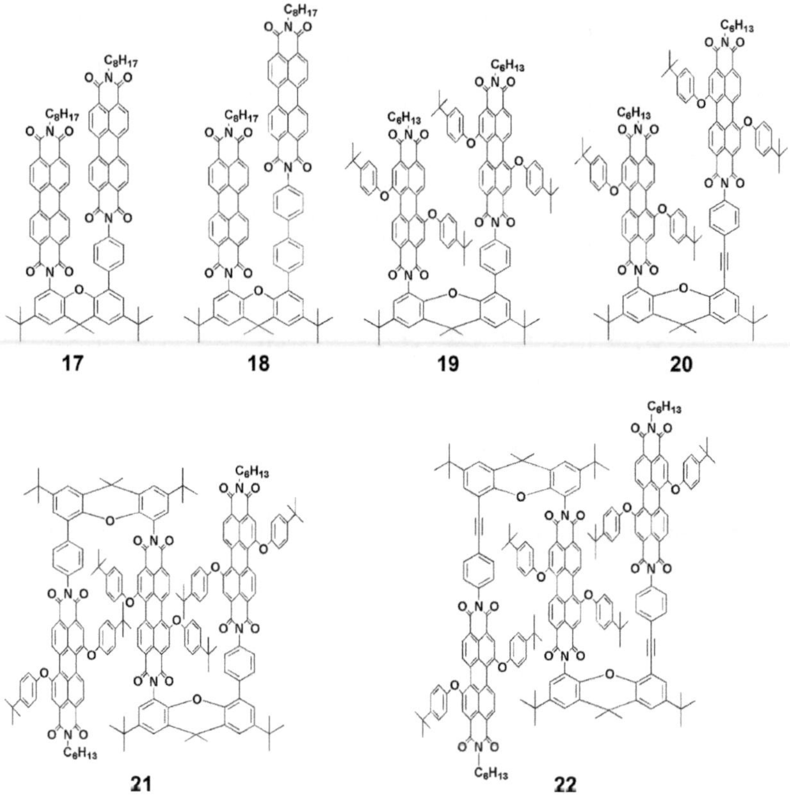

Figure 9.15 Molecular structures of "J-type" PDI aggregates.

and **20**, respectively. The angles between the center-to-center connection line and the plane of PDI subunits (θ) of these four compounds read from the energy-minimized molecular structure are smaller than 54.7°, therefore, they are classified as "J" type aggregates following exciton theory. However, the strongly red-shifted absorption band, which is typical of PDI J-aggregates,[155,156] is not observed. Instead, compounds **17** and **19** display an apparent enhancement of their 0–1 vibronic band, which is more typical of H-aggregation, while the 0–0 vibronic bands of **18** and **20** display only 5 nm and 10 nm red shifts, respectively (Figure 9.16). This is because the vibronic coupling in the exciton states of the dimer relieves the symmetry restrictions inherent to the simple model,[58] thus, the transition to the higher exciton state can be partially allowed and blue-shifted absorption bands in the electronic absorption spectra of **17** and **19** are observed. The increase of the distance between the PDI subunits can significantly reduce the interactions between the two PDI subunits and decrease the exciton splitting. Therefore, the observed small red-shifts on the absorption spectra of **18** and **20** are rational. Compounds **19** and **20** exhibit J-type absorption spectra, but their emission spectra are dominated by the excimer emission.

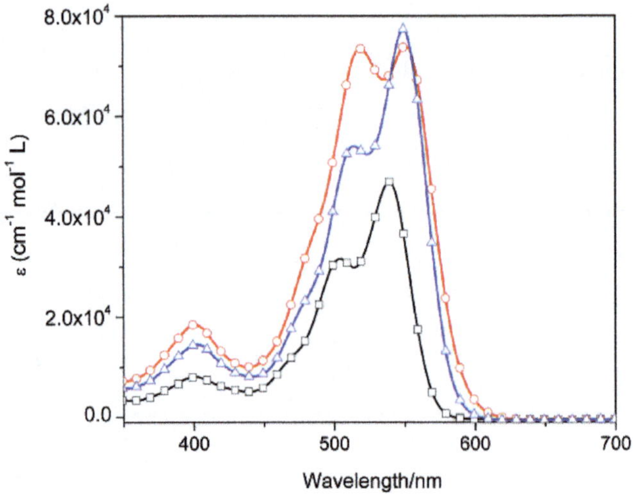

Figure 9.16 Absorption spectra of **19** (circle), **20** (triangle) and PDI monomer (square) in toluene.

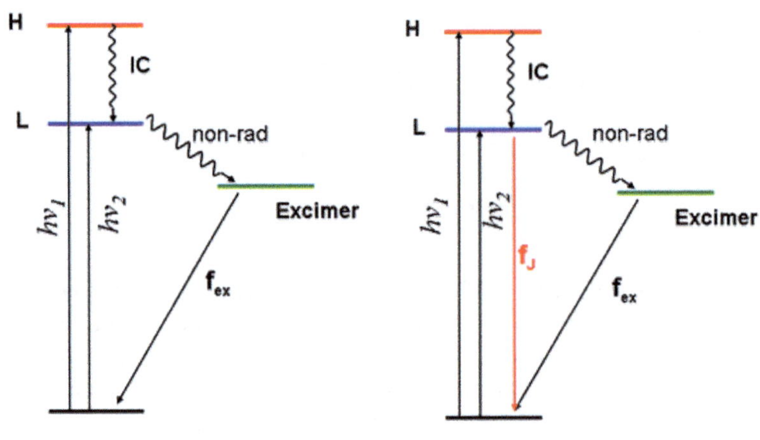

Figure 9.17 The decay pathway for the excited states of **19** (left) and **20** (right). H: high exciton state; L: low exciton state; IC: internal conversion; non-rad: nonradiative decay; f_J: "J-type" fluorescence; f_{ex}: excimer fluorescence; $h\nu_1$ and $h\nu_2$: excitation lights.
Reproduced from ref. 154 with permission from the PCCP Owner Societies and the Royal Society of Chemistry.

Only compound **20** exhibits a very weak emission besides the excimer emission in its fluorescence spectrum, which may be ascribed to the J-type emission. This can be attributed to its large displacement along the long molecular axis. The decay pathways of the excited states of **19** and **20** and the competition between the excimer-emission and J-type emission are illustrated in Figure 9.17.

In compounds **21** and **22**, three PDI subunits form "J-type" aggregates, however, because of the large steric hindrance caused by the phenoxyl groups at the bay positions, the three PDI subunits do not stack in an ideal face-to-face way. The minimized structures of these two compounds revealed that the PDI subunits are not parallel to each other in the two compounds. Large transverse displacements and rotation displacements between the neighbor PDIs are found. The absorption spectra of **21** and **22** show red-shifted bands, which is typical of J-type aggregates. Their fluorescence spectra are similarly dominated by the excimer emission. The picosecond time-resolved fluorescence spectra of **21** and **22** revealed two components in the fluorescence of both. The long lifetime (10.36 ns) component can be assigned to the emission of "excimer-like" states, while the short lifetime (1.93 ns) component is suggested to be the emission from the low-exciton states (J-type emission).

The triplet state formation in "J-type" PDI aggregates is useful for fabricating PDI-based materials, especially useful for organic photovoltaics (OPVs). Triplet state formation has been shown to limit OPV device performance when it competes with charge separation.[157,158] However, in principle, this long-lived state could be used to dramatically improve device efficiency in a single junction cell if it is formed by singlet exciton fission (SF).[159] As for "J-type" aggregates, the photophysics of the triplet state formation of compounds **17**, **18** and their corresponding PDI monomer **23** (Figure 9.18 left) has been investigated by using ultra-fast transient absorption and stimulated Raman spectroscopy. The mechanism of ^{3*}PDI (triplet state of PDI) formation is illustrated in Figure 9.18, right. The mechanism is found to depend strongly on a competition between the rate of Xan$^{\bullet+}$ – PDI$^{\bullet-}$ (charge separated state) formation involving the spacer

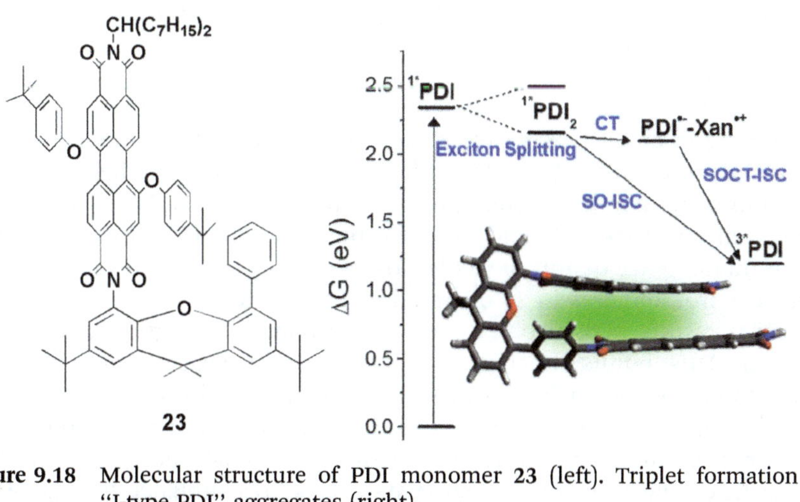

Figure 9.18 Molecular structure of PDI monomer **23** (left). Triplet formation of "J-type PDI" aggregates (right).
Reprinted with permission from ref. 61. Copyright 2013 American Chemical Society.

group and the rate of excimer-like state formation. SF to produce ^{3*}PDI does not compete kinetically with these processes. The excimer-like state decays relatively slowly with $\tau = 28$ ns to produce ^{3*}PDI, while charge recombination of $Xan^{\bullet+} - PDI^{\bullet-}$ yields ^{3*}PDI more than an order of magnitude faster. The perpendicular orientation between the π orbitals of PDI and the xanthene bridge provides a large enough orbital angular momentum change to greatly increase the intersystem crossing rate *via* $Xan^{\bullet+} - PDI^{\bullet-} \rightarrow {}^{3*}$PDI charge recombination. These results highlight the importance of understanding inter-chromophore electronic coupling in a wide range of geometries as well as the active role that molecular spacers can play in the photophysics of covalent models for self-assembled chromophore aggregates.[61]

9.4.3 Orthogonal Arrays of PDIs

Independent operation of two or more chromophores within the same molecule is of interest for the construction of complex molecular functional units characterized by a very high degree of integration. However, it is difficult to realize because exciton effects,[123] energy[160] and electron[161] transfer processes can take place easily between the chromophores. As mentioned above, there is only one electronic transition in the visible region for PDI, which is polarized in the direction of the N–N connection line. Therefore, the orthogonal orientation between neighboring PDI units may be suitable for independent operation. In compound **24**, the transition moments of the chromophores (N–N connection lines) are designed to be orthogonal, as demonstrated by the AM1 calculation, thus, are electronically decoupled (Figure 9.19, right).[162] The distances between the closest nitrogen atoms are about 8.5 Å, which is sufficiently close for exciton interactions. By contrast, there is only one pair of orthogonal chromophores in compound **25**, the two proximal chromophores include an angle of less than 90° for the transition moment with a N–N distance 6.5 Å, giving rise to a "H-type" interaction. These predictions can be demonstrated by the UV/Vis absorption spectra. The absorption spectrum of compound **24** is identical to that of the corresponding monomer except for a negligibly small bathochromic shift due to the different substitution at the nitrogen atoms. The more hypsochromic vibration band is more prominent in the spectrum of **25**. Furthermore, the fluorescence spectra of **24** and the corresponding monomer are very similar to the fluorescence quantum yields close to unity. All these results confirm that the chromophores in **24** operate independently.

Similar results were also observed in compound **26** (Figure 9.20), which has four PDI subunits attached to a spirobifluorene spacer. These four PDI subunits are restricted in the same plane and are all orthogonal to each other.[11] The absorption spectrum of **26** is identical to that of its monomeric counterparts, showing two absorption maxima at 545 and 510 nm. The fluorescence spectrum of **26** also bears a close resemblance to that of its monomeric counterparts with a strong emission band at 580 nm and a fluorescence quantum yield of 96%. Furthermore, the fluorescence lifetime

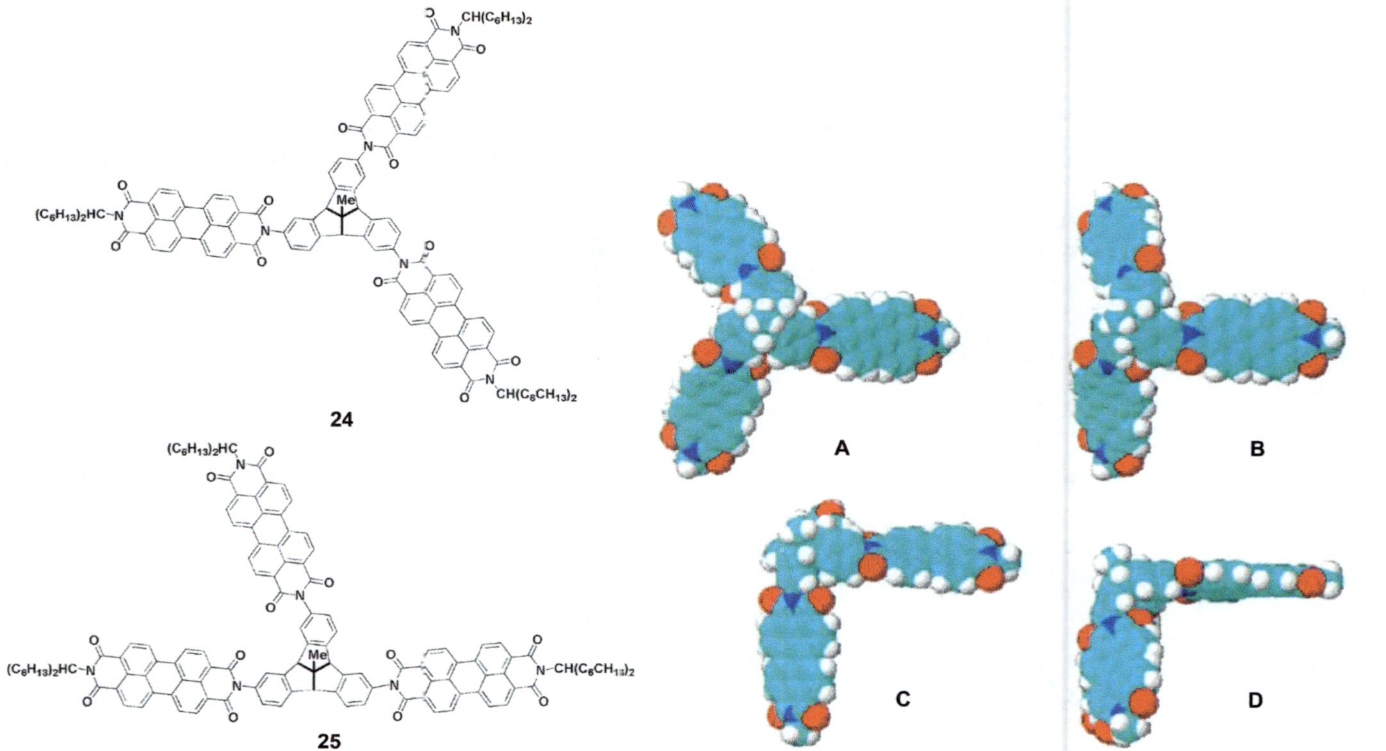

Figure 9.19 Left: molecular structures of **24** and **25**. Right: different views of the tribenzotriquinacene-based tris-(perylene) core of the C3-symmetrical compound **24**, as obtained by AM1 calculation ($R = CH_3$), (A) top view along the molecular axis; (B)–(D) three views perpendicular to the *x*-axis, the structure being turned counter-clockwise about this axis in two steps of 45°. Adapted with permission from ref. 162. Copyright 2008 American Chemical Society.

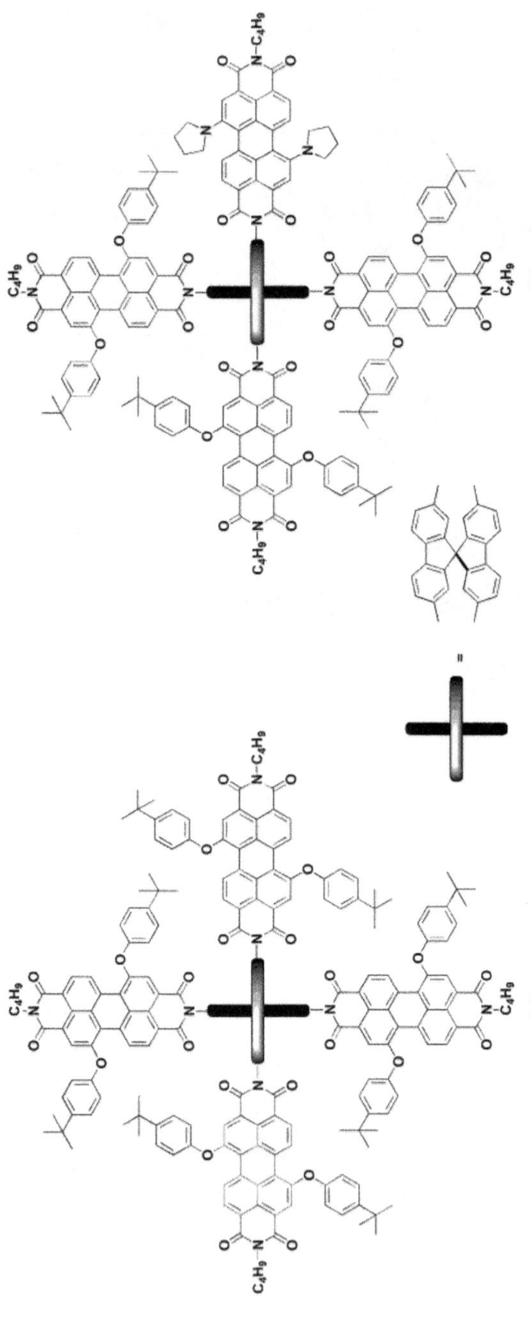

26

27

Figure 9.20 Molecular structures of orthogonal arranged PDI tetramers **26** and **27**. Adapted with permission from ref. 11. Copyright 2012 American Chemical Society.

is measured to be 4.3 ns, which is also similar to that of its monomeric counterparts. All these results demonstrate that the PDIs in **26** also operate independently from each other. When one of the PO-PDI subunits was replaced by a PY-PDI, compound **27** was obtained (Figure 9.20). There is also no significant ground state interaction between the PO-PDI and PY-PDI subunits within **27**, as indicated by the absorption spectra.[11] However, when PO-PDI was selectively excited, an efficient energy transfer from PO-PDI to PY-PDI within **27** with an efficiency *ca.* 100% was observed. In the fluorescence spectrum of **27**, only one broad band centered at 720 nm can be detected, which can be assigned to the emission of PY-PDI. The fluorescence from PO-PDI, which appears normally at 570 nm, was too weak to be detected. Obviously, the fluorescence of PO-PDI was efficiently quenched by PY-PDI. This energy transfer from PO-PDI to PY-PDI was also observed in the thin solid films of **27**. The kinetics of the energy transfer within **27** revealed three dynamic decay processes with time constants of $\tau_1 = 212$ fs (78.2%), $\tau_2 = 2.10$ ps (20.2%), and $\tau_3 = 401$ ps (1.6%). The two fastest components of the decay can be attributed to the energy transfer from PO-PDI to PY-PDI, whereas the minor slower component is attributed to another fluorescence quenching process, which was assumed to be an electron transfer between PO-PDI and PY-PDI.

A similar efficient energy transfer between a pair of orthogonally arranged transition dipole moments had also been observed in dimers **28** and **29** (Figure 9.21). It is well known that Förster resonant energy transfer (FRET)[163–166] is being generally accepted and applied as an indicator for molecular proximity of light-absorbing and fluorescent structures. Following Förster theory, the resonance energy transfer cannot happen between a pair of orthogonally arranged transition dipole moments.[11,62,167] Therefore, the efficient energy transfer within **27**, **28** and **29** is abnormal. The alternative mechanism to Förster-type energy transfer is Dexter-type energy transfer, in which the electronic coupling is dominated by electronic exchange

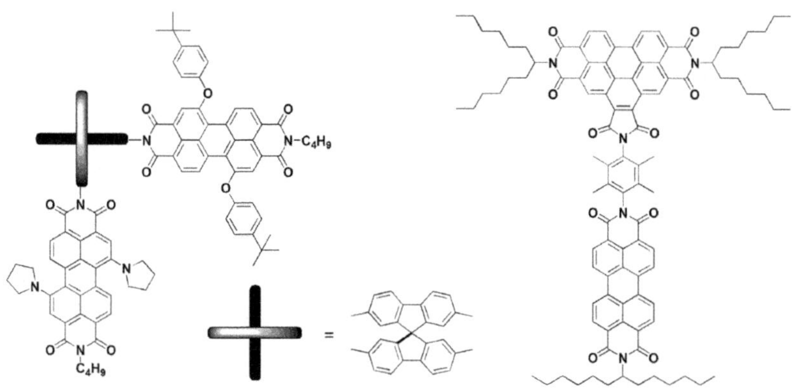

Figure 9.21 Molecular structures of orthogonal arranged PDI dimers **28** and **29**.

interactions mediated by the overlap between donor and acceptor orbitals.[62] The orbitals involved in the PDI electronic transitions are well localized on the respective chromophores, suggesting that a Dexter-type energy transfer is not at work in the dyes mentioned above. Dexter-type energy transfer can also be experimentally excluded by changing the spacer in both the length and chemical character of **29**.[62] Langhals and coworkers suggested that Förster-type energy transfer is functioning in this system, but the thermally populated ground-state vibrations are breaking the orthogonal arrangement.[62] Clearly, this kind of "unexpected" energy transfer might be a common phenomenon in orthogonally configured systems.

9.4.4 Linear Arrays of PDIs

The simplest molecule in the linear series is a "head-to-tail" dimer in which the two PDIs are covalently linked through the nitrogen head atoms (**30**). The addition of another PDI molecule using the same bonding motif results in a linear trimer (**31**, Figure 9.22).[168] In these linear complexes, oscillator strength is concentrated on the transition to the lowest energy exciton, which is nondegenerate and the energy of the lowest energy exciton red-shifts with increasing the number of PDI chromophores, consistent with what is expected for J-aggregates. Absorption and emission are polarized entirely along the long molecular axis. The absorption bands of **30** and **31** exhibit the vibration structure that is typical for PDI monomers. With an increase in the number of PDI chromophores from two to three, both the absorption and fluorescence peak are bathochromically shifted. However, the exciton effects do not interfere with the high fluorescence quantum

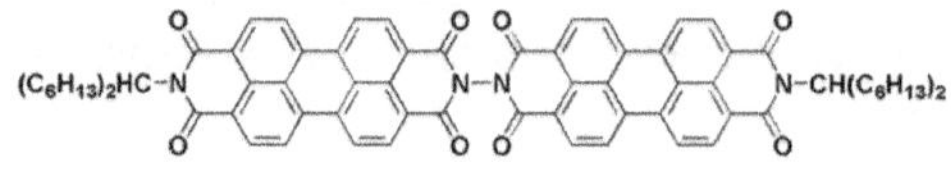

30

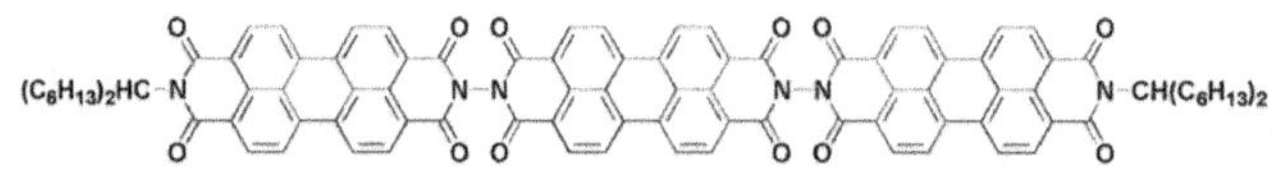

31

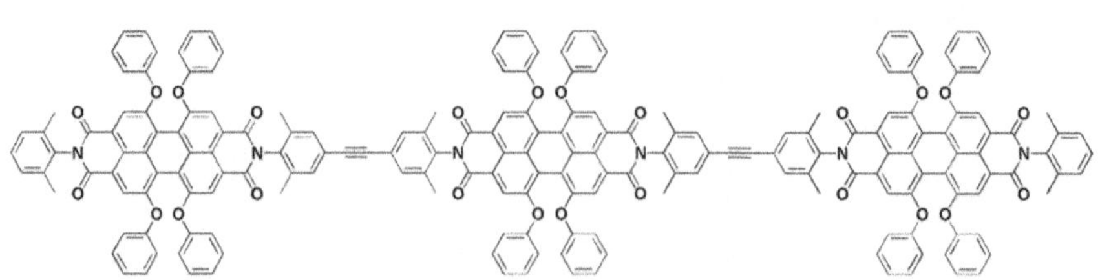

32

Figure 9.22 Molecular structures of **30**, **31** and **32**.

yields because quantum yields close to 100% were observed for both **30** and **31**.[168] In compound **32**, the distance between the PDI chromophores is increased by introducing diphenyl-acetylene units as spacers, leading to much weaker interactions between the chromophores. The center-to-center distance between the two PDI units in **32** is 2.396 nm, whereas the edge-to-edge distance (taken from the nitrogen atom) is 1.263 nm by semiempirical (MNDO) chemical calculations.[169] Minimal differences between **32** and the corresponding monomer are observed for the absorption and fluorescence spectra, which indicates that the coupling between the two PDI subunits is much weaker due to the increased distances between the PDI units, although the transition dipole moments are oriented parallel. The fluorescence quantum yield of **32** has a value of about 0.9, whereas the fluorescence quantum yields for **30** and **31** are close to 1. The fluorescence quantum yield decrease of **32** is attributed to the presence of the diphenyl acetylene bridge, which introduces some flexibility in the structure and opens up new radiationless deactivation channels. The decay time of the main fluorescence component of **32** is about 4.7 ns, shorter than that of the corresponding monomer (*ca.* 5.2 ns), in toluene. This is also due to the opening up of additional radiationless decay pathways in **32** and is in agreement with the drop in fluorescence quantum yield.

The spectral overlap between the emission spectrum and ground-state absorption or excited-state absorption spectrum of **32** is shown in Figure 9.23. The spectral overlaps indicated that in these linear PDI oligomers, two energy transfer processes: energy hopping and excited state annihilation, may be considered within the framework of Förster theory for energy transfer.[170] The energy hopping process is an energy transfer from the S_1 state toward the ground-state of a neighboring PDI. In the annihilation process, the energy is transferred toward the excited-state of the neighbor PDI, resulting in the promotion of the latter into a higher excited state. The spectral overlap integral (J) between the donor's emission (corresponding to the PDI S_1–S_n transition) and acceptor's absorption spectrum (the stationary absorption spectrum for the energy hopping and the excited-state absorption spectrum for annihilation) is a key parameter for the two processes. The critical energy transfer distance R_0 can be calculated from parameter J, Φ_D (the fluorescence quantum yield of the donor), and the relative orientation of the two dipoles expressed by the orientation factor κ^2. For the linear PDI molecules, the orientation factor is assumed to have the maximum value ($\kappa^2 = 4$) as the transition dipoles are quasi collinearly oriented and the S_0–S_1 transition is oriented along the long molecular axis.[62] The spectral overlap between the emission spectrum and ground-state absorption or excited-state absorption spectrum is shown in Figure 9.23 and the calculated overlap integrals and Förster critical distances are given in Table 9.1. A rather high value for the overlap integral is found in the case of energy hopping as compared to the value found for the singlet–singlet annihilation. The presence of energy hopping and singlet–singlet annihilation can both be found in **32**. In the time-resolved ensemble experiment results, a

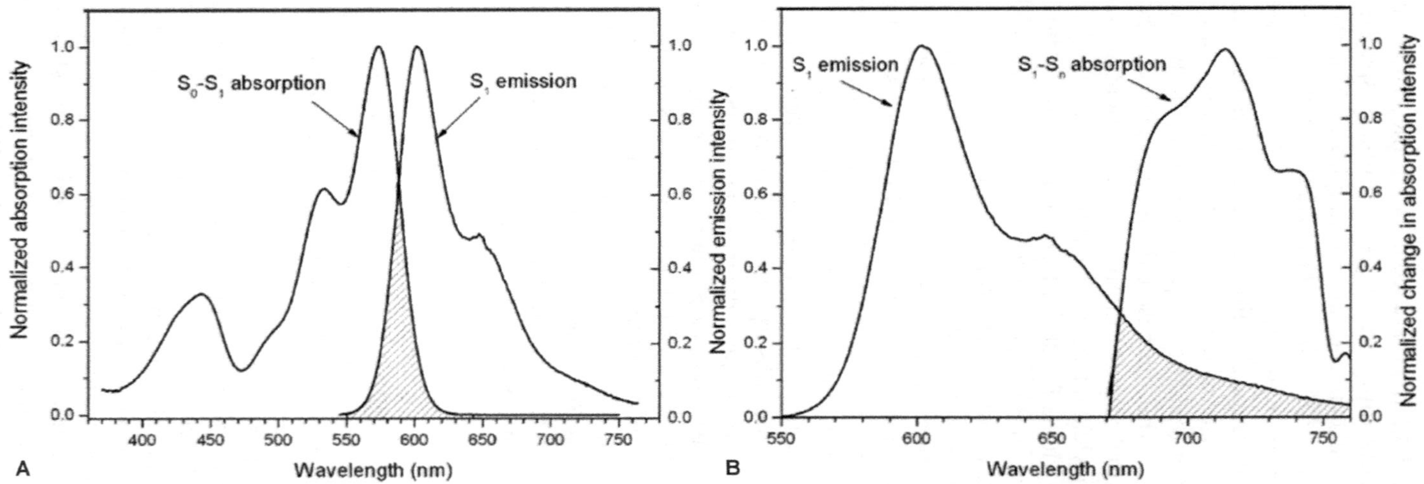

Figure 9.23 (A) Spectral overlap of the fluorescence and ground-state absorption relevant for the energy hopping process in **32** in toluene. (B) Spectral overlap of the fluorescence and excited-state absorption band relevant for the S1–S1 annihilation process in **32** in toluene.
Reprinted with permission from ref. 169. Copyright 2009 American Chemical Society.

Table 9.1 Calculated overlap integrals and Förster critical distances for energy hopping and S_1–S_1 annihilation processes for **32** in toluene.[169]

	Φ_D	κ^2	J (mol^{-1} dm^3 nm^4)	R_0 (nm)
Energy hopping	0.99	4	7.80×10^{13}	3.45
S_1–S_1 annihilation	0.99	4	1.11×10^{13}	2.49

decay component of 356 ps is retrieved in the high excitation power experiments. This decay component is attributed to singlet–singlet annihilation between two PDI chromophores in the trimer. Sequential fitting of the dipole emission pattern recorded with defocused wide field imaging of a single **32** molecule, immobilized in a PMMA polymer film, demonstrated that emission can switch *via* energy hopping between sequential emission of all the chromophores or emission from one chromophore that likely is the lowest in energy.[169]

As revealed by compound **27**, excited energy can transfer from donor PO-PDI to acceptor PY-PDI efficiently because the emission of the PO-PDI unit overlaps very well with the absorption of the PY-PDI unit. Therefore, a more efficient and quick energy transfer is expected if PO-PDI and PY-PDI are linearly linked. A series of rigid linear PDI-based donor–acceptor oligomers linked together by phenyl bridges (**33**, **34**, **35**) have been prepared for the purpose of evaluating the effects of the donor–acceptor distance on the energy transfer efficiency (Figure 9.24).[12] The absorption spectra of these three compounds indicate that no strong ground state interactions exist between the PO-PDI and PY-PDI units. Interestingly, no matter which PO-PDI subunit is excited in these linear molecules, the excitation energy is efficiently collected by the PY-PDI subunit. The fluorescence quenching efficiency in all three compounds is larger than 99% and all the PO-PDI subunits in these compounds contribute equally to the energy transfer to PY-PDI. The Förster-type energy transfer efficiencies (Φ_{EnT}) from PO-PDI to PY-PDI as well as that between the neighbouring PO subunits within these three compounds have been estimated and the results are summarized in Table 9.2. The calculated Φ_{EnT} for the energy transfer from PO-PDI to PY-PDI, which are not connected directly (**34** and **35**), is significantly smaller than that in **33**, indicating that the direct energy transfer from the second and third PO-PDI subunits within **34** and **35** to the PY-PDI unit should be less efficient. However, due to the efficient energy hopping between the PO-PDI subunits within **34** and **35**, the total energy transfer efficiencies in these three compounds are all close to 100%. The excitation energy can transfer as long as 47 Å without any decrease in efficiency. The most important characteristic of these molecules is the relative weak ground interactions between the neighboring PO subunits, which can keep the excitation energy hopping between the neighboring PDI subunits, and meanwhile, can reduce the nonradiative decay for the excited states in the PO array. Therefore, there are two kinds of energy transfer processes within

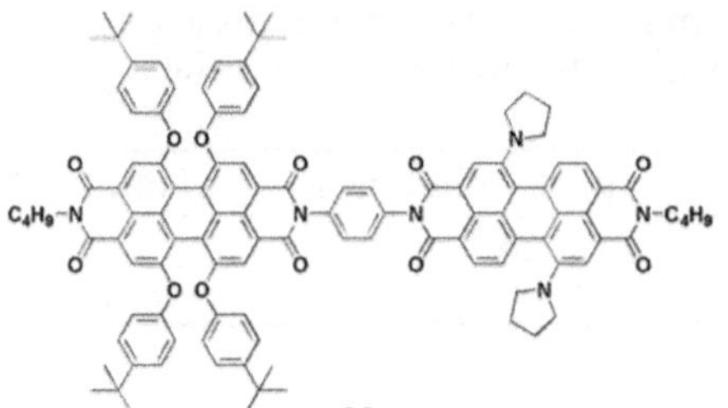

33

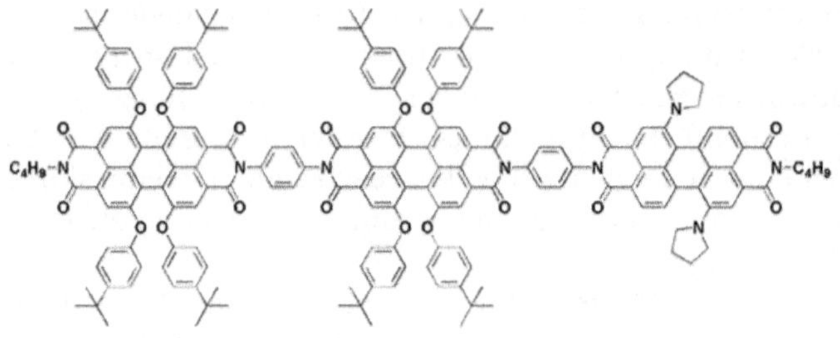

34

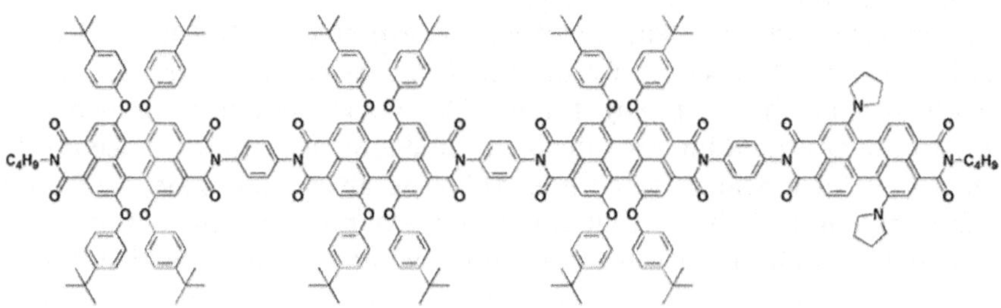

35

Figure 9.24 Molecular structures of **33**, **34** and **35**.

Table 9.2 Estimated Förster-type energy transfer parameters within compounds **33**, **34** and **35** in toluene.[12]

Donor–acceptor pair	R_0 (Å)	J (mol^{-1} dm^3 nm^4)	R^d (Å)	Φ_{EnT} (%)
PO-PO	34.5	7.8×10^{13}	16.93	98.6
PO_1-PY^a	40.7	2.1×10^{14}	16.93	99.5
PO_2-PO-PY^b	40.7	2.1×10^{14}	33.86	75.1
PO_3-PO-PO-PY^c	40.7	2.1×10^{14}	50.79	21

^{a}The energy transfer from the directly connected PO to PN.
^{b}The energy transfer from the second PO subunit to PN.
^{c}The energy transfer from the third PO subunit to PN.
^{d}The centre-to-centre distance between the donor and acceptor read from the minimized structure.

these compounds: (i) energy hopping between PO-PDI subunits, which is energy transfer from the S_1 state toward the ground-state of a neighboring PO-PDI, and (ii) the energy transfer from PO-PDI to PY-PDI. The energy transfer from PO-PDI to PY-PDI within **34** (1.78 ps and 23.6 ps) and **32** (1.39 and 9.67 ps) is slower than that in **33** (236 fs). However, even though the long distance between PO-PDI and PY-PDI in **35**, the rate constant for the energy transfer from PO-PDI to PY-PDI is very fast and close to that of the energy transfer from B800 to B850 in the light harvesting system (LH II) of photosynthesis.

9.4.5 Cyclic or Hyperbranched Arrays of PDIs

In nature, cyclic arrangements of chromophores have been found in photosynthesis systems. For example, in the bacterial photosynthetic antenna of light-harvesting complexes 1 and 2 (LH1 and LH2), photosynthetic tetrapyrrolic pigments are deliberately positioned in a wheel-like arrangement.[118,171,172] The absorbed light energy is transferred through intra- or inter-wheel excitation energy transfer (EET) to the reaction center. Special interactions of the chromophores are expected for this arrangement. The construction of PDI-based cyclic molecules can be used as model molecules to reveal the structure–property relationship and photoinduced processes in the photosynthesis system.

In dye **36** (Figure 9.25),[173] the angles between the adjacent transition moments are 120° (larger than 90°), therefore, an α transition (Figure 9.9) is more favorable. The absorption spectra resemble the spectra of the corresponding monomer except for small bathochromic shifts. Different from the linear series exhibiting canonical J-aggregate behavior, the symmetric trimer shows entirely unique photophysical behavior, neither J- nor H-like.[174] Similar to a linear series, along with the increase of the number of PDI subunits, an amplification of the absorptivity is observed for **36** in comparison with the corresponding monomer. When PDI subunits are arranged to a tetrahedral geometry, as compound **38**, whose steady-state spectra are similar as the corresponding monomer, an ultrafast intramolecular EET occurs between adjacent PDI molecules in $\tau = 6$ ns.[175] In compound **37** (Figure 9.25), the angle between the transition moments of the PDI chromophores is thereby diminished to 60°. The smaller angle induces strong exciton coupling in **37**. As a consequence, the molar absorptivity of each chromophore moiety is diminished, the main absorption band is shifted to shorter wavelengths and an effective fluorescence quenching is observed. Compound **37** exhibits canonical H-aggregate behavior. Similar results have also been observed in compound **39** (Figure 9.26), whose structure is similar to **37** but with a larger spacer hexaphenylbenzene group and bay-substituted PDI subunits.[63] A reversal in the intensity of 0–0 and 0–1 vibronic bands is observed in the absorption spectrum of **39** due to the intramolecular interactions. The relative intensity of the 0–0 and 0–1 vibronic bands correlates directly to the ratio of aggregates to the monomer in solution as well as to the strength of the interactions between the PDI subunits.[176] In compound **40** (Figure 9.26), the intensity ratio of the 0–0 and 0–1 vibronic bands clearly decreased, compared with the corresponding monomer, but was not

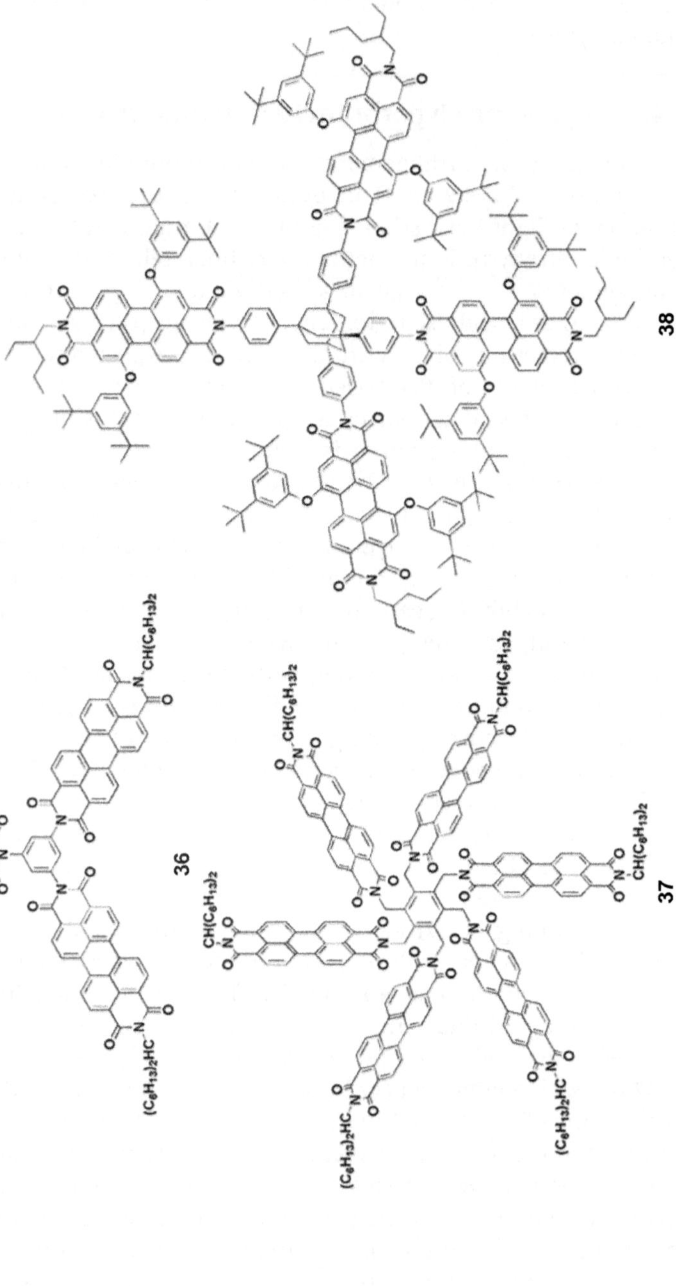

Figure 9.25 Molecular structures of 36, 37 and 38.

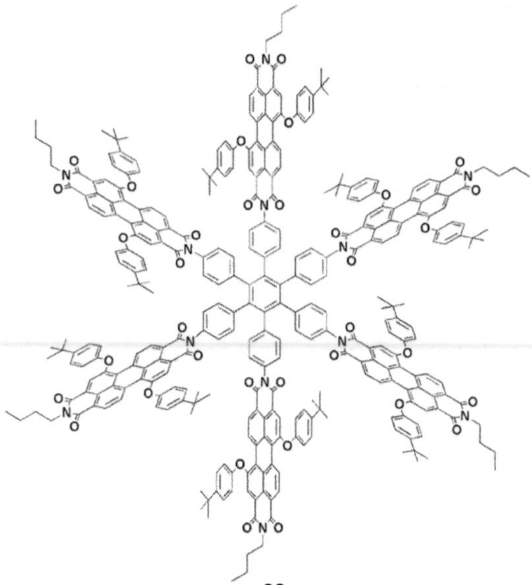

39

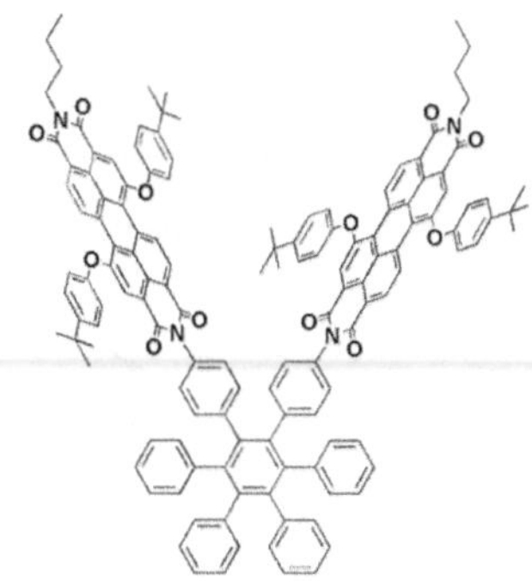

40

Figure 9.26 Molecular structures of **39** and **40**.

reversed. This is unexpected as the structure of **40** is similar to that of **39**. The relative orientation of the PDI subunits in **40** and **39** is the same and interactions between them should also be identical. From the optimized structures obtained with the AM1 method, we know that although the center-to-center distances between the neighboring PDI subunits in **40** are the same as those in **39**, the dihedral angle (α) between the PDI plane and the central phenyl ring in **40** is significantly smaller than that in **39** (25.66 *vs.* 35.77°). Moreover, because of the steric hindrance of the two neighboring subunits, the rotation of the PDI subunits along the molecular long axis in **39** is blocked. Therefore, compound **39** exhibits a more rigid "face-to-face" stacking mode. Compared with the corresponding monomer, the fluorescence spectra of both **39** and **40** are composed of monomer emission and excimer-like state emission. The fluorescence quantum yields of **39** and **40** were found to be 8% and 42%, respectively, which indicates that the interactions among the PDI subunits in **39** are stronger than those in **40**. The fluorescence lifetime of the excimer-like states of **40** (6.7 ns) is found to be shorter than that of **39** (8.3 ns); this can also be ascribed to the more rigid and strictly face-to-face stacked structure of **39** with respect to that of **40**. To further explore the relaxation dynamics of the excited-states of **39** and **40**, femtosecond transient absorption anisotropy (TAA) and femtosecond transient absorption (TA) spectra were recorded for them and their corresponding monomer.[177–181] The fitting results are summarized in Table 9.3. Two time constants were obtained for the monomer while three time constants were obtained for hexamer **39** and dimer **40**. The shortest (100 fs for both **39** and **40**) and longest (>10 ps) components are similar to those of the

Table 9.3 Fitting results of TAA decay at 710 nm for compounds **39**, **40** and the corresponding monomer.[10]

Compound	τ_1 (fs)	τ_2 (fs)	$\tau_3{}^a$ (fs)
Monomer	97	—	>10
39	100	168	>10
40	100	371	>10

^{a}Long lifetime component in the TAA decay, which was longer than the time window of the instrument.

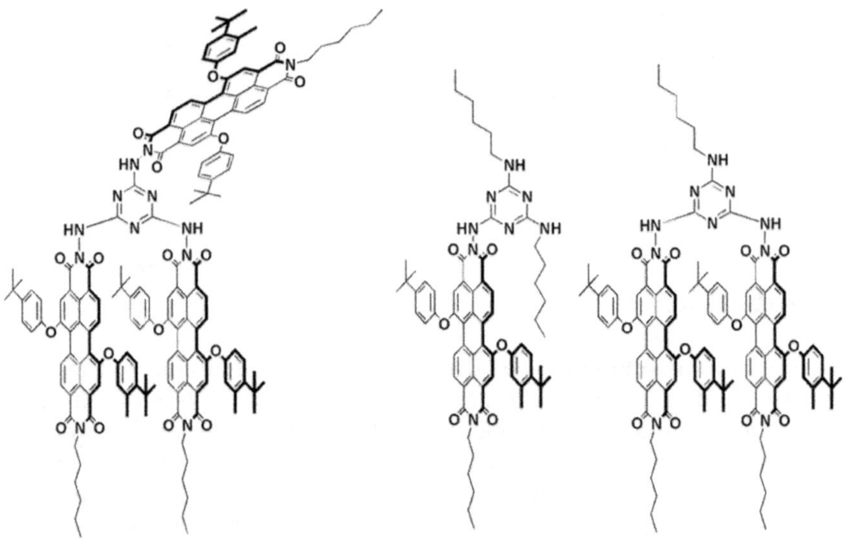

41 **42** **43**

Figure 9.27 Molecular structures of **41**, **42** and **43**.
Reprinted with permission from ref. 10. Copyright 2009 American Chemical Society.

monomer and, therefore, can be assigned to S2 to S1 internal conversion and rotation diffusion, respectively. Another fast component (371 fs for **40** and 168 fs for **39**) is attributed to the excitation delocalization between/among the different PDI branches.[182] Delocalization of the excited states in **39** (168 fs) is faster than that in **40** (371 fs), which supports the idea that the intramolecular interactions in **39** are stronger than those in **40.** This behavior is also seen in photosynthetic light-harvesting antenna pigment systems.[183]

As mentioned above, when two PDI subunits are attached to a triazine spacer, the majority of the dimers in solution present in the form of a face-to-face stacked conformation.[14] When three PDI subunits are attached to a triazine spacer, a flexible tri-branched trimer can be obtained (**41** in

Figure 9.27).[10] Interestingly, the absorption spectrum of compound **41** is almost identical to the sum of that of monomer **42** and dimer **43**. The intermolecular aggregation of **41** can be excluded by the concentration independent absorption spectrum in the range of 1×10^{-6} to 1×10^{-4} mol L^{-1}. These results indicate that compound **41** is composed of a "dimer" and a "monomer" and there is no significant interaction between them at ground state. Two of the three PDIs in **41** formed a face-to-face stacked dimeric structure and the third one appended acts as a monomer, which partially resembles the structural block in the LH2 complex, that is, B800 BChl monomer and B850 BChl dimer. The fluorescence spectrum of **41** gives a broad emission band in the range of 550–750 nm with a fluorescence lifetime of 22 ns, typical for the excimer's emission. The expected monomer emission was not observed but the excitation spectra of **41** showed clearly the contributions of the monomer's absorption at 542 nm to the emission of the dimer at 650 nm. The emission properties of **41** suggest that the excimer-like emission of the PDI trimer can be achieved by a photoinduced energy transfer between non-interacted monomer and dimer. The corresponding kinetics have been proposed as shown in Figure 9.28. The energy transfer from the monomer to the dimer is 0.8 ps, which is similar to that between B800 to B850 (0.8–0.9 ps).[184] This observation is meaningful for the design and construction of novel artificial LH systems.

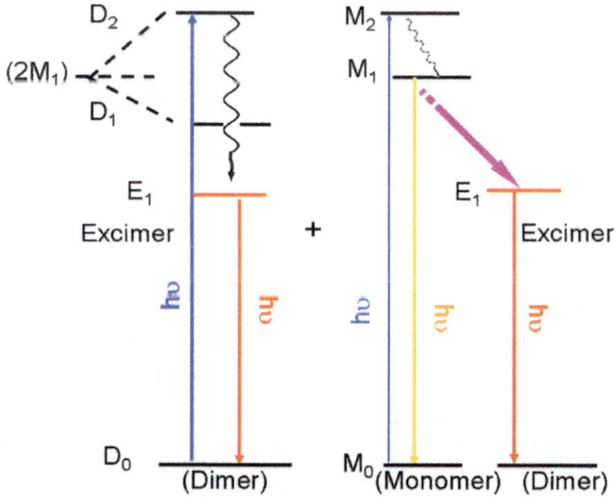

Figure 9.28 Model for energy relaxation and transfer in the trimer where the monomeric and dimeric subunits are coexcited: (left) dimeric excitation path; (right) monomeric excitation path. D_n $(n=0\text{–}2)=$ energy levels of the dimer, where D_1 and D_2 arise from the energy splitting of the monomeric level because of the formation of H-aggregates; M_n $(n=0\text{–}2)=$ energy levels of monomer; $E_1=$ the lowest excited energy level of the excimer.
Reprinted with permission from ref. 10. Copyright 2009 American Chemical Society.

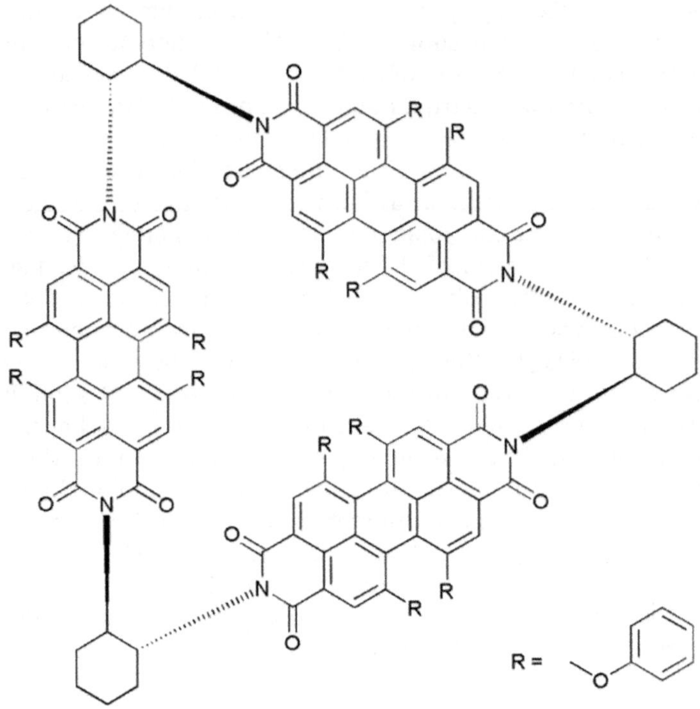

44

Figure 9.29 Molecular structure of **44**.
Reprinted with permission from ref. 185. Copyright 2015 American Chemical Society.

Recently, an interesting chiral molecular triangle composed of three covalently linked PDI units has been reported, as shown in Figure 9.29.[185] Cofacial PDI dimers usually lead to photodimerization or ultrafast formation of a lower-energy excimer state. The rigid triangle-shaped cyclic trimer **44** can reduce the electronic coupling between the PDI units and decrease the excimer formation rate. After photoexcitation with visible light, the initial singlet locally excited state (S_1) can generate four energetically degenerate symmetry-broken states depending on the relative rates of electron transfer (ET) and hole transfer (HT). Then six energetically degenerate ion-pair states may exist because of charge shifts, as shown in Figure 9.30.

9.5 Outlook and Perspective

Perylene bisimide and its derivatives have been investigated for about 100 years. Their outstanding chemical-, thermal- and photo-stabilities have attracted a lot of research interest in different fields. Various synthetic methods for PDIs have been developed recently, which makes PDIs one of

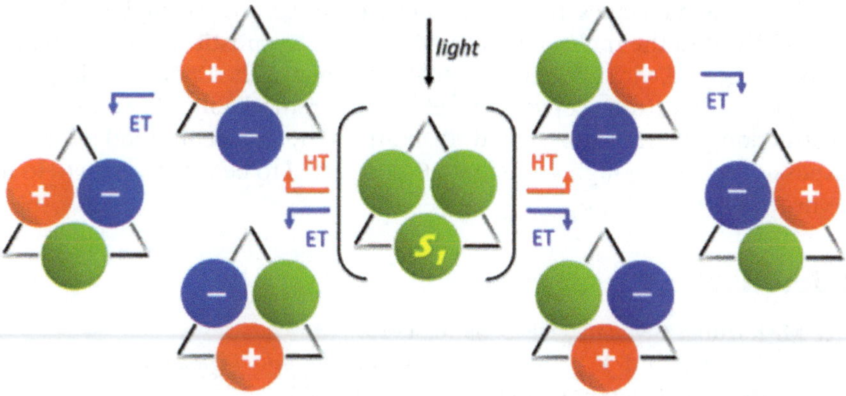

Figure 9.30 Schematic of symmetry-breaking charge separation within (−)-**44** upon photoexcitation, showing four possible charge (hole or electron) transfer processes and the subsequent charge hopping between adjacent redox centers to achieve six energetically degenerate ion-pair states. Reprinted with permission from ref. 185. Copyright 2015 American Chemical Society.

the most popular organic dyes in the area of functional organic aromatic compounds. They have great application potential as semiconductors in organic field effect transistors (OFET), light harvesting chromophores in artificial photosynthesis, fluorescence probes in biological systems and environmental monitoring, and in photocatalysis. The most popular and well developed area of PDI chemistry is the application of PDIs as model compounds to duplicate the photoinduced processes in the light-harvesting systems of artificial photosynthesis. Both supramolecular systems as well as covalently linked systems are constructed for this purpose. The covalently linked systems are particularly important for the purpose of revealing the structure–property relationship of these compounds. This area will continually be an active direction for PDI research in the future, because there are still plenty of problems that need to be solved in this area. For example, the mechanisms of the excited state decay of PDI molecular arrays are still not fully resolved. Energy transfer between different molecular arrays or molecular aggregates is rarely successful.

Singlet exciton fission (SF) has recently attracted much attention due to the possibility of it overcoming the Shockley–Queisser efficiency limit in conventional photovoltaic cells.[186] Optimal SF chromophores should satisfy the condition E(S1), E(T2) > 2E(T1), which is not easy to fulfill in most common chromophores. Moreover, the most investigated SF molecules until now have been tetracene and pentacene compounds, which are well known for their instability towards light and oxygen.[187] PDI comes close to satisfying the energetic requirements for SF. More importantly, the crystal of an *ortho*-substituted PDI derivative has recently been found to conduct SF efficiently.[188] This result suggests that *ortho*-substituted PDIs are ideal molecules for SF study because they are much more stable than the traditional SF molecules (tetracene, pentacene *etc.*). Once highly efficient SF is achieved in the molecular array of these *ortho*-substituted PDIs, then it can be integrated with OPV or dye-sensitized solar cells (DSSC) and improve the light to electron conversion efficiency remarkably. Therefore, the construction of *ortho*-substituted PDI arrays by covalent bonds for the purpose of achieving highly efficient SF is expected to be a very active field in the near future.

References

1. M. Kardos, *Ger. Pat.*, DE 276956, 1913.
2. W. Herbst and K. Hunger, *Industrial Organic Pigments*, Wiley-VCH, Weinheim, 2nd completely revised edn, 1997.
3. F. Würthner, *Chem. Commun.*, 2004, 1564–1579.
4. P. M. Kazmaier and R. Hoffmann, *J. Am. Chem. Soc.*, 1994, **116**, 9684–9691.
5. C. Huang, S. Barlow and S. R. Marder, *J. Org. Chem.*, 2011, **76**, 2386–2407.
6. G. Geissler and H. Remy, *Ger. Pat. Appl.*, DE 1130099, 1959 (*Chem. Abstr.*, 1962, **57**, P11346f).
7. M. P. O'Neil, M. P. Niemczyk, W. A. Svec, D. Gosztda, G. L. Gaines III and M. R. Wasielewski, *Science*, 1992, **257**, 63–66.
8. Y. Zhang, L. L. Zhang, H. Y. Liu, D. Sun and X. Y. Li, *CrystEngComm*, 2015, **17**, 1453–1463.
9. X. Y. Li, L. E. Sinks, B. Rybtchinski and M. R. Wasielewski, *J. Am. Chem. Soc.*, 2004, **126**, 10810–10811.
10. Y. F. Wang, H. L. Chen, H. X. Wu, X. Y. Li and Y. X. Weng, *J. Am. Chem. Soc.*, 2009, **131**, 30–31.
11. Y. N. Du, L. L. Jiang, J. Zhou, G. J. Qi, X. Y. Li and Y. Q. Yang, *Org. Lett.*, 2012, **14**, 3052–3055.
12. G. J. Qi, L. L. Jiang, Y. Y. Zhao, Y. Q. Yang and X. Y. Li, *Phys. Chem. Chem. Phys.*, 2013, **15**, 17342–17353.
13. F. Würthner, A. Sautter, D. Schmid and P. J. A. Weber, *Chem. – Eur. J.*, 2001, **7**, 894–902.
14. Y. F. Wang, Y. L. Chen, R. J. Li, S. Q. Wang, W. Su, P. Ma, M. R. Wasielewski, X. Y. Li and J. Z. Jiang, *Langmuir*, 2007, **23**, 5836–5842.
15. X. Zhan, Z. A. Tan, B. Domercq, Z. An, X. Zhang, S. Barlow, Y. Li, D. Zhu, B. Kippelen and S. R. Marder, *J. Am. Chem. Soc.*, 2007, **129**, 7246–7247.
16. X. W. Zhan, A. Facchetti, S. Barlow, T. J. Marks, M. A. Ratner, M. R. Wasielewski and S. R. Marder, *Adv. Mater.*, 2011, **23**, 268–284.
17. J. G. Mei, Y. Diao, A. L. Appleton, L. Fang and Z. N. Bao, *J. Am. Chem. Soc.*, 2013, **135**, 6724–6746.
18. C. Xiao, W. Jiang, X. Li, L. Hao, C. Liu and Z. Wang, *ACS Appl. Mater. Interfaces*, 2014, **6**, 18098–18103.

19. X. Zhan, J. Zhang, S. Tang, Y. Lin, M. Zhao, J. Yang, H. L. Zhang, Q. Peng, G. Yu and Z. Li, *Chem. Commun.*, 2015, **51**, 7156–7159.
20. L. Schmidt-Mende, A. Fechtenkötter, K. Müllen, E. Moons, R. H. Friend and J. D. MacKenzie, *Science*, 2001, **293**, 1119–1122.
21. M. Planells, F. J. Céspedes-Guirao, A. Forneli, Á. Sastre-Santos, F. Fernández-Lázaro and E. Palomares, *J. Mater. Chem.*, 2008, **18**, 5802–5808.
22. A. Sharenko, D. Gehrig, F. Laquai and T.-Q. Nguyen, *Chem. Mater.*, 2014, **26**, 4109–4118.
23. Yu. Lin, Y. Wang, J. Wang, J. Hou, Y. Li, D. Zhu and X. Zhan, *Adv. Mater.*, 2014, **26**, 5137–5142.
24. R. Singh, R. Shivanna, A. Iosifidis, H.-J. Butt, G. Floudas, K. S. Narayan and P. E. Keivanidis, *ACS Appl. Mater. Interfaces*, 2015, **7**, 24876–24886.
25. P. E. Hartnett, A. Timalsina, H. S. S. R. Matte, N. Zhou, X. Guo, W. Zhao, A. Facchetti, R. P. H. Chang, M. C. Hersam, M. R. Wasielewski and T. J. Marks, *J. Am. Chem. Soc.*, 2014, **136**, 16345–16356.
26. K. D. Field, M. V. Bondar, F. E. Hernandez and F. E. Przhonska, *J. Phys. Chem. C*, 2008, **112**, 5618–5622.
27. R. Givishi, R. Reisfeld and Z. Burshtein, *Chem. Phys. Lett.*, 1993, **213**, 338–344.
28. M. Sadrai and G. R. Bird, *Opt. Commun.*, 1984, **51**, 62–64.
29. W. E. Ford and P. V. Kamat, *J. Phys. Chem.*, 1987, **91**, 6373–6380.
30. A. K. Dwivedi, M. Pandeeswar and T. Govindaraju, *ACS Appl. Mater. Interfaces*, 2014, **6**, 21369–21379.
31. A. Han, X. Liu, G. D. Prestwich and L. Zang, *Sens. Actuators, B*, 2014, **198**, 274–277.
32. P. Spenst and F. Würthner, *Angew. Chem., Int. Ed.*, 2015, **54**, 10165–10168.
33. A. K. Dwivedi, M. Pandeeswar and T. Govindaraju, *ACS Appl. Mater. Interfaces*, 2014, **6**, 21369–21379.
34. F. J. Céspedes-Guirao, A. B. Ropero, E. Font-Sanchis, A. Nadal, F. FernándezLázaro and A. Sastre-Santos, *Chem. Commun.*, 2011, **47**, 8307–8309.
35. Y. Zhao, X. Zhang, D. Li, D. Liu, W. Jiang, C. Han *et al.*, *Luminescence*, 2009, **24**, 140–143.
36. Q. Fan, K. Cheng, Z. Yang, R. Zhang, M. Yang, X. Hu, X. Ma, L. Bu, X. Lu, X. Xiong, W. Huang, H. Zhao and Z. Cheng, *Adv. Mater.*, 2015, **27**, 843–847.
37. Z. B. Hill, D. B. Rodovsky, J. M. Leger and G. P. Bartholomew, *Chem. Commun.*, 2008, 6594–6596.
38. W. Porzio, S. Destri, M. Pasini, U. Giovanella, M. Ragazzi, G. Scavia *et al.*, *New J. Chem.*, 2010, **34**, 1961–1973.
39. F. Graser and E. Hädicke, *Liebigs Ann. Chem.*, 1980, 1994–2011.
40. F. Graser and E. Hädicke, *Liebigs Ann. Chem.*, 1984, 483–494.
41. E. Hädicke and F. Graser, *Acta Crystallogr., Sect. C: Cryst. Stuct. Commun.*, 1986, **42**, 189–195.

42. E. Hädicke and F. Graser, *Acta Crystallogr., Sect. C: Cryst. Stuct. Commun.*, 1986, **42**, 195–198.
43. Y. Zhang, L. L. Zhang, H. Y. Liu, D. Sun and X. Y. Li, *CrystEngComm*, 2015, **17**, 1453–1463.
44. P. Zugenmaier, J. Duff and T. L. Bluhm, *Cryst. Res. Technol.*, 2000, **35**, 1095–1115.
45. L. Xue, Y. F. Wang, Y. L. Chen and X. Y. Li, *J. Colloid Interface Sci.*, 2010, **350**, 523–529.
46. P. K. Sukul, P. K. Singh, S. K. Maji and S. Malik, *J. Mater. Chem. B*, 2013, **1**, 153–156.
47. S. Chen, P. Slattum, C. Y. Wang and L. Zang, *Chem. Rev.*, 2015, **115**, 11967–11998.
48. F. Würthner, *Pure Appl. Chem.*, 2006, **78**, 2341–2349.
49. T. Maeda and F. Würthner, *Chem. Commun.*, 2015, **51**, 7661–7664.
50. C. T. Zhao, Y. X. Zhang, R. J. Li, X. Y. Li and J. Z. Jiang, *J. Org. Chem.*, 2007, **72**, 2402–2410.
51. S. Nakazono, S. Easwaramoorthi, D. Kim, H. Shinokubo and A. Osuka, *Org. Lett.*, 2009, **11**, 5426–5429.
52. G. Battagliarin, Y. F. Zhao, C. Li and K. Müllen, *Org. Lett.*, 2011, **13**, 3399–3401.
53. P. E. Hartnett, A. Timalsina, H. S. S. R. Matte, N. Zhou, X. Guo, W. Zhao, A. Facchetti, R. P. H. Chang, M. C. Hersam, M. R. Wasielewski and T. J. Marks, *J. Am. Chem. Soc.*, 2014, **136**, 16345–16356.
54. T. T. Clikeman, E. V. Bukovsky, X. B. Wang, Y. S. Chen, G. Rumbles, S. H. Strauss and O. V. Boltalina, *Eur. J. Org. Chem.*, 2015, 6641–6654.
55. A. P. H. J. Schenning, J. Herrikhuyzen, P. Jonkheijm, Z. J. Chen, F. Würthner and E. W. Meije, *J. Am. Chem. Soc.*, 2002, **124**, 10252–10253.
56. C.-C. You and F. Würthner, *J. Am. Chem. Soc.*, 2003, **125**, 9716–9725.
57. Y. Guan, Y. Zakrevskyy, J. Stumpe, M. Antonietti and C. F. J. Faul, *Chem. Commun.*, 2003, 894–895.
58. J. M. Giaimo, A. V. Gusev and M. R. Wasielewski, *J. Am. Chem. Soc.*, 2002, **124**, 8530–8531.
59. J. M. Giaimo, J. V. Lockard, L. E. Sinks, A. M. Scott, T. M. Wilson and M. R. Wasielewski, *J. Phys. Chem. A*, 2008, **112**, 2322–2330.
60. J. Q. Feng, Y. X. Zhang, C. T. Zhao, R. J. Li, W. Xu, X. Y. Li and J. Z. Jiang, *Chem. – Eur. J.*, 2008, **14**, 7000–7010.
61. K. M. Lefler, K. E. Brown, W. A. Salamant, S. M. Dyar, K. E. Knowles and M. R. Wasielewski, *J. Phys. Chem. A*, 2013, **117**, 10333–10345.
62. H. Langhals, A. J. Esterbauer, A. Walter, E. Riedle and I. Pugliesi, *J. Am. Chem. Soc.*, 2010, **132**, 16777–16782.
63. L. Xue, Y. Shi, L. L. Zhang and X. Y. Li, *ChemPhysChem*, 2013, **14**, 3319–3326.
64. H. Langhals, *Heterocycles*, 1995, **40**, 477–500.
65. M. R. Wasielewski, *J. Org. Chem.*, 2006, **71**, 5051–5066.
66. Y. Nagao, *Prog. Org. Chem.*, 1997, **31**, 43–49.
67. Y. Nagao and T. Misonot, *Bull. Chem. Soc. Jpn.*, 1981, **54**, 1191–1194.

68. I. K. Iverson and S.-W. Tam-Chang, *J. Am. Chem. Soc.*, 1999, **121**, 5801–5802.
69. G. Seybold and G. Wagenblast, *Dyes Pigm.*, 1989, **11**, 303–317.
70. F. Würthner, P. Osswald, R. Schmidt, T. E. Kaiser, H. Mansikkama and M. Könemann, *Org. Lett.*, 2006, **8**, 3765–3768.
71. A. Stange, G. Wagenblast and G. Seybold, *BMFT-Bericht T 86-216*, Fachinformationszentrum Karlsruhe, Germany, 1986, vol. 3, pp. 86–216.
72. A. Böhm, H. Arms, G. Henning and P. Blaschka, *Ger. Pat. Appl.*, 1997, vol. DE 19547209 A1.
73. F. Würthner, V. Stepanenko, Z. J. Chen, C. R. Saha-Möller, N. Kocher and D. Stalke, *J. Org. Chem.*, 2004, **69**, 7933–7939.
74. L. Q. Fan, Y. P. Xu and H. Tian, *Tetrahedron Lett.*, 2005, **46**, 4443–4447.
75. P. Rajasingh, R. Cohen, E. Shirman, L. J. W. Shimon and B. Rybtchinski, *J. Org. Chem.*, 2007, **72**, 5973–5979.
76. G. Golubkov, H. Weissman, E. Shirman, S. G. Wolf, I. Pinkas and B. Rybtchinski, *Angew. Chem., Int. Ed.*, 2009, **48**, 926–930.
77. W. M. Chen, Z. Dai, H. Q. Liu, H. Y. Liu, Y. Shi and X. Y. Li, *J. Lumin.*, 2015, **168**, 192–198.
78. D. Schmidt, M. Son, J. M. Lim, M.-J. Lin, I. Krummenacher, H. Braunschweig, D. Kim and F. Würthner, *Angew. Chem., Int. Ed.*, 2015, **54**, 13980–13984.
79. W. F. Qiu, S. Y. Chen, X. B. Sun, Y. Q. Liu and D. B. Zhu, *Org. Lett.*, 2006, **8**, 867–870.
80. U. Rohr, C. Kohl, K. Müllen, A. van de Craats and J. Warman, *J. Mater. Chem.*, 2001, **11**, 1789–1799.
81. Z. S. An, S. A. Odom, R. F. Kelley, C. Huang, X. Zhang, S. Barlow, L. A. Padilha, J. Fu, S. Webster, D. J. Hagan, E. W. V. Stryland, M. R. Wasielewski and S. R. Marder, *J. Phys. Chem. A*, 2009, **113**, 5585–5593.
82. Y. Shi, H. X. Wu, L. Xue and X. Y. Li, *J. Colloid Interface Sci.*, 2012, **365**, 172–177.
83. D. L. Wang, Y. Shi, C. T. Zhao, B. L. Liang and X. Y. Li, *J. Mol. Struct.*, 2009, **938**, 245–253.
84. W. Yue, Y. Li, W. Jiang, Y. G. Zhen and Z. H. Wang, *Org. Lett.*, 2009, **11**, 5430–5433.
85. S. Nakazono, Y. Imazaki, H. Yoo, J. Yang, T. Sasamori, N. Tokitoh, T. Cédric, H. Kageyama, D. Kim, H. Shinokubo and A. Osuka, *Chem. – Eur. J.*, 2009, **15**, 7530–7533.
86. L. E. Shoer, S. W. Eaton, E. A. Margulies and M. R. Wasielewski, *J. Phys. Chem. B*, 2015, **119**, 7635–7643.
87. T. Teraoka, S. Hiroto and H. Shinokubo, *Org. Lett.*, 2011, **13**, 2532–2535.
88. G. Battagliarin, C. Li, V. Enkelmann and K. Müllen, *Org. Lett.*, 2011, **13**, 3012–3015.
89. E. Kianmehr, M. Yahyaee and K. Tabatabai, *Tetrahedron Lett.*, 2007, **48**, 2713–2715.

90. S. Seifert, D. Schmidt and F. Würthner, *Chem. Sci.*, 2015, **6**, 1663–1667.
91. M. Sadrai, L. Hadel, R. R. Sauers, S. Husain, K. Krogh-Jespersen, J. D. Westbrook and G. R. Bird, *J. Phys. Chem.*, 1992, **96**, 7988–7996.
92. M. Gsänger, J. H. Oh, M. Könemann, H. W. Höffken, A.-M. Krause, Z. Bao and F. Würthner, *Angew. Chem., Int. Ed.*, 2010, **49**, 740–743.
93. W. Yue, W. Jiang, M. Böckmann, N. L. Doltsinis and Z. Wang, *Chem. – Eur. J.*, 2014, **20**, 5209–5213.
94. J. Gao, C. Xiao, W. Jiang and Z. Wang, *Org. Lett.*, 2014, **16**, 394–397.
95. B. L. Liang, Y. X. Zhang, Y. F. Wang, W. Xu and X. Y. Li, *J. Mol. Struct.*, 2009, **917**, 133–141.
96. F. Würthner, A. Sautter and J. Schilling, *J. Org. Chem.*, 2002, **67**, 3037–3044.
97. W. Su, Y. X. Zhang, C. T. Zhao, X. Y. Li and J. Z. Jiang, *Chem. Phys. Chem.*, 2007, **8**, 1857–1862.
98. J. Q. Feng, D. L. Wang, H. L. Wang, D. P. Zhang, L. L. Zhang and X. Y. Li, *J. Phys. Org. Chem.*, 2011, **24**, 621–629.
99. K. Balakrishnan, A. Datar, T. Naddo, J. L. Huang, R. Oitker, M. Yen, J. C. Zhao and L. Zang, *J. Am. Chem. Soc.*, 2006, **128**, 7390–7398.
100. A. Rademacher, S. Märkle and H. Langhals, *Chem. Ber.*, 1982, **115**, 2927–2934.
101. F. Würthner, C. Thalacker, S. Diele and C. Tschierske, *Chem. – Eur. J.*, 2001, **7**, 2245–2253.
102. F. Würthner and C. Thalacker, *Ger. Pat. Appl.*, DE 10039232 A1, 2000 (*Chem. Abstr.*, 2002, **136**, 185323).
103. L. E. Sinks, B. Rybtchinski, M. Iimura, B. A. Jones, A. J. Goshe, X. B. Zuo, D. M. Tiede, X. Y. Li and M. R. Wasielewski, *Chem. Mater.*, 2005, **17**, 6295–6303.
104. R. Schmidt, M. M. Ling, J. H. Oh, M. Winkler, M. Könemann, Z. Bao and F. Würthner, *Adv. Mater.*, 2007, **19**, 3692–3695.
105. Z. Chen, M. G. Debije, T. Debaerdemaeker, P. Osswald and F. Würthner, *ChemPhysChem*, 2004, **5**, 137–140.
106. B. A. Jones, A. Facchetti, M. R. Wasielewski and T. J. Marks, *J. Am. Chem. Soc.*, 2007, **129**, 15259–15278.
107. M. J. Ahrens, M. J. Fuller and M. R. Wasielewski, *Chem. Mater.*, 2003, **15**, 2684–2686.
108. Y. Zhao and M. R. Wasielewski, *Tetrahedron Lett.*, 1999, **40**, 7047–7050.
109. A. S. Lukas, Y. Zhao, S. E. Miller and M. R. Wasielewski, *J. Phys. Chem. B*, 2002, **106**, 1299–1306.
110. D. Görl, X. Zhang and F. Würthner, *Angew. Chem., Int. Ed.*, 2012, **51**, 2–23.
111. L. Zang, Y. Che and J. S. Moore, *Acc. Chem. Res.*, 2008, **41**, 1596–1608.
112. P. D. Frischmann, K. Mahata and F. Würthner, *Chem. Soc. Rev.*, 2013, **42**, 1847–1870.
113. M. R. Wasielewski, *Acc. Chem. Res.*, 2009, **42**, 1910–1921.
114. S. Chen, P. Slattum, C. Y. Wang and L. Zang, *Chem. Rev.*, 2015, **115**, 11967–11998.

115. T. Pullerits and V. Sundström, *Acc. Chem. Res.*, 1996, **29**, 381–389.
116. X. Hu and K. Schulten, *Phys. Today*, 1997, **54**, 29–34.
117. K. McLuskey, S. M. Prince, R. J. Cogdell and N. W. Isaacs, *Biochemistry*, 2001, **40**, 8783–8789.
118. G. McDermott, S. M. Prince, A. A. Freer, A. M. HawthornthwaiteLawless, M. Z. Papiz, R. J. Cogdell and N. W. Isaacs, *Nature*, 1995, **374**, 517–521.
119. J.-M. Lehn, *Science*, 2002, **295**, 2400–2403.
120. J. A. A. W. Elemans, R. van Hameren, R. J. M. Nolte and A. E. Rowan, *Adv. Mater.*, 2006, **18**, 1251–1266.
121. W. Kuhn, *Trans. Faraday Soc.*, 1930, **26**, 293–308.
122. M. Kasha, H. R. Rawls and M. A. El-Bayoumi, *Pure Appl. Chem.*, 1965, **11**, 4575f.
123. M. Kasha, H. R. Rawls and M. A. El-Bayoumi, *Pure Appl. Chem.*, 1965, **11**, 371–392.
124. P. B. Walczak, A. Eisfeld and J. S. Briggs, *J. Chem. Phys.*, 2008, **128**, 044505.
125. U. Rösch, S. Yao, R. Wortmann and F. Würthner, *Angew. Chem., Int. Ed.*, 2006, **45**, 7026–7030.
126. H. Langhals, *Helv. Chim. Acta*, 2005, **88**, 1309–1343.
127. T. M. Wilson, M. J. Tauber and M. R. Wasielewski, *J. Am. Chem. Soc.*, 2009, **131**, 8952–8957.
128. B. Chiński, L. E. Sinks and M. R. Wasielewski, *J. Phys. Chem. A*, 2004, **108**, 7497–7505.
129. D. Man, S. M. A. Chopin, S. C. J. Meskers, M. M. Groeneveld, R. M. Williams and R. A. J. Janssen, *J. Phys. Chem. A*, 2008, **112**, 5846–5857.
130. F. Schlosser, J. Sung, P. Kim, D. Kim and F. Würthner, *Chem. Sci.*, 2012, **3**, 2778–2785.
131. N. A. Montgomery, G. J. Hedley, A. Ruseckas, J.-C. Denis, S. Schumacher, A. L. Kanibolotsky, P. J. Skabara, I. Galbraith, G. A. Turnbull and I. D. W. Samuel, *Phys. Chem. Chem. Phys.*, 2012, **14**, 9176–9184.
132. H. Yoo, S. Furumaki, J. Yang, J.-E. Lee, H. Chung, T. Oba, H. Kobayashi, B. Rybtchinski, T. M. Wilson, M. R. Wasielewski, M. Vacha and D. Kim, *J. Phys. Chem. B*, 2012, **116**, 12878–12886.
133. Y. Deng, C. J. Chang and D. G. Nocera, *J. Am. Chem. Soc.*, 2000, **122**, 410–411.
134. K. M. Kadish, L. Frémond, Z. Ou, J. Shao, C. Shi, F. C. Anson, F. Burdet, C. P. Gros, J.-M. Barbe and R. Guilard, *J. Am. Chem. Soc.*, 2005, **127**, 5625–5631.
135. T. van der Boom, R. T. Hayes, Y. Zhao, P. J. Bushhard, E. A. Weiss and M. R. Wasielewski, *J. Am. Chem. Soc.*, 2002, **124**, 9582–9590.
136. H. A. Staab, N. Riegler, F. Diederich, C. Krieger and D. Schweitzer, *Chem. Ber.*, 1984, **117**, 246–259.
137. H. Yoo, J. Yang, A. Yousef, M. R. Wasielewski and D. Kim, *J. Am. Chem. Soc.*, 2010, **132**, 3939–3944.

138. A. E. Clark, C. Qin and A. D. Q. Li, *J. Am. Chem. Soc.*, 2007, **129**, 7586–7595.
139. A. D. Q. Li, W. Wang and L.-Q. Wang, *Chem. – Eur. J.*, 2003, **9**, 4594–4601.
140. D. Carbonera, N. DiValentin, C. Corvaja, G. Agostini, G. Giacometti, P. A. Liddell, D. Kuciauskas, A. L. Moore, T. A. Moore and D. Gust, *J. Am. Chem. Soc.*, 1998, **120**, 4398–4405.
141. T. Okada, I. Karaki, E. Matsuzawa, N. Mataga, Y. Sakata and S. Misumi, *J. Phys. Chem.*, 1981, **85**, 3957–3960.
142. T. Kircher and H.-G. Löhmannsröben, *Phys. Chem. Chem. Phys.*, 1999, **1**, 3987–3992.
143. T. Nakano and T. Yade, *J. Am. Chem. Soc.*, 2003, **125**, 15474–15484.
144. E. A. Margulies, L. E. Shoer, S. W. Eaton and M. R. Wasielewski, *Phys. Chem. Chem. Phys.*, 2014, **16**, 23735–23742.
145. A. Nowak-Król, B. Fimmel, M. Son, D. Kim and F. Würthner, *Faraday Discuss.*, 2015, **185**, 507–527.
146. B. Fimmel, M. Son, Y. M. Sung, M. Grüne, B. Engels, D. Kim and F. Würthner, *Chem. – Eur. J.*, 2015, **21**, 615–630.
147. M. J. Ahren, L. E. Sinks, B. Rybtchinski, W. Liu, B. A. Jones, J. M. Giaimo, A. V. Gusev, A. Goshe, D. M. Tiede and M. R. Wasielewki, *J. Am. Chem. Soc.*, 2004, **126**, 8284–8294.
148. H. Langhals and R. Ismael, *Eur. J. Org. Chem.*, 1998, 1915–1917.
149. K. E. Brown, W. A. Salamant, L. E. Shoer, R. M. Young and M. R. Wasielewski, *J. Phys. Chem. Lett.*, 2014, **5**, 2588–2593.
150. M. Shellaiah, Y. C. Rajan and H.-C. Lin, *J. Mater. Chem.*, 2012, **22**, 8976–8987.
151. F. Würthner, T. E. Kaiser and C. R. Saha-Möller, *Angew. Chem., Int. Ed.*, 2011, **50**, 3376–3410.
152. S. Yagai, T. Seki, T. Karatsu, A. Kitamura and F. Würthner, *Angew. Chem., Int. Ed.*, 2008, **47**, 3367–3371.
153. T. E. Kaiser, V. Stepanenko and F. Würthner, *J. Am. Chem. Soc.*, 2009, **131**, 6719–6732.
154. H. Y. Liu, L. Shen, Z. Z. Cao and X. Y. Li, *Phys. Chem. Chem. Phys.*, 2014, **16**, 16399–16406.
155. C. Shao, M. Gruene, M. Stolte and F. Würthner, *Chem. – Eur. J.*, 2012, **18**, 13665–13677.
156. Y. Tian, V. Stepanenko, T. E. Kaiser, F. Würthner and I. G. Scheblykin, *Nanoscale*, 2012, **4**, 218–223.
157. C. Ramanan, A. L. Smeigh, J. E. Anthony, T. J. Marks and M. R. Wasielewski, *J. Am. Chem. Soc.*, 2012, **134**, 386–397.
158. T. Roland, J. Léonard, G. Hernandez Ramirez, S. Méry, O. Yurchenko, S. Ludwigs and S. Haacke, *Phys. Chem. Chem. Phys.*, 2012, **14**, 273–279.
159. M. C. Hanna and A. J. Nozik, *J. Appl. Phys.*, 2006, **100**, 074510.
160. T. Förster, *Fluoreszenz Organischer Verbindungen*, Vandenhoeck & Puprecht, Göttingen, Germany, 1st edn, 1951.

161. For an overview, see: *Photoinduced Electron Transfer*, ed. J. Mattay, Springer-Verlag, Berlin, 1990–1993, vols. I–V.
162. H. Langhals, M. Rauscher, J. Ströbe and D. Kuck, *J. Org. Chem.*, 2008, **73**, 1113–1116.
163. T. Förster, *Naturwissenschaften*, 1946, **33**, 166–175; *Chem. Abstr.*, 1947, **41**, 36668.
164. T. Förster, *Ann. Phys.*, 1948, **6**, 55–75; *Chem. Abstr.*, 1949, **43**, 31172.
165. T. Förster, *Z. Elektrochem.*, 1949, **53**, 93–99; *Chem. Abstr.*, 1949, **43**, 33629.
166. T. Förster, *Z. Naturforsch.*, 1949, **4a**, 321–327; *Chem. Abstr.*, 1950, **44**, 43074.
167. H. Langhals, S. Poxleitner, O. Krotz, T. Pust and A. Walter, *Eur. J. Org. Chem.*, 2008, 4559–4562.
168. H. Langhals and W. Jona, *Angew. Chem., Int. Ed.*, 1998, **37**, 952–955.
169. T. Vosch, E. Fron, J. Hotta, A. Deres, H. Uji-i, A. Idrissi, J. Yang, D. Kim, L. Puhl, A. Haeuseler, K. Müllen, F. C. De Schryver, M. Sliwa and J. Hofkens, *J. Phys. Chem. C*, 2009, **113**, 11773–11782.
170. T. Förster, *Discuss. Faraday Soc.*, 1959, **27**, 7–17.
171. J. Koepke, X. Hu, C. Muenke, K. Schulten and H. Michel, *Structure*, 1996, **4**, 581–597.
172. A. W. Roszak, T. D. Howard, J. Southall, A. T. Gardiner, C. J. Law, N. W. Isaacs and R. J. Cogdell, *Science*, 2003, **302**, 1969–1972.
173. H. Langhals and J. Gold, *J. Prakt. Chem.*, 1996, **338**, 654–659.
174. C. M. Pochas, K. A. Kistler, H. Yamagata, S. Matsika and F. C. Spano, *J. Am. Chem. Soc.*, 2013, **135**, 3056–3066.
175. C. Ramanan, C. H. Kim, T. J. Marks and M. R. Wasielewski, *J. Phys. Chem. C*, 2014, **118**, 16941–16950.
176. L. F. Dçssel, V. Kamm, I. A. Howard, F. Laquai, W. Pisula, X. Feng, C. Li, M. Takase, T. Kudernac, S. De Feyter and K. Müllen, *J. Am. Chem. Soc.*, 2012, **134**, 5876–5886.
177. S. E. Bradforth, R. Jimenez, F. van Mourik, R. van Grondelle and G. R. Fleming, *J. Phys. Chem.*, 1995, **99**, 16179–16191.
178. B. Brüggemann, J. L. Herek, V. Sundström, T. Pullerits and V. May, *J. Phys. Chem. B*, 2001, **105**, 11391–11394.
179. B. Brüggemann and V. May, *J. Chem. Phys.*, 2004, **120**, 2325–2336.
180. Y. Nakamura, I. W. Hwang, N. Aratani, T. K. Ahn, D. M. Ko, A. Takagi, T. Kawai, T. Matsumoto, D. Kim and A. Osuka, *J. Am. Chem. Soc.*, 2005, **127**, 236–246.
181. H. S. Cho, H. Rhee, J. K. Song, C. K. Min, M. Takase, N. Aratani, S. Cho, A. Osuka, T. Joo and D. Kim, *J. Am. Chem. Soc.*, 2003, **125**, 5849–5860.
182. F. Zhao, X. Zheng, J. Zhang, H. Wang, Z. Yu, J. Zhao and L. Jiang, *J. Photochem. Photobiol., B*, 1998, **45**, 144–149.
183. T. Goodson III, *Acc. Chem. Res.*, 2005, **38**, 99–107.
184. X. Chen, L. Zhang, Y. Weng, L. Du, M. Ye, G. Yang, R. Fujii, F. S. Rondonuwu, Y. Koyama, S. Wu and J. Zhang, *Biophys. J.*, 2005, **88**, 4262–4273.

185. Y. Wu, R. M. Young, M. Frasconi, S. T. Schneebeli, P. Spenst, D. M. Gardner, K. E. Brown, F. Würthner, J. F. Stoddart and M. R. Wasielewski, *J. Am. Chem. Soc.*, 2015, **137**, 13236–13239.
186. M. B. Smith and J. Michl, *Chem. Rev.*, 2010, **110**, 6891–6936.
187. A. Vollmer, H. Weiss, S. Rentenberger, I. Salzmann, J. P. Rabe and N. Koch, *Surf. Sci.*, 2006, **600**, 4004–4007.
188. S. W. Eaton, L. E. Shoer, S. D. Karlen, S. M. Dyar, E. A. Margulies, B. S. Veldkamp, C. Ramanan, D. A. Hartzler, S. Savikhin, T. J. Marks and M. R. Wasielewski, *J. Am. Chem. Soc.*, 2013, **135**, 14701–14712.

Subject Index